Understanding and Governing Sustainable Tourism Mobility

Despite a growing contribution to climate change, tourist and traveller behaviour is currently not acknowledged as an important sector within the development of climate policy. Whilst tourists may be increasingly aware of potential impacts on climate change there is evidence that most are unwilling to modify their actual behaviours. Influencing individual behaviour in tourism and informing effective governance is therefore an essential part of climate change mitigation.

This significant volume is the first to explore the psychological and social factors that may contribute to and inhibit sustainable change in the context of tourist and traveller behaviour. It draws on a range of disciplines to offer a critical review of the psychological understandings and behavioural aspects of climate change and tourism mobilities, in addition to governance and policies based upon psychological, behavioural and social mechanisms. It therefore provides a more informed understanding of how technology, infrastructure and cost distribution can be developed in order to reach stronger mitigation goals whilst ensuring that resistance from consumers for socio-psychological reasons are minimized.

Written by leading academics from a range of disciplinary backgrounds and regions this ground-breaking volume is essential reading for all those interested in the effective governance of tourism's contribution to climate change now and in the future.

Scott A. Cohen is a senior lecturer in the School of Hospitality and Tourism Management at the University of Surrey, UK. He primarily researches sociological and consumer behaviour issues in tourism, leisure and mobility contexts, with a particular interest in the impacts of air travel on climate change.

James E.S. Higham holds the position of Professor, Department of Tourism, University of Otago, New Zealand, and Visiting Professor of sustainable tourism, Norwegian School of Hotel Management, Norway. His research interests address tourism and global environmental change across global-local scales of analysis, with a specific focus at present on global climate change, personal aeromobility and behaviour change.

Paul Peeters is an associate professor at NHTV Breda University of Applied Sciences, the Netherlands. His research specializes in the impacts of tourism on the environment in general and climate change in particular.

Stefan Gössling is a professor at the Department of Service Management, Lund University, and the School of Business and Economics, Linnaeus University, both in Sweden. He is also the research co-ordinator at the Western Norway Research Institute, Sogndal, Norway. His current main research interests include transport systems, energy use, greenhouse gas emissions and mobility consumption.

Contemporary geographies of leisure, tourism and mobility

Series Editor:

C. Michael Hall

Professor at the Department of Management, College of Business and Economics, University of Canterbury, Christchurch, New Zealand

The aim of this series is to explore and communicate the intersections and relationships between leisure, tourism and human mobility within the social sciences.

It will incorporate both traditional and new perspectives on leisure and tourism from contemporary geography, e.g. notions of identity, representation and culture, while also providing for perspectives from cognate areas such as anthropology, cultural studies, gastronomy and food studies, marketing, policy studies and political economy, regional and urban planning, and sociology, within the development of an integrated field of leisure and tourism studies.

Also, increasingly, tourism and leisure are regarded as steps in a continuum of human mobility. Inclusion of mobility in the series offers the prospect to examine the relationship between tourism and migration, the sojourner, educational travel, and second home and retirement travel phenomena.

The series comprises two strands:

Contemporary Geographies of Leisure, Tourism and Mobility aims to address the needs of students and academics, and the titles will be published in hardback and paperback. Titles include:

Routledge Studies in Contemporary Geographies of Leisure, Tourism and Mobility is a forum for innovative new research intended for research students and academics, and the titles will be available in hardback only. Titles include:

Understanding and Governing Sustainable Tourism Mobility

Psychological and behavioural approaches

**Edited by Scott A. Cohen,
James E.S. Higham, Paul Peeters and
Stefan Gössling**

Routledge
Taylor & Francis Group

LONDON AND NEW YORK

First published 2014
by Routledge
2 Park Square, Milton Park, Abingdon, Oxon OX14 4RN

and by Routledge
711 Third Avenue, New York, NY 10017

Routledge is an imprint of the Taylor & Francis Group, an informa business

British Library Cataloguing in Publication Data
A catalogue record for this book is available from the British Library

Library of Congress Cataloging in Publication Data
A catalog record for this book has been requested

ISBN: 978-0-415-83937-2 (hbk)
ISBN: 978-0-203-77150-1 (ebk)

Typeset in Times New Roman
by Wearset Ltd, Boldon, Tyne and Wear

Contents

Figures

Tables

Contributors

Editors

Scott A. Cohen is a senior lecturer in the School of Hospitality and Tourism Management at the University of Surrey, UK. He primarily researches sociological and consumer behaviour issues in tourism, leisure and mobility contexts, with a particular interest in the impacts of air travel on climate change.

Stefan Gössling is a professor at the Department of Service Management, Lund University, and the School of Business and Economics, Linnaeus University, both in Sweden. His current main research interests include transport systems, energy use, greenhouse gas emissions and mobility consumption.

James E.S. Higham holds the position of Professor, Department of Tourism, University of Otago (New Zealand) and Visiting Professor of sustainable tourism, Norwegian School of Hotel Management (Norway). His research interests address tourism and global environmental change across global-local scales of analysis, with a specific focus at present on global climate change, personal aeromobility and behaviour change.

Paul Peeters is Associate Professor at NHTV Breda University of Applied Sciences, The Netherlands. His research specialises in the impacts of tourism on the environment in general and climate change in particular.

Authors

Lindemberg Medeiros de Araujo has a PhD in Tourism Management from Sheffield Hallam University, UK. He teaches Geography of Tourism and Environmental Analysis at the Universidade Federal de Alagoas (UFAL) in Brazil. His current tourism research activities focus on coastal zones, environmental management, territory dynamics, internationalisation of developing regions, participation, partnerships and sustainable development.

Tom Cherrett is a senior lecturer in the Transportation Research Group at the University of Southampton, UK. His research interests focus around new technologies and operating practices to improve the efficiency of both forward and reverse supply chains in the retail sector.

Susanna Curtin is a senior lecturer at Bournemouth University, UK. Her research interests include the emotional and psychological benefits of taking a wildlife holiday, memorable wildlife encounters, how wildlife tourists attend to and perceive wildlife and the importance of tour leaders in the responsible management of wildlife tourism.

Nigel Davies is a professor of ubiquitous computing in the Computing Department at Lancaster University, UK. His research interests include mobile computing and transport.

Janet E. Dickinson is a senior lecturer at Bournemouth University, UK. Her research interests focus on sustainable travel practice and include studies on slow travel, social representations, mobile social networking and time.

Eke Eijgelaar has been a researcher at the Centre for Sustainable Tourism and Transport of NHTV Breda University of Applied Sciences, The Netherlands since 2008. His primary research interest is in climate change related issues such as tourism emissions, low carbon transport modes, carbon management and carbon offsetting.

Viachaslau Filimonau is a research fellow at Bournemouth University, UK. His research interests include sustainable mobility, information communication technology in tourism and tourism impact assessment.

Regine Gerike chairs the Institute for Transport Studies at the University of Natural Resources and Life Sciences (BOKU) in Vienna, Austria. Her research interests include all aspects of transport ecology and economy, including modelling of environmental effects, empirical research and holistic approaches to evaluating sustainable mobility.

Werner Gronau is Professor for Tourism, Travel & Transport at the University of Stralsund, Germany, Director of the "Tourism & Transport Research Center" of the University of Nicosia, Cyprus and is affiliated to the University of Bergamo, Italy as Visiting Professor. He holds a German degree in Human Geography from the Technical University of Munich and a PhD in Tourism Studies ("Leisure mobility and leisure style") from the University of Paderborn. He is a member of several research groups in the field of tourism and transport and is a reviewer for several academic journals. Furthermore he is chief editor of the transport journal *Studies on Mobility and Transport Research*. His research interests focus on sustainable destination management and tourism related transport issues.

Jo W. Guiver is a researcher and lecturer in the Institute of Transport and Tourism, School of Sports, Tourism and the Outdoors at the University of Central Lancashire, UK, where she is also the School Sustainability Lead. She is currently conducting research under the theme "disrupted mobility".

C. Michael Hall has long-standing interests in sustainable environmental behaviour and has published widely in the areas of tourism, gastronomy and environmental history. He has posts at the University of Canterbury, Christchurch, New Zealand; the Freiburg Institute of Advanced Studies, Germany; the University of Oulu, Finland; and the University of Johannesburg, South Africa.

Julia F. Hibbert is a recent PhD graduate from a partnership programme at Bournemouth University, UK and Linnaeus University, Sweden. Her current research explores the role played by identity in tourism mobility.

Eljas Johansson graduated from Lund University with an MSc in Service Management and holds a BSc with honours in Hospitality Management from Manchester Metropolitan University, UK. Prior to graduating, Eljas held management positions in the hospitality industry in Finland and Canada.

Catheryn Khoo-Lattimore holds a PhD in consumer behaviour from the University of Otago, New Zealand and is now an associate professor in the School of Hospitality, Tourism and Culinary Arts, Taylor's University, Malaysia. Her work is predominantly focused on the needs of consumer and tourist groups utilising qualitative research techniques.

Danny de Kinderen graduated in International Tourism and Travel Industry at NHTV Breda University of Applied Sciences, The Netherlands in 2012. His final thesis focused on carbon offsetting and travel behaviour.

Gunvor Riber Larsen is a recent PhD graduate from the School of Business and Economics at Linnaeus University in Sweden and the School of Sports, Tourism and the Outdoors, University of Central Lancashire, UK. She is currently undertaking postdoctoral research at the Centre for Mobilities and Urban Studies, Aalborg University, Denmark. Her prime research interest is tourism mobility.

Diem-Trinh Le-Klähn is a doctoral student at the School of Management, Technische Universität München, Germany. She is also a member of a research group at the Department of Urban Structure and Transport Planning, working on the project "sustainable mobility in the metropolitan region of Munich, Germany". Her research focuses on the mobility of tourists and their use of public transport in destinations.

Acácia Cristina Mendes Malhado has studied tourism and sustainable resource management. She holds a doctorate in Geography at the University of Tübingen, Germany. Since 2003 she has been a post-doctoral researcher at the Federal University of Alagoas (UFAL) in Brazil. Her current research focus is multi- and inter-disciplinary reflecting her interests in sustainable tourism and development, travel behaviour, transport and mega-events.

Jeroen Nawijn is a senior lecturer in tourism at NHTV Breda University of Applied Sciences, The Netherlands. His main interest is to study consumption emotions in various leisure and tourism contexts, setting and stages of destination development.

Sarah Norgate is a senior lecturer in psychology at the University of Salford, UK. Her research interests include user engagement with digital technology across the lifespan, particularly around temporalities.

Bruce Prideaux is Professor of Marketing and Tourism Management at James Cook University, Australia, based in Cairns. He heads a team of eight PhD researchers and a number of visiting scholars and is actively engaged in climatic change research with a particular interest in its impacts on coral reef systems and rainforests. Other interests include transport, crisis management and issues related to tourism development in remote areas.

Yael Ram is a lecturer at the Department of Tourism and Leisure Studies at Ashkelon Academic College, Israel. Her research interests centre around the linkage between psychology and tourism.

Arianne C. Reis is a research fellow with the School of Tourism and Hospitality Management, at Southern Cross University, Australia. Her research interests lie within the broader theme of sustainability of tourism and leisure practices, with a particular focus on social and environmental justice.

Rainer Rothfuss has studied Geography, Political Sciences and Spatial Planning at the universities of Tübingen, Stuttgart (Germany) and Mérida (Venezuela). He holds a doctorate in Geography with focus on transnational city network cooperation in the field of sustainable urban mobility. Since 2009 he has been Professor for Human Geography at the Institute of Geography at Tübingen University. Professor Rothfuss leads a research team dealing with the issue of sustainable mobility with specialization in mobility management, electric mobility and light electric vehicles.

Chris Speed is a reader in digital spaces at the University of Edinburgh, UK. His research interests include the Internet of things and time.

Sander van der Linden is currently pursuing a doctoral degree in environmental psychology at the London School of Economics and Political Science (Department of Geography and the Environment). He is also a research scholar with the Yale Project on Climate Change Communication at Yale University. Sander has conducted national studies on climate change beliefs, perceptions, attitudes and behaviours. His research interests include modelling the psychosocial determinants of pro-environmental behaviour, theories of persuasive communication and informational processing, behavioural change and the construction of human risk perception.

Chris Winstanley is a research assistant in the Computing Department at Lancaster University, UK.

Acknowledgements

The convenors of the *Freiburg 2012* workshop (3–6 July, 2012) gratefully acknowledge Tim Freytag (Freiburg Institute for Advanced Studies, Germany), Eke Eijgelaar, Jelmer Jeuring and Sandra Janssen (NHTV Breda, The Netherlands) and other members of the Scientific Advisory Committee: Bas Amelung (Wageningen University and Research, The Netherlands), Jean-Paul Ceron (Limoges University, France), Janet Dickinson (Bournemouth University, United Kingdom), Ghislain Dubois (TEC Conseil, France), Michael Hall (Freiburg Institute for Advanced Studies, Germany), Jeroen Nawijn (NHTV CSTT, Breda, The Netherlands), Daniel Scott (University of Waterloo, Canada), Gert Spaargaren (Wageningen University and Research, The Netherlands), John Urry (Freiburg Institute for Advanced Studies, Germany), all of whom contributed to the success of *Freiburg 2012*.

The existence of this book owes much to the initiative, guidance and support of Bernard Lane (Editor, *Journal of Sustainable Tourism*), who offered advice and encouragement for us to consider the option of building upon the special issue of the *Journal of Sustainable Tourism* 21(7) to develop a full edited volume of published work arising from the *Freiburg 2012* workshop. Indeed, eight of the chapters in this book (listed below) are reproduced from the 2013 special issue in the *Journal of Sustainable Tourism* 21(7) on "Psychological and behavioural approaches to understanding and governing sustainable mobility".

We are also grateful to Dr Adam Doering (University of Otago, New Zealand), who provided invaluable research assistance in reviewing and copy-editing chapters for this book, and in liaising with its contributors to bring the manuscript to completion, as well as to Sam Spector (University of Otago, New Zealand) for his contribution in indexing the book.

References from associated special issue

Cohen, S.A., Higham, J.E.S. and Reis, A.C. (2013). Sociological barriers to developing sustainable discretionary air travel behaviour. *Journal of Sustainable Tourism*, *21*(7), 982–998.

Dickinson, J.E., Filimonau, V., Cherrett, T., Davies, N., Norgate, S., Speed, C. and Winstanley, C. (2013). Understanding temporal rhythms and travel behaviour at destinations:

Potential ways to achieve more sustainable travel. *Journal of Sustainable Tourism, 21*(7), 1070–1090.

Hall, C.M. (2013). Framing behavioural approaches to understanding and governing sustainable tourism consumption: Beyond neoliberalism, "nudging" and "green growth". *Journal of Sustainable Tourism, 21*(7), 1091–1109.

Hibbert, J.F., Dickinson, J.E., Gössling, S. and Curtin, S. (2013). Identity and tourism mobility: An exploration of the attitude-behaviour gap. *Journal of Sustainable Tourism, 21*(7), 999–1016.

Khoo-Lattimore, C. and Prideaux, B. (2013). ZMET: A psychological approach to understanding unsustainable tourism mobility. *Journal of Sustainable Tourism, 21*(7), 1036–1048.

Larsen, G.R. and Guiver, J.W. (2013). Understanding tourists' perceptions of distance: A key to reducing the environmental impacts of tourism mobility. *Journal of Sustainable Tourism, 21*(7), 968–991.

Peeters, P.M. (2013). Developing a long-term global tourism transport model using a behavioural approach: Implications for sustainable tourism policy making. *Journal of Sustainable Tourism, 21*(7), 1049–1069.

Ram, Y., Nawijn, J. and Peeters, P.M. (2013). Happiness and limits to sustainable tourism mobility: A new conceptual model. *Journal of Sustainable Tourism, 21*(7), 1017–1035.

1 Why tourism mobility behaviours must change

Scott A. Cohen, James E.S. Higham, Paul Peeters and Stefan Gössling

Introduction

There now exists a general scientific consensus that anthropogenic climate change is an inescapable reality (IPCC, 2007). The climate science has been subject to, and withstood, "withering scrutiny" (Garnaut, 2008). The consequences of climate change – social, economic, environmental – will be far reaching (Stern, 2007). The critical challenge that must be taken up without delay is to achieve "radical emission reductions" in all sectors of the economy, and across all aspects of society. The climate crisis, which demands the transformation of our lives and societies (Monbiot, 2007), raises difficult questions for consumer-based neoliberal western societies (Harvey, 2011; Stern, 2007). One important but problematic aspect of the required transformation relates to contemporary western mobility (Gössling *et al.*, 2010). In singling out transport, Cuenot (2013, p. 22) of The International Energy Agency suggests that "Transport offers the easiest path for reducing oil dependency in theory: simple readily available solutions promise a 30% to 50% improvement in fuel economy, depending on the country, while reducing carbon emissions by several giga-tonnes of CO_2 each year". Wheeller (2012, p. 39), however, focusing on tourist transport, unpacks a simple paradox: "All tourism involves travel: all travel involves transport: no form of transport is sustainable: so how on earth can we have sustainable tourism?" While some modes of transport (e.g. human, electrical, solar powered) are more sustainable than others, the sustainability of high volume, high velocity, long distance transportation is clearly coming under increasing scrutiny (Peeters and Dubois, 2010).

The situation is particularly acute in the case of discretionary air travel (Cohen *et al.*, 2011; Gössling *et al.*, 2010). Monbiot (2007) highlights the considerable challenge associated with mitigating aviation greenhouse gas (GHG) emissions, given high current and projected growth in demand for air travel, and the absence of significant scope for further technical gains in aircraft efficiency (Scott *et al.*, 2010). In the absence of "game-changing" innovations in transport technology, it is clearly evident that the United Nations World Tourism Organization (UNWTO) *Tourism Barometer 2012* forecast of 1.8 billion international travellers by 2030 is incompatible with carbon mitigation. Western governments and the industry have to date been unwilling – or unable – to make meaningful responses to the tourism transport emissions challenge.

The continuing inability to bring aviation into emission trading schemes (ETS) is indicative of this impasse (Duval, 2013). As many other sectors actively respond to the call for radical emissions reduction (Scott, 2011; Scott *et al.*, 2012), tourism could find itself generating up to 40 per cent of global carbon emissions by 2050 (Dubois and Ceron, 2006; Gössling and Peeters, 2007). This failure of response is producing an industry of environmental disregard and neglect, with contemporary tourism that may be considered profligate and dissolute.

It is clearly evident that "technology and management will not be sufficient to achieve even modest absolute emission reductions" (Gössling *et al.*, 2010, p. 119). This, according to Gössling *et al.* (2010), confirms that social and behavioural change is necessary to achieve climatically sustainable tourism. Indeed the UNWTO concedes that climatically sustainable tourism requires fundamental shifts in consumer behaviour (UNWTO-UNEP-WMO, 2008). However, reliance upon shifts in behaviour raises its own issues and challenges (Semenza *et al.*, 2008). Despite evidence of growing public awareness of the impacts of air transport on climate change (Hares *et al.*, 2010; Higham and Cohen, 2011) there remains an alarming disconnection between attitudes and (tourist) behaviour (Miller *et al.*, 2010). Thus, an increasingly informed and concerned public, which is beginning to internalise the realities of the climate crisis (Cohen and Higham, 2011), displays few signs of behaviour change (Barr *et al.*, 2010; Higham *et al.*, 2014; McKercher *et al.*, 2010). The efficacy of individual consumers bearing the costs (social, economic) and responsibilities (psychological, behavioural) of a profoundly (environmentally) unsustainable industry is clearly open to question.

From this overall context, the *Freiburg 2012* workshop, held in Freiburg im Breisgau in southern Germany (3–5 July, 2012) set out to explore the psychological and social factors that both contribute to and inhibit behaviour change vis-à-vis sustainable (tourist) mobility. The workshop provided an opportunity to advance a rigorous and theoretically informed knowledge base and research agenda for effective policy interventions to address tourism's contribution to climate change. Such insights are of importance to policy makers, as policy interventions will be less effective if not based on a rigorous understanding of tourist behaviour and psychology. These understandings are needed to negotiate or remove barriers that policy makers may perceive in implementing stronger mitigation measures by signalling how such measures can be made palatable to the public. The psychological and behavioural insights achieved during the workshop informed discussion of government approaches and policy measures that are required to both (a) support the efforts of individuals/consumers to respond to the emission reduction challenge, and (b) conflate the onus of responsibility (and the anxieties of consumption fuelled climate change) from the level of the individual, to the collective levels of government, industry and economy.

The chapters in this edited book arise from the presentations at the *Freiburg 2012* workshop, which explored psychological and behavioural approaches to understanding and governing sustainable tourism mobility. In addition to this volume, the *Journal of Sustainable Tourism* published in 2013 (vol. 21, issue 7) a collection of peer reviewed papers from the workshop. This book includes the papers from this journal special issue (see Acknowledgements), as well as a number of additional contributions. Our intention here is to provide critical insights into psychological and behavioural approaches to understanding (un) sustainable mobility and, in doing so, to inform policy measures that may be required to achieve emission mitigation. We consequently now turn back to the question of why tourism mobility behaviours must change, addressing the issue from a technical perspective, before introducing further the structure and content of this edited book.

Why behaviour must change – no technical solution!

The critical question is whether tourism is able to mitigate its greenhouse gas (GHG) emissions in a way that supports political consensus to avoid dangerous climate change, i.e. stabilising temperature rise below 2°C, compared to pre-industrial levels. Evidence currently suggests otherwise, with projections of emissions from tourism to increase by more than 130 per cent over 2005 levels by 2035 (UNWTO-UNEP-WMO, 2008). Of all elements of tourism, air transport causes up to 75 per cent of the contribution to tourism related climate change (Gössling *et al.*, 2010). Sustainable transportation is now established as *the* critical issue confronting a global tourism industry that is palpably unsustainable, and aviation lies at the heart of this issue (Gössling *et al.*, 2010). Based on a resolution accepted by all International Air Transport Association (IATA) members (IATA, 2013), aviation envisages to achieve "climate neutral growth" up to 2020 and will "aspire" to continue to do so up to 2050. "Climate neutral" is considered keeping emissions at current levels (IATA, 2013, p. 3).

However, even keeping the current contribution of aviation to carbon emissions equal, means the global impact of aviation on climate change will continue to grow. "Safe" levels of emissions require a reduction of current emissions by at least 3 per cent per year (Parry *et al.*, 2008). Tourism's current global contribution to climate change ranges between approximately 5 per cent if measured as CO_2 emissions and up to 14 per cent of global GHG emissions if considering the impact of all GHG in a given year (Scott *et al.*, 2010). Tourism-related CO_2 emissions (leisure, business and VFR – visiting friends and relatives) are mainly a result of transport (72 per cent), followed by accommodation (24 per cent) and local tourism activities (4 per cent; Peeters and Dubois, 2010).

Achieving a sustainable path for climate change would require a very significant reduction of emissions by 2050 (as outlined below). While IATA considers an aviation fuel efficiency improvement of 2 per cent per annum, across all 38 years between 2012 and 2050 to be a realistic goal, the industry expects passenger-kilometers volume growth of between 4.7 per cent per year (Airbus,

2012) and 5.0 per cent per year (Boeing, 2012) over the next 20 years. Despite efficiency gains, emissions from aviation will thus continue to grow. Moreover, historically seen, efficiency gains in aviation have fallen, and always remained lower than growth in the sector (Mayor and Tol, 2010; Penner *et al.*, 1999).

Tourism transport will, in all probability, continue to grow strongly both to 2050 and beyond; growth that only in some moderate economic and population growth scenarios may slow down by the end of the twenty-first century (see Chapter 11 this volume). The main cause is an increase of average travel distance as tourism transport increases at a faster pace than the number of trips in tourism (Peeters and Dubois, 2010; UNWTO-UNEP-WMO, 2008). IATA (2013) acknowledges a gap between their carbon neutral growth vision and the failure of technology to compensate for demand growth, suggesting by way of a solution that aviation purchases credits in the global carbon market. However, as tourism accounts for a growing share of global emissions (Scott *et al.*, 2010), offsetting cannot be feasible in the long term. In sustainable emission scenarios, the cost of carbon may rise to as much as US$250–300 per ton of CO_2 (Edenhofer and Kalkuhl, 2011). It is unlikely that such costs could be sustained by the aviation sector in the absence of very significant demand reductions.

In light of this, the large-scale introduction of biofuels is the only other major innovation suggested by industry (ATAG, 2011; IATA, 2009; WTTC, 2010). The global potential of bio-energy is estimated to be between 50 and 500 exajoules (EJ; Edenhofer *et al.*, 2011), compared with energy requirements of approximately 15 EJ for aviation in 2007 (Rye *et al.*, 2010). With an expected growth in tourism transport to 10–15 times its current volume by 2100, even the most optimistic estimates for biofuel use will remain insufficient for air transport. Furthermore, the overall sustainability of biofuels is contested (Dray *et al.*, 2012; Melillo *et al.*, 2009), with first-generation biofuels causing conflicts with food production, and large-scale production of second-generation biofuels failing because of economic and technological constraints (Timilsina and Shrestha, 2011). Microalgae and other third-generation biofuels are presented to have high yield promises and relatively low costs, but remain, given a wide range of technical obstacles, little more than a future option at this point (Singh and Gu, 2010; Waltz, 2009). It is also important to note that biofuels will not reduce the non-CO_2 radiative forcing of aviation, which are several times aviation's cumulative CO_2 impact (Lee *et al.*, 2010; Owen *et al.*, 2010). The potential of biofuels to reduce the contribution of transport energy to climate change is thus less than 30 per cent, compared to fossil fuels.

It is clear that the current UNWTO (2012) tourism growth scenarios to 2030 are fundamentally incompatible with significant and sufficient reduction of greenhouse gas emissions. Peeters and Dubois (2010) have presented an economically optimised (i.e. maximum net revenues) tourism system to 2050, with the objective of emitting 70 per cent less than the current system. It shows two major outcomes based on current transport technology. Either the current volume of air transport can be maintained (with no further growth), requiring that the majority of trips by car be shifted to rail/coach. The only viable alternative to this is that commercial air transport be reduced to the level of the 1970s (with no further growth in current car use; Peeters and Dubois, 2010). Both scenarios demand fundamental changes in travel behaviour. For tourists, this implies behavioural changes towards less flying, a shift from long haul to medium and short haul travel (i.e. reduction in distance), a modal shift from car to rail and coach, and less frequent travel (with longer length of stay if a reduction in total number of nights is to be avoided; see Chapter 3 this volume). Changes within the sector may include rather complex shifts in the social and psychological valuation of long haul travel (Chapter 13 this volume), and, on the side of tour operators and airlines, diversification of airlines into other modes of transport (railways, buses), logistics, information communication technologies and, for destinations, a change in perspective towards revaluing closer markets. Under such scenarios, the idea of distance decay, i.e. close markets having far stronger relations with destinations than geographically distant markets (Peeters and Landré, 2012), seems undervalued by most destination managers and the tourism industry.

Psychological, behavioural and governance insights into sustainable mobility

This edited book presents 15 further chapters that explore psychological, behavioural and governance dimensions of sustainable mobility as it relates to the global climate crisis. The book is divided into three thematic sections, which mirror the main themes of the *Freiburg 2012* workshop: (1) psychological understandings of climate change and tourism mobilities; (2) behavioural aspects of climate change and tourism mobilities; and (3) governance and policies based upon psychological, behavioural and social mechanisms. The book concludes with the development of a research agenda by the co-editors on governing behaviour change in tourism mobilities.

The chapters responded to the aims of the workshop, which were to explore a number of dimensions surrounding psychological understandings of climate change and tourism mobilities: e.g. climate change attitudes and perceptions, the psychological benefits of tourism mobilities, the attitude-behaviour gap (i.e. cognitive dissonance between understandings of, and responses to, climate change), discrepancies between environmental behaviour at home and away and the psychology of modal shifts. We further aimed to gain behavioural insights into: the factors influencing travel behaviour, social and behavioural practices that entrench patterns of contemporary mobility, interventions for sustained behaviour change, mechanisms for encouraging modal shifts, explanations for the perceived importance of long distance travel, explorations of travel speed/time and the potential for new information technologies (e.g. social media, persuasive technologies) to influence behaviour change. We lastly aimed to identify effective policy mechanisms, as informed by insights from psychology and the social (behavioural) sciences, by considering issues such as governing the travel psyche, targeting policy interventions (e.g. in terms of demography), emission mitigation policies, social marketing/de-marketing, promoting the importance of climatically sustainable mobility in tourism strategy and planning, attending to mechanisms that foster hypermobility (e.g. low cost carrier (LCC) promotions) and addressing institutionally fostered mobilities (e.g. frequent flier and air travel loyalty/reward programmes). The 15 chapters that follow this introduction chapter responded to these aims by addressing many of these issues, and more. Their contributions are outlined in Table 1.1 by their respective part in the book, and the primary issues that they cover.

The first part of the book focuses on psychological understandings of climate change and tourism mobilities. In Chapter 2 Hibbert, Dickinson, Gössling and Curtin provide insights into the "attitude-behaviour gap" in the context of transport's CO_2 emissions. They do so through an *identity* lens, providing an in depth empirical analysis that highlights the likelihood that identity overrides other factors, including environmental impact, in the vast majority of individual travel decision-making processes. The next chapter, by Ram, Nawijn and Peeters considers distance, modal shifts and the "attitude-behaviour gap" in proposing a conceptual "three-gear" model of unsustainable tourist behaviour. Their discussion centres on the notion of *happiness*, which is implicated in number of trips, the consumption of new and novel places, speed, time and distance. They show that happiness, which is central to all elements of tourist experiences, serves as a fundamental barrier to behavioural change. Chapter 4 by Johansson and Gössling investigates the willingness of travellers to donate frequent flyer points for charitable purposes, and in particular to environmental projects. The findings show that travellers prefer to personally use their points and are reluctant to donate, except in cases where they have more miles than they can consume and they would otherwise expire. The authors conclude that addressing the climate impacts of aviation through donations has very limited potential.

In Chapter 5 by Cohen, Higham and Reis, the authors use modern and postmodern sociological theory to interpret the "home" and "away" environmental behaviours of consumers in Australia, Norway and the United Kingdom. They provide empirical insights that demonstrate the alarming degree to which much tourism decision-making is absolved of environmental concern or responsibility. This chapter concludes that significant voluntary behavioural change, in the absence of strong government intervention, in the context of sustainable air travel practices is unlikely. Khoo-Lattimore and Prideaux introduce in Chapter 6 a specific research technique intended to facilitate psychological approaches to understanding sustainable tourism mobility. The authors describe the Zaltman Metaphor Elicitation Technique (ZMET), which is informed by *Freudian psychology* and employs *photography and photo elicitation* to explore thoughts rarely expressed in verbal social exchange. The technique allows access to deep-seated psychological factors, many subconscious or repressed, that are important determinants of complex (tourist) behaviour. In Chapter 7, Malhado, Araujo and Rothfuss take us to South Africa where they surveyed tourists and residents attending the 2010 FIFA World Cup on their transport choices. They bring together research on transport behaviour and mega-events to provide insights to planners for developing measures for mega-event mobility management. The contribution points to the

Table 1.1 Overview of part themes, primary issues covered and associated chapters

Part theme	Primary issues covered	Associated chapters
Part I – Psychological understandings of climate change and tourism mobilities	• Attitude-behaviour gap • Behaviour change • Multiple identities and mobility • The role of happiness • Distance and modal shifts • Willingness to donate frequent flyer miles • Tourism as liminoid space • Home-away gap • Zaltman Metaphor Elicitation Technique (ZMET) • Shaping mega-event mobility choices	Chapter 2 – Hibbert, Dickinson, Gössling and Curtin Chapter 3 – Ram, Nawijn and Peeters Chapter 4 – Johansson and Gössling Chapter 5 – Cohen, Higham and Reis Chapter 6 – Khoo-Lattimore and Prideaux Chapter 7 – Malhado, Araujo and Rothfuss
Part II – Behavioural aspects of climate change and tourism mobilities	• Motives for and effects of carbon offsetting • Redefining *time* through information technologies • The influence of lifestyles on transport demand models • Longer term planning horizons • System dynamics simulation models • Promoting public transport to visitors • Understandings of *distance*	Chapter 8 – Eijgelaar and de Kinderen Chapter 9 – Dickinson, Filimonau, Cherrett, Davies, Norgate, Speed and Winstanley Chapter 10 – Gronau Chapter 11 – Peeters Chapter 12 – Le-Klähn, Hall and Gerike Chapter 13 – Larsen and Guiver
Part III – Governance and policies based upon psychological, behavioural and social mechanisms	• Persuasive communication • Integrative frameworks • Market-based mechanisms • Systems of provision • Structural change • Expanding governance • Future research agenda	Chapter 14 – van der Linden Chapter 15 – Hall Chapter 16 – Gössling, Peeters, Higham and Cohen

power of reliable and efficient information, along with factors such as convenience and safety, in shaping travel mode choices for mega-events.

Part II of the book turns to behavioural aspects of climate change and tourism mobilities, and begins with Eijgelaar and de Kinderen's chapter on motivations for carbon offsetting, and the effects of (non)offsetting, on the travel behaviour of Dutch tourists. The authors find through survey-based data that although off-setters are greener in daily life and feel more responsible than non-offsetters for mitigating their travel emissions, offsetters ironically fly more often and further away. They conclude that rather than leading to a reduction in travel through heightened awareness of travel's impact on carbon footprints, offsets conversely appear as an incentive to travel more. Chapter 9 by Dickinson, Filimonau, Cherrett, Davies, Norgate, Speed and Winstanley explores how personalised travel information, in real time, offers possibilities that may redefine important aspects of travel behaviour. Employing data generated via a range of traditional and emerging approaches (including a purpose built smartphone app) their analysis illuminates the heavy influence that *time*, and competing forms of time, exert upon (un)sustainable travel choices. In Chapter 10, Gronau focuses on lifestyle as a psychological determinant of mobility behaviour. He merges the strengths of quantitative transport forecasting methods with those of qualitative approaches that take account of individualised behaviour, to develop a more integrative approach to modelling transport demand. Such a "subject-oriented" transport demand model is aimed at targeting groups with specific transport products that foster more sustainable leisure and tourism mobility.

Chapter 11 by Peeters highlights the need for wide *planning time horizons* to mirror the period of climate change projections which typically extend to the year 2100. He presents a system dynamics model (SDM), which is specially intended to model insights that extend beyond normal economic equilibrium modelling. His chapter demonstrates that mitigation of tourism CO_2 emissions is extremely unlikely in the absence of strong policy intervention. In Chapter 12, Le-Klähn, Hall and Gerike examine the importance of public transport use in the sustainable development of urban destinations, by focusing on the case of Munich. The authors show how tourists use public transport in the city differently from residents, and draw out the importance of these differences for public transport management and marketing. In the following chapter by Larsen and Guiver, the authors address (tourist) mobility and *distance*. They directly attend to the fact that with increasing speed, particularly on the coat tails of aviation, the average distances consumed by tourists has increased dramatically. Larsen and Guiver conceptualise distance and then, using discourse analysis, provide empirical insights into perceptions and performances of distance among members of the Dutch travelling public.

The final part of the book focuses on governance and policies based upon psychological, behavioural and social mechanisms. This part begins with van der Linden's chapter on public communication about climate change, where he reviews the cognitive, experiential and normative approaches on which past and present climate change communication campaigns have been based. He uses this evidence to propose a new integrated conceptual framework that can guide the design of future public interventions which narrow the gap between climate change communication and individual behaviour change. Many of the threads of discussion presented in the preceding chapters are usefully tied together in Chapter 15, where Hall confronts the need for expanding governance for sustainable mobility. Hall observes that social/psychological approaches to sustainable consumer behaviour, in isolation, do not question the systems of provision that give rise to the social practices of tourist travel consumption. Hall does not deny the importance of mechanisms that may influence consumer decision-making: nudging, social marketing, education and other market-based mechanisms. He argues though, that these need to be viewed as part of a suite of approaches that include structural change. Together, these 16 chapters, closed by our conclusion that develops a research agenda on governing behaviour change in tourism mobilities, provoke and encourage further critical contemplation of the psychological and behavioural complexities of climate change, tourism and sustainability mobility at both the individual and sectorial/institutional levels.

References

Airbus. (2012). *Navigating the future. Global market forecast 2012–2031*. Paris: Airbus S.A.S.

ATAG. (2011). *Powering the future of flight. The six easy steps to growing a viable aviation biofuels industry* (No. web version). Geneva: Air Transport Action Group (ATAG).

Barr, S., Shaw, G., Coles, T. and Prillwitz, J. (2010). "A holiday is a holiday": Practicing sustainability, home and away. *Journal of Transport Geography*, *18*(3), 474–481.

Boeing. (2012). *Current market outlook 2012–2031*. Seattle: Boeing Commercial Airplanes. Marketing.

Cohen, S.A. and Higham, J.E.S. (2011). Eyes wide shut? UK consumer perceptions on aviation climate impacts and travel decisions to New Zealand. *Current Issues in Tourism*, *14*(4), 323–335.

Cohen, S.A., Higham, J.E.S. and Cavaliere, C.T. (2011). Binge flying: Behavioural addiction and climate change. *Annals of Tourism Research*, *38*(3), 1070–1089.

Cuenot, F. (2013). Driving need for fuel economy: Improve my ride. *International Energy Agency*, *4*(22).

Dray, L.M., Schäfer, A. and Ben-Akiva, M.E. (2012). Technology limits for reducing EU transport sector CO_2 emissions. *Environmental Science & Technology*, *46*(9), 4734–4741.

Dubois, G. and Ceron, J.P. (2006). Tourism/leisure greenhouse gas emission forecasts for 2050: Factors for change in France. *Journal of Sustainable Tourism*, *14*(2), 172–191.

Duval, D.T. (2013). Critical issues in air transport and tourism. *Tourism Geographies*, *15*(3), 494–510.

Edenhofer, O. and Kalkuhl, M. (2011). When do increasing carbon taxes accelerate global warming? A note on the green paradox. *Energy Policy*, *39*(4), 2208–2212.

Edenhofer, O., Pichs-Madruga, R., Sokona, Y., Seyboth, K., Matschoss, P., Kadner, S., ... and von Stechow, C. (eds). (2011). *IPCC special report on renewable energy sources and climate change mitigation*. Cambridge, UK: Cambridge University Press.

Garnaut, R. (2008). *The Garnaut climate change review*. Cambridge, UK: Cambridge University Press.

Gössling, S., Hall, C.M., Peeters, P. and Scott, D. (2010). The future of tourism: Can tourism growth and climate policy be reconciled? A climate change mitigation perspective. *Tourism Recreation Research*, *35*(2), 119–130.

Gössling, S. and Peeters, P. (2007). It does not harm the environment! An analysis of industry discourses on tourism, air travel and the environment. *Journal of Sustainable Tourism*, *15*(4), 402–417.

Hares, A., Dickinson, J. and Wilkes, K. (2010). Climate change and the air travel decisions of UK tourists. *Journal of Transport Geography*, *18*(3), 466–473.

Harvey, D. (2011). *The enigma of capital and the crises of capitalism*. Oxford: Oxford University Press.

Higham, J.E.S. and Cohen, S.A. (2011). Canary in the coalmine: Norwegian attitudes towards climate change and extreme long-haul air travel to Aotearoa/New Zealand. *Tourism Management*, *32*(1), 98–105.

Higham, J.E.S., Cohen, S.A. and Cavaliere, C.T. (2014). Climate change, discretionary air travel and the "flyers" dilemma. *Journal of Travel Research*, doi: 10.1177/004728 7513500393.

IATA. (2009). *The IATA technology roadmap report*. Montreal: IATA.

IATA. (2013). *Resolution on the implementation of the aviation "CNG 2020" strategy*. Retrieved from http://ec.europa.eu/clima/consultations/0020/organisation/iata_en.pdf.

Intergovernmental Panel on Climate Change (IPCC) (2007). *Climate change 2007: Synthesis report. An assessment of the Intergovernmental Panel on Climate Change*. Retrieved from www.ipcc.ch/.

Lee, D., Pitari, G., Grewe, V., Gierens, K., Penner, J., Petzold, A., ... and Sausen, R. (2010). Transport impacts on atmosphere and climate: Aviation. *Atmospheric Environment*, *44*(37), 4678–4734.

Mayor, K. and Tol, R.S.J. (2010). Scenarios of carbon dioxide emissions from aviation. *Global Environmental Change*, *20*(1), 65–73.

McKercher, B., Prideaux, B., Cheung, C. and Law, R. (2010). Achieving voluntary reductions in the carbon footprint of tourism and climate change. *Journal of Sustainable Tourism*, *18*(3), 297–317.

Melillo, J.M., Reilly, J.M., Kicklighter, D.W., Gurgel, A.C., Cronin, T.W., Paltsev, S., ... and Schlosser, C.A (2009). Indirect emissions from biofuels: How important? *Science*, *326*(5958), 1397–1399.

Miller, G., Rathouse, K., Scarles, C., Holmes, K. and Tribe, J. (2010). Public understanding of sustainable tourism. *Annals of Tourism Research*, *37*(3), 627–645.

Monbiot, G. (2007). *Heat: How to stop the planet burning*. London: Penguin Books.

Owen, B., Lee, D.S. and Lim, L. (2010). Flying into the future: Aviation emissions scenarios to 2050. *Environmental Science & Technology*, *44*(7), 2255–2260.

Parry, M., Palutikof, J., Hanson, C. and Lowe, J. (2008). Squaring up to reality. *Nature Reports Climate Change*, 68.

Peeters, P. and Dubois, G. (2010). Tourism travel under climate change mitigation constraints. *Journal of Transport Geography*, *18*(3), 447–457.

Peeters, P. and Landré, M. (2012). The emerging global tourism geography – an environmental sustainability perspective. *Sustainability*, *4*(1), 42–71.

Penner, J.E., Lister, D.H., Griggs, D.J., Dokken, D.J. and McFarland, M. (eds). (1999). *Aviation and the global atmosphere: A special report of IPCC working groups I and III*. Cambridge, UK: Cambridge University Press.

Rye, L., Blakey, S. and Wilson, C. (2010). Sustainability of supply or the planet: a review of potential drop-in alternative aviation fuels. *Energy & Environmental Science, 3*(1), 17–27.

Scott, D. (2011). Why sustainable tourism must address climate change. *Journal of Sustainable Tourism, 19*(1), 17–34.

Scott, D., Hall, C.M. and Gössling, S. (2012). *Tourism and climate change: Impacts, adaptation and mitigation.* London: Routledge.

Scott, D., Peeters, P. and Gössling, S. (2010). Can tourism deliver its "aspirational" greenhouse gas emission reduction targets? *Journal of Sustainable Tourism, 18*(3), 393–408.

Semenza, J.C., Hall, D.E., Wilson, D.J., Bontempo, B.D., Sailor, D.J. and George, L.A. (2008). Public perception of climate change: Voluntary mitigation and barriers to behavior change. *American Journal of Preventative Medicine, 35*(5), 479–487.

Singh, J. and Gu, S. (2010). Commercialization potential of microalgae for biofuels production. *Renewable and Sustainable Energy Reviews, 14*(9), 2596–2610.

Stern, N.H. (2007). *The economics of climate change. The Stern review.* Cambridge, UK: Cambridge University Press.

Timilsina, G.R. and Shrestha, A. (2011). How much hope should we have for biofuels? *Energy, 36*(4), 2055–2069.

Waltz, E. (2009). Biotech's green gold? *Nat Biotech, 27*(1), 15–18.

Wheeller, B. (2012). Sustainable mass tourism: More smudge than nudge – the canard continues. In T.V. Singh (ed.), *Critical debates in tourism* (pp. 39–43). Bristol: Channel View Publications.

United Nations World Tourism Organization. (2012). *UNWTO tourism highlights 2012 edition.* Retrieved from http://dtxtq4w60xqpw.cloudfront.net/sites/all/files/docpdf/unwtohighlights12enlr_1.pdf.

UNWTO-UNEP-WMO. (2008). *Climate change and tourism: Responding to global challenges.* Madrid: UNWTO.

WTTC. (2010). *Climate change: A joint approach to addressing the challenge.* London: World Travel & Tourism Council.

Part I

Psychological understandings of climate change and tourism mobilities

2 Identity and tourism mobility

An exploration of the attitude–behaviour gap

Julia F. Hibbert, Janet E. Dickinson,
Stefan Gössling and Susanna Curtin

Introduction

Tourists are becoming increasingly mobile. As physical tourism mobility is largely based on the use of fossil fuels, tourism has become a significant contributor to global climate change (UNWTO, UNEP and WMO, 2008). Options to reduce emissions from tourism through technology are promising, but unlikely to achieve absolute emission reductions in line with international efforts (Scott *et al.*, 2010). It has consequently been argued that climatically sustainable tourism mobility demands behavioural change, which appears difficult to achieve (McKercher *et al.*, 2010). Why do we continue to participate in environmentally harmful activities, even though consumers are aware of these interlinkages (Giddens, 2009)?

From a sociological point of view, it is suggested that identity issues lie at the heart of our desire for greater tourism mobility (e.g. Desforges, 2000). It is well documented that travel is important in shaping the perception of self through experiences of other people and places (Bruner, 1991; Crompton, 1979; Desforges, 2000; Noy, 2004; Urry, 2000). However, very little has been written about how the perception of self influences travel choices and drives our desire for travel. This chapter helps fulfil that research gap.

Currently, society views highly mobile lifestyles in a positive light: high mobility is associated with a high degree of "meetingness", i.e. an individual's standing in society is reflected in mobility patterns, ultimately necessitating air travel (Urry, 2011). This is also demonstrated by airlines' frequent flyer programmes which "reward and thus increase interest in mobility" (Gössling and Nilsson, 2010, p. 242). It could be argued that such marketing strategies hold some responsibility for the status implied in highly mobile lifestyles through their inclusion of VIP lounges for some members and added status attached to long-haul travel and exotic international tourism, a positive identity marker for most people (Gössling and Nilsson, 2010). Through tourism choices people seek to reinforce or develop particular identity markers and, therefore, a desired identity appears to affect their decisions and behaviour (Markus and Nurius, 1986). Given that individuals can be persuaded to choose low carbon tourism products, it is vital to examine the underlying identity processes at work. There is a general assumption that individuals predisposed to environmental concern will modify their behaviour accordingly, however, this has not proved to be a potent force in other areas of life, such as car use (Dickinson and Dickinson, 2006; Schwanen and Lucas, 2011; Steg and Vlek, 2009).

Steg and Vlek (2009) suggest that to change behaviour, it is necessary to understand the factors underlying the behaviour. Factors determining behaviour include (1) perceived costs and benefits, (2) moral and normative concerns, (3) affect, (4) contextual factors and (5) habits. For instance, with regard to costs and benefits, travel mode choice is dependent on variables such as money, effort and the perceived benefits of tourism (Hares *et al.*, 2010).While higher moral and normative concern for the environment is associated with more pro-environmental behaviour, in tourism, climate change awareness appears to have little effect on tourism consumption (e.g. Anable *et al.*, 2006; Dickinson *et al.*, 2009; Eijgelaar *et al.*, 2010; Hares *et al.*, 2010; McKercher *et al.*, 2010). Here contextual factors, such as lack of alternatives to air travel, and habitual travel choices, seem to steer people with environmental concern to unsustainable choices (Hares *et al.*, 2010).

This chapter focuses on the gap between environmental awareness and lack of behavioural change. In addition to the factors outlined above, it argues that identity plays a significant role in explaining the attitude–behaviour gap (Stets and Biga, 2003). It therefore questions the assumption that behaviour change can be effectively managed given individuals' needs to manage a variety of identity interests. Thus, the chapter analyses the assumption that behaviour can change, based on an identity perspective.

Identity, tourism and environmental behaviours

While there is a large body of knowledge on identity based in psychology and sociology, this chapter draws predominantly on the social psychology literature and some strands within sociology. It, therefore, focuses predominantly on the individual and the construction of "self" relative to others. "Self concept" is how individuals see and describe themselves with reference to others in society. It is defined through group membership and the acknowledgement of necessary characteristics to belong (Hogg and Terry, 2000; Onkvisit and Shaw, 1987). The human need for connectedness is deeply enrooted in our social evolution (Cacioppo and Patrick, 2008; Miller, 2007). Self-image is hard to distinguish from self-concept; Graeff (1996) uses the terms interchangeably and self-image is generally used in more recent studies (Fein and Spencer, 1997). Hogg and Abrams (1988) propose that self-conception could be placed on a continuum ranging from exclusively social to exclusively personal identity and that the social setting in which the individual is placed in at any moment in time will dictate which self-conception is most prominent. Self-concept is, therefore, multi-faceted (Gergen, 1971). Individuals may also define themselves through who they aspire to be: such "possible selves" act as "incentives toward future behaviours, representing the individual's significant hopes, fears, aspiration and fantasies ... possible selves may be seen as acting in the role of a powerful motivational force" (Morgan, 1993, p. 430). Crucially for this chapter, Morgan acknowledges that "certain disposition behaviours, such as environmentally friendly activities, may be motivated by a desire to avoid or approach possible selves, rather than being motivated by perceptions of the current self" (Morgan, 1993, p. 431).

Linked to understanding tourists' motivations for travel is the work of Pearce and Lee (2005). Their study aims to understand a tourist's travel career pattern, a development of Pearce's work on the travel career ladder (see: 1988, 1991, 1993). Understanding the travel career pattern (Pearce and Lee, 2005) allows attention to be paid to overlapping and multiple motivations for travel. Pearce and Lee (2005) found that self-actualisation and self-development was a major motivator for travel and that travel to a culturally different destination was one way of achieving this.

Self-concepts can conflict, resulting in dissonance or an identity crisis, where an individual is uncomfortable with contrasting self-concepts (Hogg and Abrams, 1988) and seeks to justify their behaviour to themselves. Studies have found evidence of conflicts between sustainable "at home" and tourist lifestyles with individuals using sustainable practices at home to justify unsustainable flying (Randles and Mander, 2009) and indicating a willingness to pay higher taxes as a "penance" (Barr *et al.*, 2009).

Self-concept evolves through the process of self-categorisation which involves making comparisons between self and others. The theory suggests that through categorisation of "ingroups" and "out-groups" and the normative behaviour necessary for group membership, it is possible to replicate the norms or to avoid them in order to associate or disassociate with a particular group (Hogg and Terry, 2000). For instance, a tourist may choose a particular style of holiday to reflect the norms of a group they seek association with. Self-presentation is related to how others may influence an individual's behaviour (Ellemers *et al.*, 1999). When a person is aware of an "audience" they can choose to accentuate or subdue certain elements of their identity in order to present the identity that they feel most appropriate for this group or audience (Goffman, 1959).

Social identity more explicitly relates to the various groups to which an individual belongs. It is not simply about relationships with other group members but also involves the relationships and distinctions to those not in the group (Hogg and Terry, 2000). Social identity is constructed through self-descriptions that relate to the individual's membership in various social categories. These categories are wide ranging and include enforced categories, such as nationality, sex and race, and chosen ones, such as occupation and membership of sports teams (Hogg and Abrams, 1988). Within sociology, Giddens (1991) refers to "self identity" as a reflexive project involving appraisal of others and oneself. "A person's identity is not to be found in behaviour, nor – important though this is – in the reactions of others, but in the capacity *to keep a particular narrative going*" (Giddens, 1991, p. 54, original emphasis). Smith and Sparkes (2006, p. 169) support Giddens' statement and present the argument that identity is created through narratives, "the stories that people tell and hear from others form the warp and weft of who they are and what they do". Both Burke (2001) and Giddens (1991) acknowledge that identity is relative to those around us, therefore, identity is created and maintained through reflexively appraising oneself and those around. Reference groups could be family, friends, those who share the same leisure interests, or wider society. Kitzinger and Wilkinson (1996, p. 8) state that "'we' use the 'Other' to define ourselves: 'we' understand ourselves in relation to what 'we' are not". This means that identities are culturally specific and will have different meanings across cultures. In this chapter, identity is understood to be the idiosyncratic descriptions of self which are expressed through narratives either told to oneself or to others (Giddens, 1991; Hogg and Abrams, 1988; Smith and Sparkes, 2006).

In contemporary society, an individual has identity choices: these are "not only about how to act but who to be" (Giddens, 1991, p. 81). Tourism provides a site of consumption in which to perform our identity choices (Gram, 2005). It provides a platform to accentuate chosen elements of identity and avoid or develop possible selves. At present, literature relating to identity and tourism tends to focus on identity issues such as "finding yourself" through tourism (see Bruner, 1991; Fullagar, 2002; Noy, 2004). Noy's (2004) study of Israeli backpackers' narratives demonstrates how these travellers used tourism to construct their identity, returning from their trips as "changed" people. Bruner (1991) suggests that this is used as a selling point in tourist brochures but that it is inevitably the host populations in non-western cultures that emerge "changed". Tourism is also described as affirming identity: Thurlow and Jaworski (2006, p. 100) discuss how frequent flyer programmes create or reinforce status and desire to travel and "reward" travelling: "regular customers are declared 'elite' and ... this status is then fabricated and regulated". Identity is, therefore, thought to be important to tourism research. In contrast, research undertaken by Cohen (2010) proposes that it might be unrealistic to use travel to "find yourself" because the self is constantly evolving and changing: the lifestyle travellers of Cohen's research are simply acting out the expected discourse of finding themselves.

For many, holidays play an important part of who they are. Memories are not always just stored away, they can shape the future self of the traveller (Bruner, 1991; Desforges, 2000; Fullagar, 2002; Noy, 2004; Thurlow and Jaworski, 2006). Desforges (2000) suggests that understanding identity can give insight into tourism consumption because, by understanding the person and their needs and desires, it could be possible to predict their future travel patterns. If the tourism identity process that an individual goes through could be understood, it might be possible to influence desired identities and consequently travel behaviour.

Given that identities are an intrinsic part of who we are, they are linked to every part of our lives. This means that identity issues are at play when it comes to environmental concern. Becoming "green" can be a lifestyle choice and green identities are presented through behaviour and consumption choices (e.g. clothes, food, travel). According to Horton (2003), for identities to be performed there needs to be a stage and props. The stage can be social settings and props consumption choices. The gap in environmental attitudes and behaviours comes about when people have multiple identities requiring differing performances. Identities are contextual, with some dormant while others come to the fore.

Palmer (2005) suggests that in various situations different identities may be drawn upon to present oneself in a certain way. In situations where actions do not meet the requirements of the identity "script", the individual will need to reason with themselves about why their behaviour is acceptable. Thus while a long-haul flight may break green cultural codes, a tourist may draw on other identity resources to justify this, such as the need to maintain close family bonds. Frew and Winter (2010) observed that tourists' concerns about time and cost of travel, family commitments and the simple desire to "see the world" can outweigh any consideration of the environmental impact of their travel.

There is considerable literature about the gap between attitude and behaviour linked to environmental awareness, but little success in bridging this gap. Stets and Biga (2003) suggest that previous research does not take the role of identity into account, arguing that psychology-based attitude theory demonstrates how individuals make choices based specifically on the object or situation they are in. On the other hand, identity theory rooted in sociology and social psychology is based on how choices are embedded in wider social settings and, given that individuals operate in wide-ranging social settings, there are multiple identities thus making choices multi-faceted. They summarise the deficiency, stating that "identity theory links individuals to the larger social structure in ways that attitude theory neglects" (Stets and Biga, 2003, p. 399). Bond and Falk (2012) introduce a model of identity-related tourism motivation. They even suggest that "ALL tourist experiences are in some way motivated by the individual's self-perceived identity related needs" (Bond and Falk, 2012, p. 10). This suggests that an understanding of identity-related tourism decisions could aid the understanding of the attitude–behaviour gap. This chapter stimulates that debate and opens the door for further research.

Method

The earlier discussion demonstrated the complexity and fluid nature of identity. That, and the role of context in identity presentation call for an exploratory research approach. Given the focus on the individual, rather than group identity, interviews were considered the primary means to generate data. A narrative interview approach had the capacity to enable participants to give an account of historic events, allowing participants to reflect on their tourist travel by recounting their stories, an approach much used in identity research (Holloway and Wheeler, 2010; Kraus, 2006; Smith and Sparkes, 2006; Van De Mieroop, 2009). This raises various issues about the identity presented. For example, is this the participant's "real" identity or one formulated in response to the interview context? This question, though vexing, is largely irrelevant: all our accounts of past events are, to some extent, managed for the audience (Chase, 2005). It would be unrealistic to think you could capture a "real" identity and the act of presenting a story reveals many identity markers (McAdams, 1993) that are useful to understand the role of identity in tourist travel.

A two-stage process was undertaken. In a first interview, participants recounted their tourist travel history from childhood to present day. A true narrative approach engages participants with one main question, in this case, "tell me about all the holidays you have been on throughout your life course". However, it is unrealistic to expect participants to talk at length without researcher intervention (Bryman and Bell, 2011; Reissman, 1993). To aid the flow of the interview, participants were encouraged to recount their travel history chronologically from childhood, through teens, early adulthood and adult life, with the interviewer providing relevant prompts to capture the required detail. Following analysis of the first interview, a second interview was conducted. This returned to key themes from the first interview and asked the participants to reflect on climate change issues in the context of their travel.

Participants were recruited in Dorset, United Kingdom, selecting adults who had achieved a degree of stability in both their identity and their adult travel behaviour. Therefore, all participants were aged over 25, and British in order to avoid cultural comparisons (see profiles in Table 2.1). A minimum age for recruitment was set: it was thought that interviewees over 25 would have been through multiple instances of identity formation. Participants were initially selected using a snowballing technique but it was rapidly clear that this strategy was leading to a relatively homogenous sample with little variability with respect to key research themes. Therefore, sampling became more purposeful and sought out individuals who would bring new knowledge to the study, particularly in relation to the theme of identity with respect to tourism impacts on climate change. This was achieved by contacting local environmental groups and requesting potential participants make contact with the researcher. In total, 24 participants took part in two interviews which ranged in length from approximately 30 minutes up to two hours.

All interviews were recorded following participant consent and transcribed. A narrative analysis approach initially employed thematic analysis followed by dialogic/performance analysis (Reissman, 2008). In dialogic/performance analysis meaning is created through the interaction of the participant with the interviewer with a recognition that the interviewee is "performing" for the interviewer. While thematic analysis focuses on "what" is said, dialogic/performance analysis focuses on "how" it is said and pays close attention to the use of phrases, repetition, pauses and use of tense (Reissman, 2008). Attention is, therefore, paid to how the participant attempts to persuade the listener through their story (Reissman, 2008). Results presented here come primarily from the thematic analysis. Initially each participant was treated as an individual case, with interviews analysed as a whole. Subsequent analysis sought to identify commonalities and discordances across cases.

Table 2.1 Profile of participants

	Name (pseudonym)	Gender	Age	Occupation
1	Martin	Male	67	Retired RAF pilot
2	Trisha	Female	47	Hair salon owner
3	Penny	Female	61	Retired teacher
4	James	Male	56	Retired banking industry
5	June	Female	59	University lecturer
6	Stephanie	Female	38	University administrator
7	Jill	Female	62	Semi-retired teacher
8	Claire	Female	28	Building surveyor
9	Heather	Female	65	Housewife
10	Paul	Male	65	Retired from IT industry
11	Susie	Female	31	Unemployed teacher
12	Mark	Male	29	Goods driver
13	Katherine	Female	66	Retired teacher
14	Simon	Male	33	Journalist
15	Liz	Female	61	Retired civil servant
16	Dennis	Male	59	Civil servant
17	Stuart	Male	58	Retired transport planner
18	Tom	Male	47	Owner of local tour company
19	Michael	Male	64	Retired school inspector
20	Beth	Female	31	Researcher
21	Richard	Male	63	Travel writer
22	Samantha	Female	34	Journalist/writer
23	Reece	Male	46	Town planner
24	Daniel	Male	63	Retired local government officer

Findings and discussion

Many interviewees spoke of holidays to visit loved ones. For example, Susie talks about the holiday with her (now ex) husband:

> … his dad lived in Milan so we went to Italy and we went to Holland to see my aunt and we went to Spain to see my uncle, funny how it was always family…
>
> (Susie)

The data reinforced existing studies (see, for example, Moscardo *et al.*, 2000; Seaton and Palmer, 1997) demonstrating mobility patterns shaped by the desire to visit friends or relatives.

An increasingly globalised world has resulted in increased global networks meaning it is possible to have friends and family spread across thousands of miles (Axhausen, 2005); visiting friends and relatives (VFR) is now the second most important motive for travel after leisure and before business travel (UNWTO, 2012). While global networks can evolve without travel (Hannam *et al.*, 2006, p. 2), face-to-face contact has remained important for maintaining relationships and trust (Axhausen, 2005; Urry, 2003). While the frequency of VFR travel is not surprising, it would be pertinent to ask why this is such an important form of mobility. Identity-related factors may explain this. Arguably, the strongest bonds in our social networks consist of family. Visiting family is consequently an important motive for travel: we usually strive to have social standing in family networks. Physical meetings with family may thus be a precondition for presenting specific identities, and to receive reaffirmation of our role and standing within these networks (Cacioppo and Patrick, 2008).

Social identity performance and tourism decisions

Holidays are a way of performing certain selves and allow us to nurture or maintain relationships which are important to self. For example, Penny's eldest daughter Rachel has travelled extensively and Penny shows pride when talking about this: "Rachel, she's travelled the world you know." Being well travelled is a signifier of success and hence status (Urry, 2011), which parents deem desirable for their children, signalling success as parents and providing a positive identity. This would be particularly important for Penny whose identity is heavily influenced by relationships with significant others, demonstrating this through constant reference to her family, often by name, and even stating family members' opinions before her own. Rachel's travel stories and photos have opened Penny's eyes and created a desire for travel. Rachel may emigrate to Australia in the future and this will play a part in Penny's future tourism mobility:

> Rachel was more or less offered a job back there any time she wanted to go ... I would only go if they went back there. It's not a destination I particularly want to see but, obviously, if they were there, I would make the effort to go because it is such a long way.

While there is nothing new in the notion of an individual's tourism mobility being influenced by the desire to visit friends and relatives, the possibility that this travel is fuelled by the need to reinforce an identity is important. Using Stets and Biga's (2003, p. 401) definition of identity as "a set of meanings attached to the self that serves as a standard or reference that guides behaviour" we can suggest that Penny, and other interviewees, are creating meaning related to relationships. Their identities stem from being connected physically and emotionally to significant others and this inevitably involves travel.

VFR travel also provides a framework for identity performance. Claire has undertaken many holidays to visit family and friends, even spending three weeks in Malaysia visiting an ex-boyfriend who was travelling for a year. Claire presents her trips as exotic, dangerous and adventurous. However, a thread running through her stories is one of seeking a "home away from home", a place where she feels as comfortable and safe as if she were at home. Her explanation for visiting Australia was: "I think because Ian had an uncle out there, that was a good starting block to go and do somewhere like that."

Claire was going a long way from home and it was important for her to have a security blanket should things go wrong. For Claire, friends and relatives were often a seed for a bigger trip. The relative knew where to go, where to avoid and provided other sound advice. Knowing "a local" also provides a base to return to if things should go wrong – as things sometimes do when on holiday (Löfgren, 2008). In Claire's case the relative met her at the airport, providing an extra security measure and equipping her for the trip before venturing out alone: "So we got in to Australia and Ian's uncle picked us up and we actually stayed with him for two weeks and then he set us up with a car, bought us a car and got us sorted."

Claire's strongest memory was a week in a caravan during her Australia trip:

> ... it was my best week, because you just felt like you were at home, it was nice and chilled, we actually stayed in a stationary caravan for the week, which when you've been in and out of hostels and everything else, ok, they're probably nicer but when you've got a stationary caravan and you've got your little kitchen area and everything else you've got a chance to make it feel like a home for a week and I think that's a part of it.... And I think, I don't know, it's strange that we go away to find somewhere different but when you find somewhere that's home from home it always seems to be one of your favourite places, so yeah. That's probably one of my strongest holiday memories.

This admission causes Claire some identity conflict. She wants to be seen as adventurous but the reality is more ordinary. She plays with an identity that goes beyond her "true" identity. Here VFR travel bridges a gap between the need for security (emphasised by her constant reference to home – a place of security) and the desire to project an adventurous identity. Given global networks, this generates both opportunities and desire for long-haul travel.

Throughout both her interviews June referred to the idea of togetherness as being central to her holidays, with her family as a whole or just with her husband. She highlights the importance of this through the use of repetition while she is telling her story. For example:

> I do remember the idea of being away as a family as being something nice and certainly that did probably influence the holidays we took when I had a young family, things like caravan holidays, holidays where we were all *together* in the same place. It didn't have to be grand. We did have some holidays abroad but even when we were abroad, it was all about being *together* and swimming *together* and walking out *together* so I think they are influential but you don't know it at the time. [Emphasis added by author]

June suggests that it did not matter where they were or what they were doing, as long as they were together. This is supported by Small (2002) and Haldrup

(2004, p. 433), who suggests "family based vacationing is more concerned with the extra ordinary ordinariness of personal social relations than with the documenting and gazing at spectacular sights". Stephanie emphasises the importance of family togetherness over place in the following:

> We'd be going ... somewhere in a tent but they [friends] went on a tour of the west coast of the States so you know as far as I was concerned there was no comparison, but you know I didn't think "oh no that's not fair I wish I was doing that" because I actually knew I would rather be with my family than with her family so it wasn't necessarily where she was going. Yeah sure I might have liked to have tried it but I wouldn't want to swap it for what I was doing so looking back I certainly wouldn't have swapped it.

Trauer and Ryan (2005) suggest that place significance is created through the meaningful relationships that take place in the location and also the enduring memories from the experiences. June also acknowledges the importance of the journey in assisting the feeling of "togetherness": "First of all there is the travel down, travelling down in the car, wherever we're going." This is similar to the example in Dickinson *et al.* (2011) and also by Sheller (2004, p. 44) who proposes that "private cars are now also becoming mobile leisure spaces..."

Togetherness is not just physical but also psychological. Holidays can bring us back in touch with basic needs and commonalities that are often lost in modern life where we are isolated in big houses and through technology brought about by greater prosperity. Holidays "help create or re-create a common feeling of unity" (Shaw, 2001, p. 128). This unity while on holiday presents a powerful motivation for continuing to travel. The association of the idea of togetherness with the self and tourism is through the creation of a "family identity" and the security that brings.

Jill appears to use travel as a way of engaging in various communities or subcultures, something that has carried on from her childhood family holidays following motorcycle scrambling rallies.

> They [her childhood holidays] were all based on motorcycling scrambling and motorcycle rallies.... So the smell of the diesel from the motorbikes was really nice, they used to have campfires where everybody would come and congregate and ummmmm sort of wooden huts which people would sit in and eat so it was all very basic but errrmmm yeah I quite enjoyed that, that was the purpose of it [the holiday] really. I think they used to follow those rallies.

This passage illustrates an intimacy, sharing and closeness with the other community members. Jill's subsequent holidays demonstrate her desire to be part of a collective, including going on art retreat holidays, festivals and gatherings for social movements such as transition groups (local community groups responding to issues of peak oil and climate change).

Membership of various social networks creates social capital. Social capital can take one of two forms (Putnam, 2000), bonding social capital or bridging social capital. Strong ties reinforce bonding capital and are maintained through social networks of people "like us" where we maintain strong relationships such as in families. Weak ties reinforce bridging capital which "can generate broader identities and reciprocity, whereas bonding social capital bolsters our narrower selves" (Putnam, 2000, p. 23). The use of tourism to reinforce bonding capital has a direct impact on sense of self through the close proximity and extended time spent with like-minded people. However, in the temporary communities generated by tourism, bridging capital is significant (Filimonau *et al.*, 2013) and extends opportunities for identity formation. For example:

> When I got off at the station, there was supposed to be a bus to the campsite, because it was 7 miles and errrmmm so yes there was somebody sitting down on their big army bag reading a book so basically the four of us had got talking and this girl had come all the way from Oslo for this festival so we shared a taxi and yeah we ended up killing ourselves laughing because we both had big hair and both had daughters who are wild and various other things.
>
> (Jill)

In Jill's example she is using her travel to create new relationships: the destination is used as a "centre for physical and emotional exchange" (Trauer and Ryan, 2005, p. 482). Talking about shared holiday experiences helps maintain relationships (Heimtum, 2007), even new relationships as in Jill's case. Jill has remained in touch with the woman she met at the train station and they have formed a strong bond with plans to holiday together in the future. According to Heimtum (2007, p. 284), the creation and recall of memories is significant given that "highly mobile and fleeting friendships in today's society need repetition of pleasurable memories in order to renew themselves".

The influence of significant others (family, family friends, work colleagues or wider reference groups) is not just through their physical presence but also "psychological presence", i.e. the ways in which we "mentally represent" them (Shah, 2006). Here we adopt Anderson and Chen's (2002, p. 619) definition that significant others are those who have been deeply influential in a person's life or people to whom someone has given a significant emotional investment. According to Anderson and Chen (2002, p. 619), a significant other may be able to influence an individual's sense of self through "thoughts, feelings, motives, and self-regulatory strategies". Thus other people drive our motives, set standards and regulate our behaviour. Tourism decisions can be made to please a significant other in order to maintain an established relationship position and identity. For instance, Beth holds strong opinions regarding environmental behaviour and would like to give up flying, however, she has recently married and her husband does not share her strength of feeling:

I mean I feel really bad because this year I have taken flights to Ibiza and Marrakech … especially because there's my husband, although he kind of supports and understands that, I suppose has less of a personal commitment and he sees it as being a bigger society issue, so it's always tricky.

Shah (2006) suggests that significant others can guide our behaviour and experiences through our own views of the goals and expectations that the significant other holds for us. In Beth's case her relationship with her husband is stronger than that with the goal, therefore, the significant other is driving her behaviour. Significant others can provide "motives" that cause us to act in a certain way in order to meet their standards (Anderson and Chen, 2002). As a continuation of Goffman's (1959, p. 46) proposal that it is possible to perform in a "favourable social style", Rosenblatt and Russell (1975) suggest that our behaviour while on holiday with significant others is censored by their presence because of a need to present a favourable self.

Possible selves and travel behaviour

A concept evident through many of our interviews was that people undertook travel because of future visions of themselves. They held various images of the person they could become, either positive or negative, and undertook travel to either become or avoid becoming that person. For example, Tom explains his hitchhiking trips:

I think I had a bit of a chip on my shoulder about being a carrot cruncher from a village because even when you went to Purbeck School you had people from Swanage who seem really worldly wise and street [interviewer laughs] you know compared to Wool they were. [Wool has approximately 4,000 inhabitants whereas Swanage has 10,000. Tom currently lives in a town with a population of 180,000.]

Tom's possible self here is of a "carrot cruncher", someone rural, and thought to be a bit backward. Here it could be interpreted that by travelling and seeing the world he is avoiding the negative possible self of the carrot cruncher or he is trying to reach the positive possible self of being worldly wise (like Swanage people).When Tom left home he engaged in extended long-haul travel. Though this cannot be directly attributed to his view of his possible self, his narrative suggests it played a role.

Several other interviewees spoke of instances where there was the potential for their possible selves to influence future behaviour. Heather and Paul (a couple who wanted to be interviewed together) spoke of their friend who died suddenly at the age of 50:

… And I think this friend dying made us realise that, you know, he was our age. [Heather] Seize the day. He was our age, we'd known him for 30 odd years and it was a bit of a wakeup call and you start to think, you know, there were lots of things we were saying "oh when we retire we'll do these things" and it made you realise that you might not actually retire, you know, so if you want to do things, do them when you can.

(Paul)

The possible self here is of someone who had died young, had missed out on certain experiences, an unfulfilled self. This possible self prompted them to leave work for a year and go travelling, thus making the most of life and not delaying experiences. Their friend's death made them realise that there might not be a "later".

Possible selves occur when an individual considers their potential and their future (Markus and Nurius, 1986). There are three types: the ideal self – the self that we would very much like to become; the self that we could become, which includes all the positive opportunities open to us; and the selves we are afraid of becoming. Possible selves are thought to be a means in which an individual evaluates their current self-concept. They may represent goals, aspirations, motives, fears and threats and can act as a motivation for behaviour change to achieve or avoid a possible self (Markus and Nurius, 1986). Hoyle and Sherrill (2006) distinguish between "hoped-for" or feared selves. Tourism provides a unique platform on which people can act out a hoped-for self, making tourism highly relevant and sought after with respect to self-concept development. Giddens (1991) indicates that identities are no longer prescribed by society; the individual is in control of their identity and future identities which include possible selves (Desforges, 2000). Desforges (2000, p. 934) provides a similar example in Molly who "didn't want to end up like some of her husband's peers who have recently retired". Possible selves are, therefore, acknowledged to be a catalyst for motivation and ultimately behavioural change (Markus and Nurius, 1986), however, this phenomenon has not been extensively discussed in terms of tourism behaviour but mainly in educational or sporting settings (e.g. Cross and Markus, 1994; Oyserman *et al.*, 1995; Phoenix and Sparkes, 2007).

In the examples presented here, identity is driving tourism consumption because of the desire to evade or approach a certain possible self. The individual feels that by making a certain trip – or trips – they will achieve the desired or avoid the undesired possible self as the outcome. That possible self could travel more – or less. Tourism could be used to generate a certain possible self by undertaking a particular type of travel, e.g. more sustainable travel behaviours. This is supported by Morgan (1993) who suggests, specifically about environmental behaviour, that it is not the current self-concept which drives behaviour but the imagined future self. There is also evidence that once behaviour has been altered by the possible self, there is the opportunity for validation or social feedback. This results in identity control, as "individuals will adjust their behaviour until social feedback aligns with the identity standard" (Grandberg, 2006, p. 111).

Presentation of self through tourism

Identities are rooted in the stories that we tell ourselves and others in order to make meaning of our existence and place in the world (Bagnoli, 2009). Many interviewees recounted stories presenting themselves in a certain light, choosing what to tell, and what to keep to themselves.

Martin, a retired RAF pilot, who appeared to be a sensible, golf-loving father and husband wanted to display that he also had a wilder side to him. Through his narratives, Martin demonstrated a rebellious streak when he undertook things he knew to be wrong: however there was a certain pride in telling these stories. It is a type of "performance" (Goffman, 1959) and may not be entirely grounded in reality. Martin mentions a 1,400-mile car journey with his young family across Europe that lasted 23 hours without stopping. He begins by saying "And then I did something that I am ashamed of to this day..." but this episode is mentioned three times during the interview and, since it resulted in no ill harm, offers Martin a dangerous and exciting identity marker. Another example comes from a childhood holiday with a group of boys and a schoolmaster:

> We went across on the ferry to Palermo in Sicily and down to the south coast of Sicily, Agrigento, where I was hospitalised for two days because he told me not to go on the hot sand in bare feet and I did. I burnt my feet so the whole group had to wait while I recovered and then that was another little episode.

The use of the phrase "another little episode" shows how Martin embellishes his story by brushing this incident aside as insignificant. This is also demonstrated by his use of the word "another", i.e. he was always doing things like this. Mark was another interviewee presenting a reckless side; he recounted an episode on a family holiday where he got arrested:

> I think it was the Greek holiday, it was a bit more bizarre, because I got drunk with my sister and ended up drunkenly going around in this taxi for ages and I got arrested sort of, so yeah that was a very bizarre one.

Later on in the interview it became apparent that Mark was not "sort of" arrested, he was drunk and could not remember where he was staying so slept it off in the police station. He describes another holiday as "crazy and irresponsible". Here both Martin and Mark use tourism as a site for excess where the normal rules of behaviour do not apply.

Research has found that people's behaviour changes in tourism settings and people who otherwise maintain fairly sustainable lifestyles abandon this in the tourism domain (Barr *et al.*, 2011). The findings here are consistent with Barr *et al.* (2011) in that even those who were fully committed at home and willing to acknowledge responsibility do not transfer their home commitment and responsibility to their "away" behaviour. For instance, Mark's best holiday was an all-inclusive package holiday to a 5* resort in Mexico where a highlight was swimming with dolphins. However, in Mark's case, his current belief is that he will not go on holiday unless there is some "greater good" that comes from making the trip as he reflects that, given his "current moral standards and outlook on life I would probably tut and be disgusted with what I had got up to". This is an example of identity evolving. As people move through life and social positions they "identify with identity positions initially rejected" (Gillespie, 2007, p. 591). Cohen (2010) found that some of his interviewees were aware the self was always evolving and that they could perform or present certain chosen selves at any one time. While Mark did not have this same level of insight, he was aware of the contrast in his stories and acknowledged that he had significantly changed, even if he did not realise the multiplicity of his current self-concept.

Interviewees often did not want to be identified as a tourist and spoke negatively of the presence of large numbers of tourists. Mark criticised the behaviour of British tourists as drunken louts: "we met up with some English people who said raaaaaay you're English as well, let's get really, really drunk because that's what we do". Taken at face value this sentence does not seem that condemning, however, it was said with a particularly sarcastic and condescending tone. His next sentence was very matter-of-fact: "so I got ridiculously drunk".

Richard specifically sought out a destination that was off the beaten track: "We looked up all the islands, all the Greek islands that had brochures about them; all that were in the guide book, we crossed off and we got to one that wasn't mentioned in any guide book or anything." For his identity, it was important that he was not seen as a mass tourist. Michael also chooses to differentiate himself from typical tourists, speaking about the embarrassment caused by being identified by the type of holiday he has been on, a Saga holiday, designed for an older age group and perhaps the "shame" reflects that being old is not part of his self-concept:

> Our friends said the only way to get to Cuba was with, I am so ashamed I can hardly say it, Saga. So we were now 50, aged 50, and we went with Saga and there were people on this trip who were in their 80's, and we and a few others were known as the Saga louts, because we were out for a good time, but it wasn't what you are meant to do with Saga.

Michael, although conceding that he took a Saga holiday, is trying to differentiate himself from the traditional Saga tourist. He states his (much younger) age compared to the other tourists. He also states that he was known as a "Saga lout". Although being a lout is not normally seen as desirable behaviour, here Michael is presenting it as preferable to being a "normal" Saga tourist, thus presenting himself in a fun and youthful way. Gillespie (2007) noted a similar occurrence in his research about self-identification through differentiation. His interviewees derogated certain tourist behaviours when actually they conducted themselves in a similar manner. Given the growth of the global tourism industry, it becomes increasingly difficult for the "traveller" to seek out places that mark out a different identity. Richard specifically sought out a destination that was off the beaten track: "We looked up all the islands, all the Greek islands that had brochures about them, all that were in the guide book we crossed off and we got to one that wasn't mentioned in any guide book or anything". For his identity it was important that he was not just part of the masses. This encourages long-haul flights: people seek out the new.

Conclusion

Our findings indicate that identities can play a significant role in tourism mobility decisions. Identities are present in all elements of our lives. They are the stories that make us who we are. It is possible to have multiple identities at the same time (Palmer, 1999), however, some identities will be more dominant at certain times depending on many factors, such as the context of the situation, who we are with and even the person that we might become (Morgan, 1993). Tourism provides a unique site of consumption separate to everyday life that brings certain identities to the fore and in some cases necessitates the performance of particular identities to reinforce self-concept (Goffman, 1959). This performance of identity intervenes in behaviour that might otherwise lead to sustainable mobility choices and highlights the significant role of self in behaviour. At least three different mechanisms influencing identity formation with consequences for travel behaviour have been identified.

First, VFR travel is used to affirm identities based on relationships with significant others. VFR is also used as a "safety blanket", enabling more adventurous travel and presenting a desired identity. Similarly, family holidays allow for the enactment of a family identity (Haldrup and Larsen, 2003), that due to the complexities of modern day living is not possible at home. It brings families emotionally and physically closer together. Tourism is inherently social (Dickinson *et al.*, 2011; Rickly-Boyd, 2010) providing opportunities to be with friends and family. Urry (1990) describes the "collective gaze" and Haldrup and Larsen (2003) discuss the "family gaze". Trauer and Ryan (2005) argue that the holiday is not the purchase of "place", but of "time" in which to create an intimacy with significant others. The destination acts as the "scene" for the family performance. Based on this perspective these authors suggest the purpose of the holiday is not the search for "other" but an endeavour to make sense of their own relationships through the shared experience of the holiday. Larsen *et al.* (2006, p. 45) suggest "families are most at home when away from home". However, Rosenblatt and Russell (1975) propose that holidays are the perfect opportunity for conflict through increased proximity, change of routines and the removal of agreed territories and boundaries.

Crompton (1979) produced one of the earliest studies indicating that holidays could be used as ways to demonstrate, refine or modify identities. He found that holidays could be used to strengthen family relations as members were brought closer together. Similarly, Haldrup and Larsen (2003, p. 24) move on from the notion of tourism as a way of "finding yourself", to tourism as a platform to display and create new identities and "ways of being together", even if only for the duration of the holiday. This is because there might be few situations at home where we can focus on each other to the same degree as during a holiday; the result is particularly intense feelings of togetherness. In addition, the shared holiday experience can also be used as a way of maintaining, reinforcing or even creating a specific identity. This is proposed by Gram (2005, p. 17) who states: "Identity is today built up carefully by a number of choices, which are not necessarily stable. Holidays (as other forms of consumption) are one brick in the identity building process."

The significance of existing home identities should not go un-noted. These identities can be so strong that they can lead people to undertake travel that they do not desire, particularly when significant others are involved (Anderson and Chen, 2002). The example presented here was of Beth, a recently married woman who did not want to limit her husband's travel by imposing her views on him. To be a "good wife" she took several flights, contrary to her beliefs on sustainable travel. Her behaviour was in contrast to her attitudes on environmental issues and was a direct result of her acting consistently with an identity of a good wife. For Beth, that was more important than that of someone who was environmentally concerned and she regulated her behaviour in accordance (Shah, 2006). There may be other cases where peer pressure can be important.

Second, possible selves are a catalyst for behaviour change (Markus and Nurius, 1986) and for many of our participants this meant travel behaviour. Undertaking certain forms of travel allowed them to avoid or approach their future self they had imagined. This was a powerful motivating force well documented in other fields (Cross and Markus, 1994; Oyserman *et al.*, 1995; Phoenix and Sparkes, 2007). We feel that possible selves impact travel behaviour. There is scope for future research on how industry and policy-makers could creatively utilise this concept to instigate desired behaviour changes.

Finally, the narratives of travel allowed the interviewees to present themselves in a certain light (Goffman, 1959). The recollected tourist experience provides people with a stage to embellish stories or recount stories that are not congruent to their current identity, however, in recalling the stories they demonstrate how multiple identities can and do coexist (Palmer, 2005). Tourism can be a site for excess and provides a canvas to construct narratives of self. As such, attempts to get tourists to voluntarily restrain behaviour seem unlikely to be successful. However, there is an opportunity for industry and policy-makers to provide new narratives that are less dependent on the exotic or long-haul travel but provide tourists with scope to construct personal stories.

These findings help to explain why previous policies promoting more environmentally sustainable behaviour have not worked. Identities play such a large part in the decision-making process of tourists, overriding other factors such as environmental impact and in some cases personal choice (when people travel for the benefit of a significant other), in addition to pragmatic decisions such as time available and cost (Frew and Winter, 2010). Given that decisions are deeply rooted in identity issues, instigating a behavioural change in tourism mobility will be very difficult to achieve. Policy-makers and marketers advocating climatically sustainable holidays need to pay more attention to this issue. It should be possible to address identities either by promoting positive identities or showcasing negative identities, and much more could be done to encourage the use of social media to connect virtually rather than physically with significant others.

Specifically, taking into account the social aspects of identity, there needs to be a shift in the way wider society views highly mobile lifestyles, sought after and desired because of the implied status of exotic travel. There is a possibility to introduce a counter identity, one that suggests a positive status for those who travel sustainably. There could even be the possibility to create a negative social identity by placing a stigma on those who are highly mobile.

This chapter has some limitations. It is a small study, restricted to one relatively affluent part of a small western country. All identity research has to admit that interviewees can embellish stories or leave out vital details to present a desired or unrealistic identity to their audience. Due to the multiplicity of identities and the contextual nature in which they are presented, there will never be one "true" identity. Nevertheless, the results reveal interesting and telling findings, pointing to further research in order to make recommendations that would be able to be tested by the industry and by policy-makers.

References

Anable, J., Lane, B. and Kelay, T. (2006). *An evidence base review of public attitudes to climate change and transport*. London: The Department for Transport.

Anderson, S.M. and Chen, S. (2002). The relational self: An interpersonal social-cognitive theory. *Psychological Review*, *104*(4), 619–645.

Axhausen, K.W. (2005). *A dynamic understanding of travel demand: A sketch*. Zurich: Institut für Verkehrsplanung, Transporttechnik, Strassen und Eisenbahnbau (IVT).

Bagnoli, A. (2009). On "an introspective journey": Identities and travel in young people's lives. *European Societies*, *11*(3), 325–345.

Barr, S., Gilg, A. and Shaw, G. (2011). "Helping people make better choices": Exploring the behaviour change agenda for environmental sustainability. *Applied Geography*, *31*, 712–720.

Barr, S., Shaw, G., Coles, T. and Prillwitz, J. (2009). "A holiday is a holiday": Practicing sustainability, home and away. *Journal of Transport Geography*, *18*(3), 474–481.

Bond, N. and Falk, J. (2012). Tourism and identity-related motivations: Why am I here (and not there)?. *International Journal of Tourism Research*, *15*(5), 430–442.

Bruner, E.M. (1991). Transformation of self in tourism. *Annals of Tourism Research*, *18*, 238–250.

Bryman, A. and Bell, E. (2011). *Business research methods* (3rd edn). Oxford: Oxford University Press.

Burke, P.J. (2001, 27–29 April). *Relationships among multiple identities*. Paper presented at the conference on Advances in Identity Theory, Bloomington, IN.

Cacioppo, J.T. and Patrick, W. (2008). *Loneliness. Human nature and the need for social connection*. New York: W.W. Norton.

Chase, S.E. (2005). Narrative inquiry: Multiple lenses, approaches, voices. In N.K.

Denzin and Y.S. Lincoln (eds), *The Sage handbook of qualitative research* (3rd edn) (pp. 651–681). Thousand Oaks, CA: Sage.

Cohen, S.A. (2010). Chasing a myth? Searching for "self" through lifestyle travel. *Tourist Studies*, *10*(2), 117–133.

Crompton, J.L. (1979). Motivations for pleasure vacations. *Annals of Tourism Research*, *1*(4), 408–424.

Cross, S.E. and Markus, H.R. (1994). Self-schemas, possible selves, and competent performance. *Journal of Education Psychology*, *86*(3), 423–438.

Desforges, L. (2000). Travelling the world: Identity and travel biography. *Annals of Tourism Research*, *27*(4), 926–945.

Dickinson, J.E. and Dickinson, J.A. (2006). Local transport and social representations: Challenging the assumptions for sustainable tourism. *Journal of Sustainable Tourism*, *14*(2), 192–208.

Dickinson, J.E., Lumsden, L.M. and Robbins, D. (2011). Slow travel: Issues for tourism and climate change. *Journal of Sustainable Tourism*, *19*(3), 281–300.

Dickinson, J.E., Robbins, D. and Fletcher, J. (2009). Representation of transport: A rural destination analysis. *Annals of Tourism Research*, *36*, 103–123.

Eijgelaar, E., Thaper, C. and Peeters, P. (2010). Antarctic cruise tourism: The paradoxes of ambassadorship, "last chance tourism" and GHG emissions. *Journal of Sustainable Tourism*, *18*(3), 337–354.

Ellemers, N., Spears, R. and Doosje, B. (1999). *Social identity: Context, commitment, content*. Oxford: Blackwell.

Fein, S. and Spencer, S.J. (1997). Prejudice as self-image maintenance: Affirming the self through derogating others. *Journal of Personality and Social Psychology*, *73*(1), 31–44.

Filimonau, V., Dickinson, J., Cherrett, T., Davis, N., Norgate, S. and Speed, C. (2013, January). *Rethinking travel networks: Mobile media and collaborative travel in the tourism domain*. Paper presented at the UTSG Conference, Oxford.

Frew, E.A. and Winter, C. (2010). Tourist response to climate change: Regional and metropolitan diversity. *Tourism Review International*, *13*(4), 237–246.

Fullagar, S. (2002). Narratives of travel: Desire and the movement of feminine subjectivity. *Leisure Studies*, *21*, 57–74.

Gergen, K.J. (1971). *The concept of self*. New York: Holt, Rinehart & Winston.

Giddens, A. (1991). *Modernity and self-identity: Self and society in the late modern age*. Cambridge: Polity Press.

Giddens, A. (2009). *The politics of climate change*. Cambridge: Polity Press.

Gillespie, A. (2007). Collapsing self/other positions: Identification through differentiation. *British Journal of Social Psychology*, *46*, 579–595.

Goffman, E. (1959). *The presentation of self in everyday life*. London: Penguin.

Gössling, S. and Nilsson, J.H. (2010). Frequent flyer programmes and the reproduction of aeromobility. *Environment and Planning A*, *42*, 241–252.

Graeff, T.R. (1996). Image congruence effects on product evaluations: The role of self-monitoring and public/private consumption. *Psychology and Marketing*, *13*(5), 481–499.

Gram, M. (2005). Family holidays. A qualitative analysis of family holiday experiences. *Scandinavian Journal of Hospitality and Tourism*, *5*(1), 2–22.

Grandberg, E. (2006). "Is that all there is?" Possible selves, self-change, and weight loss. *Social Psychology Quarterly*, *69*(2), 109–126.

Haldrup, M. (2004). Laid-back mobilities: Second-home holidays in time and space. *Tourism Geographies*, *6*(4), 434–454.

Haldrup, M. and Larsen, J. (2003). The family gaze. *Tourist Studies*, *3*(1), 23–45.

Hannam, K., Sheller, M. and Urry, J. (2006). Mobilities, immobilities and moorings. *Mobilities*, *1*(1), 1–22.

Hares, A., Dickinson, J. and Wilkes, K. (2010). Climate change and the air travel decisions of UK tourists. *Journal of Transport Geography*, *18*, 466–473.

Heimtum, B. (2007). Depathologizing the tourist syndrome: Tourism as social capital production. *Tourist Studies*, *7*(3), 271–293.

Hogg, M.A. and Abrams, D. (1988). *Social identifications*. London: Routledge.

Hogg, M.A. and Terry, D.J. (2000). Social identity and self-categorization process in organizational contexts. *The Academy of Management Review*, *25*(1), 121–140.

Holloway, I. and Wheeler, S. (2010). *Qualitative research in nursing and healthcare*. Oxford: Blackwell.

Horton, D. (2003). Green distinctions: The performance of identity among environmental activists. *The Sociological Review*, *51*, 63–77.

Hoyle, R.H. and Sherrill, M.R. (2006). Future orientation in the self system: Possible selves, self regulation, and behaviour. *Journal of Personality*, *7*(6), 1673–1696.

Kitzinger, C. and Wilkinson, S. (1996). Theorizing representing the other. In S. Wilkinson and C. Kitzinger (eds), *Representing the other: A feminism and psychology reader* (pp. 1–32). London: Sage.

Kraus, W. (2006). The narrative negotiation of identity and belonging. *Narrative Inquiry*, *16*(1), 103–111.

Larsen, J., Urry, J. and Axhausen, K. (2006). *Mobilities, networks, geographies*. Aldershot: Ashgate.

Löfgren, O. (2008). The secret lives of tourists: Delays, disappointments and daydreams. *Scandinavian Journal of Hospitality and Tourism*, *8*(1), 85–101.

Markus, H. and Nurius, P. (1986). Possible selves. *American Psychologist*, *41*(9), 954–969.

McAdams, D.P. (1993). *The stories we live by*. New York: The Guildford Press.

McKercher, B., Prideaux, B., Cheung, C. and Law, R. (2010). Achieving voluntary reductions in the carbon footprint of tourism and climate change. *Journal of Sustainable Tourism*, *18*(3), 297–317.

Miller, A. (2007). *The drama of the gifted child: The search for the true self*. New York: Basic Books.

Morgan, A.J. (1993). The evolving self in consumer behaviour: Exploring possible selves. *Advances in Consumer Research*, *20*, 429–432.

Moscardo, G., Pearce, P., Morrison, A., Green, D. and O'Leary, J. (2000). Developing a typology for understanding visiting friends and relatives markets. *Journal of Travel Research*, *38*, 251–259.

Noy, C. (2004). Performing identity: Touristic narratives of self-change. *Text and Performance Quarterly*, *24*(2), 115–138.

Onkvisit, S. and Shaw, J. (1987). Self-concept and image congruence: Some research and managerial implications. *The Journal of Consumer Marketing*, *4*(1), 13–23.

Oyserman, D., Gant, L. and Ager, J. (1995). A socially contextualized model of African American identity: Possible selves and school persistence. *Journal of Personality and Social Psychology*, *69*(6), 1216–1232.

Palmer, C. (1999). Tourism and the symbols of identity. *Tourism Management*, *20*, 313–321.

Palmer, C. (2005). An ethnography of Englishness: Experiencing identity through tourism. *Annals of Tourism Research, 32*(1), 7–27.

Pearce, P.L. (1988). *The Ulysses factor: Evaluating visitors in tourist settings*. New York: Springer-Verlag.

Pearce, P.L. (1991). *Dreamworld: A report on public reactions to dreamworld and proposed developments at dreamworld*. A Report to Ernst and Young on behalf of the IOOF in conjunction with Brian Dermott and Associates. Townsville: Department of Tourism, James Cook University.

Pearce, P.L. (1993). Fundamentals of tourist motivation. In D. Pearce and R. Butler (eds), *Tourism research: Critiques and challenges* (pp. 85–105). London: Routledge and Kegan Paul.

Pearce, P.L. and Lee, U.I. (2005). Developing the travel career approach to tourist motivation. *Journal of Travel Research, 43*, 226–237.

Phoenix, C. and Sparkes, A.C. (2007). Sporting bodies, ageing, narrative mapping and young team athletes: An analysis of possible selves. *Sport, Education and Society, 12*(1), 1–17.

Putnam, R.D. (2000). *Bowling alone: The collapse and revival of American community*. New York: Simon & Schuster.

Randles, S. and Mander, S. (2009). Aviation consumption and the climate change debate: "Are you going to tell me off for flying?". *Technology Analysis & Strategic Management, 21*(1), 93–113.

Reissman, C.K. (1993). *Narrative analysis*. California: Sage.

Reissman, C.K. (2008). *Narrative methods for the human sciences*. London: Sage.

Rickly-Boyd, J.M. (2010). The tourist narrative. *Tourist Studies, 9*(3), 259–280.

Rosenblatt, P.C. and Russell, M.G. (1975). The social psychology of potential problems in family vacation travel. *Family Coordinator, 24*(2), 209–215.

Schwanen, T. and Lucas, K. (2011). Understanding auto motives. In K. Lucas, E. Blumenberg and R. Weinberger (eds), *Auto motives. Understanding car use behaviours* (pp. 3–39). Bingley: Emerald.

Scott, D., Peeters, P. and Gössling, S. (2010). Can tourism deliver its "aspirational" emission reduction targets?. *Journal of Sustainable Tourism, 18*(3), 393–408.

Seaton, A.V. and Palmer, C. (1997). Understanding VFR tourism behaviour: The first five years of the United Kingdom tourism survey. *Tourism Management, 18*(6), 345–355.

Shah, J. (2006). When your wish is my desire: A triangular model of self-regulatory relationships. In K.D. Vohs and E.J. Finkel (eds), *Self and relationships: Connecting intrapersonal and interpersonal processes* (pp. 387–406). New York: Guilford Press.

Shaw, J. (2001). "Winning territory": Changing place to changing pace. In J. May and N. Thrift (eds), *Timespace: Geographies of temporality* (pp. 120–132). London: Routledge.

Sheller, M. (2004). Mobile publics: Beyond the network perspective. *Environment and Planning D: Society and Space, 22*, 39–52.

Small, J. (2002). Good and bad holiday experiences: Women's and girls' perspective. In M.B. Swain and J.H. Momsen (eds), *Gender/tourism/fun(?)* (pp. 24–38). New York: Cognizant Communication Corporation.

Smith, B. and Sparkes, A. (2006). Narrative inquiry in psychology: Exploring the tensions within. *Qualitative Research in Psychology, 3*, 169–192.

Steg, L. and Vlek, C. (2009). Encouraging pro-environmental behaviour: An integrative review and research agenda. *Journal of Environmental Psychology, 29*, 309–317.

Stets, J.E. and Biga, C.F. (2003). Bringing identity theory into environmental sociology. *Sociological Theory, 21*(4), 398–423.

Thurlow, C. and Jaworski, A. (2006). The alchemy of the upwardly mobile: Symbolic capital and the stylization of elites in frequent-flyer programmes. *Discourse and Society, 17*(1), 99–135.

Trauer, B. and Ryan, C. (2005). Destination image, romance and place experience – an application of intimacy theory in tourism. *Tourism Management, 26*, 481–491.

UNWTO (World Tourism Organization). (2012). *Tourism highlights, 2012 edition.* Retrieved from World Tourism Organization: http://mkt.unwto.org/en/publication/unwto-tourism-highlights-2012-edition.

UNWTO, UNEP and WMO. (2008). *Climate change and tourism: Responding to global challenges.* Madrid: UNWTO, UNEP and WMO.

Urry, J. (1990). *The tourist gaze.* London: Sage.

Urry, J. (2000). *Sociology beyond societies: Mobilities for the twenty-first century.* London: Routledge.

Urry, J. (2003). Social networks, travel and talk. *British Journal of Sociology, 54*(2), 155–175.

Urry, J. (2011). Social networks, mobile lives and social inequalities. *Journal of Transport Geography, 21*, 24–30.

Van De Mieroop, D. (2009). A good story or a good identity? The reportability of stories interfering with the construction of a morally acceptable identity. *Narrative Inquiry, 19*(1), 69–90.

3 Happiness and limits to sustainable tourism mobility

A new conceptual model

Yael Ram, Jeroen Nawijn and Paul Peeters

Introduction

Tourism is both a victim and a vector of climate change (Cabrini *et al.*, 2009). Although the impact of climate change on tourism will lead to considerable behavioural change for entrepreneurs, governments and tourists, this chapter examines behavioural change required to mitigate the impact of tourism on climate. Global tourism is responsible for 5 per cent of current global CO_2 emissions (UNWTO-UNEP-WMO, 2008). The impact of tourism on radiative forcing – the main parameter determining the changes in temperature – is greater. Up to 12.5 per cent of radiative forcing is caused by non-carbon effects such as contrails and contrail-induced cirrus clouds that are caused by aviation (Scott *et al.*, 2010). Approximately, 75 per cent of all CO_2 emissions are caused by transportation, of which 40 per cent is caused by air transportation alone (UNWTO-UNEP-WMO, 2008). Adding the effects of radiative forcing, these shares could be as high as 91 per cent and 74 per cent, respectively (UNWTO-UNEP-WMO, 2008). Although air transportation's share was only 17 per cent of all transportation emissions in 2005, multiple studies (Dubois *et al.*, 2011; Peeters and Dubois, 2010; Scott *et al.*, 2010) explained that current behavioural trends towards a higher share of air transportation and concomitant increase of distances have caused a significant increase in emissions.

This growth is found not only in business as usual (BAU) scenarios but also in scenarios that take account of proposed pro-environmental measures (UNWTO-UNEP-WMO, 2008). Furthermore, Peeters and Dubois (2010) and Dubois *et al.* (2011) demonstrate that it is not feasible to maintain current growth trends for trips and achieve the required emission reductions for sustainable development relating to climate change. The required emission reductions are approximately 80 per cent of 1990 global CO_2 emissions and other greenhouse gases by 2050 (Hansen, 2008; Hansen *et al.*, 2008; Parry *et al.*, 2009; Rockstrom *et al.*, 2009). BAU emissions are expected to quadruple over this time period. To accommodate a reduction of 80 per cent, overall travel efficiency must improve by a factor of 16, or almost 94 per cent. Although the industry suggests such a reduction is achievable (ATAG, 2010), science cannot confirm this (Lee *et al.*, 2013; Owen *et al.*, 2010). Therefore, even though current aviation emissions are low, measured as a share of global emissions, the main problem is the strong future growth rate.

To attain such drastic reductions, the tourism system itself needs to change. Theoretically, opportunities for such a change are evident. Long-haul trips and air transportation cause most emissions. This is partly due to shorter lengths of stay, causing more transport kilometres per guest-night (Peeters *et al.*, 2009). Approximately 80 per cent of trips in the world are domestic and involve surface transportation (83 per cent in 2005; UNWTO-UNEP-WMO, 2008). Hence, a reduction of the share of both air transportation and long-haul trips is a very effective method to reduce emissions, with relatively few tourists affected by such measures. Using scenario methods, a recent study (Peeters and Dubois, 2010) indicates that two main "solutions" to the emission reduction problem exist. The first is a significant reduction of total air transportation volume to 1970s levels, with the modal split between cars and other surface transportation modes remaining the same (assuming the same number of trips as in the BAU scenario). The second option is to maintain 2005 air transportation volumes (i.e. no growth until 2050 in passenger-kilometres), but to transfer 80 per cent of all car use to other transportation modes, particularly rail and coach (Peeters and Dubois, 2010). Changes are needed in leisure mobility behaviour because the current situation is not sustainable. There is an increase in distances, frequency of transportation in planes and cars, and trip numbers, all bringing larger trip volumes and emissions per trip. For instance, the number of trips by Dutch tourists between 2002 and 2010 increased by less than 2 per cent, whereas the total distance travelled increased by 36 per cent (de Bruijn *et al.*, 2012). The share of air transportation increased from 12 per cent in 2002 to 17 per cent in 2010, and all Dutch tourism-related CO_2 emissions increased by almost 20 per cent.

This study has two main goals: (1) to explore the psychological causes for this development towards unsustainable tourist mobility behaviour, by focusing on the notion of happiness and (2) to demonstrate how happiness hinders desirable behavioural change towards more sustainable tourist mobility behaviour. Although tourism is a global phenomenon, the chapter specifically addresses the leisure travel of western tourists. According to UNWTO (2012) data, this segment creates the majority of tourism; more than half of global trips are made by western tourists for leisure purposes. Additionally, most studies of leisure mobility involve western tourists (Antimova *et al.*, 2012; Cohen *et al.*, 2011; Hares *et al.*, 2010; McKercher *et al.*, 2010). Therefore, the ability to make inferences about non-western tourists is limited. This study aims to explain the gap between the broad understandings of the importance of changing western tourists' leisure mobility behaviours and the opposition of these tourists to behave in a sustainable way. It seeks to eventually bridge this gap.

The chapter is structured as follows. First, the shortcomings of the existing literature explaining insufficient behavioural change in leisure mobility are reviewed. Then, the conceptual method of analysis (Xin *et al.*, 2013) is used to explain the unsustainable foundations of leisure mobility. Third, a synthesis of theories derived from psychology and tourism (e.g. Diekmann and Preisendörfer, 2003; Fridgen, 1984; Nawijn, 2011b; Nawijn *et al.*, 2013; Tung and Ritchie, 2011a; Uriely *et al.*, 2011) is described to present a new conceptual model that explains why leisure tourist mobility is difficult to change. The conclusion portrays the practical and theoretical contributions of the suggested conceptual model as well as its limitations.

Unsustainable leisure travel and behavioural change

Most previous studies that addressed the issue of behavioural change in a leisure mobility context (e.g. Antimova *et al.*, 2012; Cohen *et al.*, 2011; Hares *et al.*, 2010; Higham and Cohen, 2011; McKercher *et al.*, 2010; Miller *et al.*, 2010) reported two main gaps that hinder the likelihood of changing western tourists' leisure mobility behaviours. The first was described by Hares *et al.* (2010, p. 472) as "an awareness attitude gap rather than an attitude-behavior gap", indicating that tourists' awareness of their unsustainable behaviours does not produce corresponding support for more sustainable tourist behaviour or enhance behavioural change. The second gap is a contextual gap, labelled by Barr *et al.* (2010, p. 474) as "a holiday is a holiday" problem, describing resistance to adopt environmentally friendly mobility practices during vacations, although they apply environmentally friendly practices in their everyday lives (Barr *et al.*, 2010; Cohen *et al.*, 2011; McKercher *et al.*, 2010; Miller *et al.*, 2010). Cohen *et al.* (2011) suggested that this gap produces tension, causing negative emotions of guilt and denial among specific consumers. A defence mechanism of rationalisation was revealed in a study describing how the opportunity to recycle paper causes an increase in paper consumption (Catlin and Wang, 2013). Based on this study, it can be assumed that people may exhibit environmentally friendly behaviour, such as recycling or "eco tours," as an "alibi" or rationalization to consume more. These defence mechanisms, together with the human tendency for cognitive dissonance, may hinder behavioural change.

Interestingly, the comprehensive review of Pearce and Packer (2013) about new links between psychology and tourism highlights topics such as memory, satisfaction, and personal growth but overlooks the growing debate about the need for behavioural change with regard to sustainable tourism requirements. Our discussion about behavioural change is, therefore, also based on literature outside the tourism domain, and refers to behavioural change in general, and in environmental context in particular.

Festinger's (1957) cognitive dissonance theory posits that individuals need to avoid inconsistencies in beliefs, attitudes, and behaviours because inconsistencies are unpleasant. In environmentally responsible behaviour, this entails individuals displaying consistent behaviour across various life domains. According to Thøgersen (2004), empirical evidence traditionally points in different directions. Some studies found consistent behaviour across life domains (Berger, 1997), whereas others did not (Stern and Oskamp, 1987). In the latter cases, environmentally responsible behaviour does not appear to spill over to other life domains. Rashid and Mohammad's (2012) conceptual framework explains that attitude formation within a certain life domain must take place before cognitive dissonance could even play a role. Additionally, Diekmann and Preisendörfer (2003) find that, regarding environmental concerns, the degree of influence such an attitude has on behaviour diminishes with increasing behavioural costs. In other words, if it is difficult for tourists to change their behaviour to avoid inconsistencies, then behavioural change will not occur. Literature on tourism transportation supported this logic, arguing that cognitive dissonance would be more severe with respect to tourist travel than for other aspects of behaviour (Barr *et al.*, 2010).

These two gaps – the awareness-attitude gap and the contextual gap – imply that the issue of leisure mobility does not correspond to existing theories about behavioural change, such as the theory of planned behaviour (Ajzen, 1991) or the motivation-ability opportunity-behaviour model (Ölander and Thøgersen, 1995). According to these two models, attitudes, norms and perceived controllability (Ajzen, 1991) together with opportunity and ability (Ölander and Thøgersen, 1995) may influence behaviours. However, these models do not refer to different contexts and do not answer the question why norms and attitudes relevant in one context are overlooked by consumers in another context that provides similar opportunities and abilities. Additionally, the models mentioned above cannot resolve the contradiction between awareness of environmental issues and attitudes that support flying as an acceptable mode of leisure mobility (the awareness-attitude gap). The "low cost hypothesis" of Diekmann and Preisendörfer (2003) referred to these two gaps, and related them to the high cost of avoiding vacations. However, their study did not define the components of the cost or why people perceive vacations as so important.

This situation calls for a new model to explore the specific context of leisure mobility and distinguish it from everyday practices. This new theoretical model builds on the conclusions of Barr *et al.* (2011) and Verbeek and Mommaas (2008) about the importance of contextually analysing sustainable practices. These studies suggested distinguishing everyday practices from leisure and tourism practices when addressing sustainable issues. In other words, the analysis of tourism mobility as a special case of sustainable practice would not need to focus solely on mobility characteristics, such as modes of transportation, information about greenhouse gas emissions and people's preferences, but would instead address the context of leisure vacations.

The conceptual method

In their recent work, Xin *et al.* (2013) defined the conceptual method of tourism studies as

> a set of activities that focus on the systematic analysis and profound understanding of tourism concepts.... Its major outcomes include the clarification of a concept, the proposing of a new concept, the modification of an existing one (re-conceptualization) or ideological or other critique.
>
> (p. 84)

Xin *et al.* (2013) noted that this method is not popular in tourism studies and that empirical methods (both quantitative and qualitative) are much more prevalent.

However, they also demonstrated that the conceptual method is superior to the empirical methods in cases with "big, holistic questions" (p. 73) and when new approaches are suggested. Furthermore, Xin *et al.* (2013) argued that conceptual analysis enables creativity, adds new insights to old problems and builds bridges between different disciplines.

The conceptual method was determined as appropriate for the current research because it corresponds well with the goal of bringing new insight to the old issues of a lack of behavioural change in tourist mobility by bridging two different contexts: the leisure vacation and mobility. Furthermore, it follows Wolf and Moser's (2011) conclusion about the diversity of findings of empirical studies on environmentally responsible behaviour, which led them to conclude that there is no single empirical theory that explains these variations in behaviour.

The current study adopts two themes for conceptual analysis from Xin *et al.*'s (2013) review. The first is a synthesis of concepts from different disciplines using a literature review; the second applies an existing concept to a new context. Thus, the next section of the chapter includes a literature review of different fields and subjects, including the tourist experience, motivations for vacation, happiness, perception of distance and mobility. A model synthesising these fields is then presented in two steps. First, the model refers to the tourist experience domain (the "happiness loop" model), and will then be stretched to include motivations and mobility aspects (the "three-gear model"). The latter will demonstrate not only that the tourist experience and distance are interrelated in the context of leisure mobility but *also* that happiness affects all factors of the model – the tourist experience, the motivations for vacations, the perception of distance and the mobility outcomes. Thus, the conceptual analysis extends the concept of happiness to the context of mobility. However, it is worth mentioning here that even though a model is an effective way to present a synthesis of theories (Pearce, 2008), it is a simplification of reality and not reality itself. According to Haggett and Chorley (1967), the function of a model is to highlight important notions or relationships in general terms. Following this logic, the current model does not aim to present the rich range of tourist motivations or to present solutions for unsustainable mobility behaviour, but to offer an explanation for the two previously mentioned gaps, the awareness-attitude gap and the "holiday is a holiday" gap, which together hinder behavioural change towards more sustainable tourism mobility behaviour.

The conceptual model: happiness limits sustainable tourism mobility

Literature review: the tourist experience, happiness, perception of distance and mobility

The tourist experience

The Fridgen (1984) model of the "tourist experience" is based on Clawson and Knetsch's (1966) "recreational experience" theory that describes the five phases a tourist experiences: anticipation, travelling to the destination, on-site experience, travelling from the destination and recollection. The "tourist experience" concept includes elements from before, during and after the actual experience, which implies that thoughts, plans and memories are part of that experience. Furthermore, both travel to and from the destination are viewed as integral parts of the experience and not as inescapable costs or marginal events. This suggests that tourists may anticipate and recollect not only the on-site experience but also their travel to and from the destination. Recently, Tung and Ritchie (2011b) emphasised the importance of the recollection phase. In their study, they implemented the ideas of Pine II and Gilmore (1998) about memorable experiences by describing conditions that would support tourists remembering their travel experiences.

The "tourist experience" model portrays behaviours and feeling, but does not answer questions such as why people travel and why different tourists engage in different behaviours. It cannot provide ways to promote behavioural change to improve the sustainability of tourism mobility. Therefore, a closer look at the psychological foundations of the tourist experience is needed. The next section is dedicated to understanding the psychological foundations of the tourist experience and, more precisely, tourists' motivations.

Tourists' motivations

Plog (1974, 2002) was the pioneer in developing a tourists' motivation theory. According to Plog's "venturesomeness theory", tourists can be described along a conceptual continuum, ranging from allocentric to psychocentric. Allocentric tourists are characterised by leadership, curiosity and risk-taking motivations, whereas psychocentric tourists are indecisive and risk-averse. Although this theory was defined as the most accepted personality theory in the tourism context (McGuiggan and Foo, 2004), it did not receive much empirical support (Madrigal, 1995; Smith, 1990).

Another approach for analysing interpersonal differences in tourists' motivations adopted well-known and established personality theories rather than developing new constructs. The work of Pearce and his colleagues (Moscardo and Pearce, 1986; Pearce, 1988, 1993; Pearce and Lee, 2005) was partly based on Maslow's hierarchy of needs (Maslow, 1968, 1970) and expanded this theory to two dimensions of analyses. The first dimension addressed travellers' needs (such as relaxation, safety and fulfilment), whereas the second dimension referred to past experiences, or the tourist's travel career. However, Pearce and Lee (2005) indicated that the majority of tourists shared a common pattern of motivations or needs, such as novelty, escape and enhancing relationships. This pattern of motivations is congruent with the theory of motivation for recreational travel (Iso-Ahola, 1983). According to Iso-Ahola (1983), the motivations of novelty and change are the foundations of the social activity of recreational travel. However, Iso-Ahola indicated that the combination of the desire to escape everyday life and the need to maintain interpersonal relationships might create a contradiction that the tourist must solve to obtain an optimal solution.

Both Pearce and Lee (2005) and Iso-Ahola (1983) suggest that the motivations for change, novelty and social relations affect the behaviours and experiences of different people in different trips, vacations and locations. Thus, although the "tourist experience" may have different expressions (e.g. backpacking, luxury travels or family vacations), its foundations and motivations are universal. Moreover, it may be reasoned that the motivations of novelty, escaping and enhancing relationships drive the choice of destinations farther from home within constraints posed by travel costs, time and, to some extent, the relationship element. Interestingly, Pearce and Lee (2005) found that more experienced travellers tend to seek social relations with locals in different cultures. This motivation may enhance the importance of distance especially among experienced travellers, instead of focusing on social relations with friends and relatives in known and close destinations. Nicolau (2008, p. 50) noted that

> a greater willingness to travel longer distances is associated with high income (although with a saturation point); with large cities ... with the use of intermediaries (as their use allows a reduction in the inherent uncertainty of long-distance destinations and to save time in the organization of multi-component trips) ... with the interest of discovering new places (as people with this interest may be willing to cover long distances to satisfy this intellectual need, according to the Ulysses factor); with the variety-seeking behavior...

The importance of distance

According to the sociological approach, tourism settings allow tourists to have experiences that differ from daily life. Goffman (1963) defined the tourism context as "action spaces" where tourists are allowed – sometimes even encouraged – to experience extraordinary adventures. Psychoanalytical sociology explained the permissiveness of tourism settings by exploring the mental mechanisms that cause tourists to behave in unusual ways. Specifically, it emphasised the idea that perceived distance enables tourists to utilise defence mechanisms such as sublimation, projection and omnipotence, which convert unconscious needs to unusual normative or alternatively deviant experiences. Examples of this idea are the normative participation in safari tours (normative sublimation of unconscious aggression needs) or, alternatively, a deviant participation in elephant hunts (deviant projection of unconscious aggression needs; Uriely *et al.*, 2011).

The relationship between the perception of distance and opportunities to express unconscious needs through unusual experiences affects all five phases of the "tourist experience". It shapes pre-trip expectations, stresses the importance of long-haul travel, explains various behaviours and experiences on-site and is involved in the mental process of recollection. Distance is an integral part of the tourist experience. Without the element of distance, the experience will not be fulfilled and completed, negatively affecting the happiness of the tourists.

However, the important point is "perceived" distance and the role that it plays in tourists' fantasies. Exotic destinations often "epitomize the dream of the average traveler" (Buhalis, 2000, p. 103). Tourists believe that vacationing at far away or exotic destinations makes them happier, although no empirical evidence suggests that. Burns and Bibbings (2009) suggested that the positive image of far away or exotic destinations is related to Maslow's concept of "peak experience", which is a social desire that can be changed. This social desire for distance may be translated into an intensity bias in affective forecasting. According to Buehler and McFarland (2001), this bias causes individuals to overestimate the effect that an event has on their happiness. Similarly, Wilson and Gilbert (2005) named this phenomenon the impact bias in affective forecasting. According to their research, focalism is a major cause of this bias; people focus on one aspect, the effect of which they consequently overestimate. Kahneman *et al.* (2006) observed the same phenomenon and refer to it as a focusing illusion.

In this context, it is worth mentioning that the concepts of time and space are also socially constructed (Molz, 2009). Miller *et al.* (2010) argued that longer vacations are more difficult to arrange because of work and family obligations, but Dickinson and Peeters (in press) demonstrated that this does not decrease travel distances. Interestingly, the opposite was observed when increase in travel speed almost automatically increased distances instead of saving travel time. The importance of distance was found even for tourists who return to the same destinations every year. Selwood *et al.* (2004) demonstrated that in the 1920s, second homes were built less than 20 km from their owners' urban primary homes, whereas two-thirds of people in the same area (Perth) had second homes 400 km away in the 2000s.

Subjective well-being and happiness

Happiness and subjective well-being are umbrella terms used for several types of cognitive well-being measures and affective indicators of well-being. When people assess their lives, they draw on two sources of information: how well they feel and to what extent their lives meet their expectations (Veenhoven, 2009). The former is a more affective component of happiness and consists of moods, affect and emotions (Diener, 1984). The latter is more cognitive in nature and concerns life satisfaction and domain satisfactions (e.g. job satisfaction; Veenhoven, 1984). Affect is a term used to describe general feelings or an entire category of emotions (e.g. positive emotions). Emotions are more intense than mood. Emotions are more strongly felt but briefer than moods; emotions are caused by events, rather than occurring within a person, as is the case with moods (Beedie *et al.*, 2005).

Nawijn's studies (e.g. Nawijn, 2011b; Nawijn *et al.*, 2013) suggest that individuals become happier by vacationing, but only temporarily. Vacations have no lasting effect on tourists' life satisfaction (Nawijn, 2011b). Life satisfaction scores are elevated briefly after returning home (Lounsbury and Hoopes, 1986) or when tourists are asked to recall past vacations and immediately assess their life satisfaction after this instruction (Sirgy *et al.*, 2011). Post-trip effects are present for a minority of tourists and wear off quickly (Nawijn *et al.*, 2010). Vacations do have an effect on emotion. Tourists experience a peak in positive emotions during their vacation (Mitas *et al.*, 2012). Even after returning home, approximately 40 per cent of tourists experience increased levels of positive to negative emotions for about two weeks (Nawijn *et al.*, 2010). In everyday life, vacationers' emotional balance (the difference between positive and negative emotions) is slightly higher than that of non-vacationers (Nawijn, 2011b). Thus, positive feelings are important not only during the on-site phase of the "tourist experience" but also during the fifth phase of recollection. Negative emotions, conversely, were not related to memories of vacations (Tung and Ritchie, 2011a, 2011b). An underlying explanatory factor of this increase in positive emotions is a sense of relatedness and autonomy (Ryan and Deci, 2000).

Mobility

Transport physically moves people or goods "from A to B as quickly and as smoothly as possible" (Peters, 2006, p. 3). For tourism transport, the idea of transportation always being a cost to be minimised might not be always true because "it is also appropriate to differentiate transport as a means to an end and transport which is integral to the tourism experience" (Lumsdon and Page, 2004, p. 2). Mobility is a characteristic of the persons (or goods) to be transported. As we discuss tourists' psychology, mobility seems a more appropriate term than transport. Mobility can be expressed in many terms, but common ones are trips and travel distance, time, speed and cost. These properties are connected, and the connections form an important background to our model. The main connections are budgets for travel time, money and a constancy of number of trips per person per day (Schäfer, 2000). The travel time budget is found to be constant at approximately 70 minutes on average per day per person, the money budget is approximately 10 per cent of personal income (Schäfer, 2000), and the number of trips per day is constant at approximately 4–5 (Hupkes, 1982) for a very wide range of populations (cities and countries) worldwide and with varying income and transportation infrastructure.

The constants (trips, travel time and money budget) are statistical artefacts, meant to be used at population level only, but they result from averaging all individual mobility patterns. An important consequence of these constants is that increases in the average speed of transport systems will increase the distance people travel rather than saving time (Banister, 2011; Metz, 2008). In terms of average distances travelled, mobility increases with wealth because increased income gives access to faster transport modes such as car (Arentze and Timmermans, 2011) and air transport (Schäfer *et al.*, 2009). Therefore, the common policy to follow demand with infrastructure improvements extends not only capacity but also network density, efficiency quality and overall transport speed, thus creating additional travel demand (Metz, 2008). Theoretically, it can be demonstrated that a positive policy–demand–infrastructure feedback loop exists and naturally will lead to ever increasing mobility and distances travelled (Peeters, 2010).

The "happiness loop" and unsustainable tourist mobility behaviour

One possible conclusion from the literature review is that a relationship may exist between tourist experience, happiness, perception of distance and the concept of mobility. In light of Xin *et al.*'s (2013) analysis, these different concepts combined may provide new insights on an old problem. Therefore, these concepts were integrated into models, seeking to shed light on the two goals of the current study: enhancing the understanding of the psychological causes for unsustainable tourist mobility behaviour and demonstrating how happiness hinders desirable behavioural change towards more sustainable tourist mobility behaviour. The first model, the "happiness loop" (presented in Figure 3.1) demonstrates how the tourist experience theory is integrated with the concept of happiness. According to this model, the positive emotions of happiness are not only linked exclusively to on-site experiences but also affect expectations and recollection. In that sense, every good experience may translate to pleasant memories that produce anticipation for the next trip. However, the phases of travelling to and from the destination are placed outside the happiness loop, because according to the happiness theories, mobility, alone, is not considered a source of happiness. This model suggests that tourists may be aware of the negative aspects of mobility because mobility aspects threaten sustainability, but they may perceive the mobility aspects as a "necessary evil" that enables their tourism-related happiness. In other words, this model offers a possible explanation for the awareness-attitude gap. Tourists may ignore the negative aspects of mobility to preserve their happiness. Therefore, any attempt to promote behavioural change in relation to sustainable tourist mobility behaviour would face resistance because it may reduce happiness.

The happiness loop shows only part of the whole system of tourist behaviour and has multiple relations, direct and indirect, with tourist motivations, perception of distance and mobility. Satisfying tourist motivations, and specifically motivations for novelty and change, will accelerate the happiness loop. Furthermore, satisfying the perception of distance, mainly by means of mobility, will enhance the motivations for novelty and change, and as a result would strengthen the happiness loop. These interrelations are presented in Table 3.1. Given the psychological motivation of tourists for novelty and change, there is a continuous demand for more exciting experiences that may require travelling greater distances (cf. Jang and Feng, 2007; Raghunathan and Irwin, 2001). This process may be accelerated through the components of the happiness loop (i.e. the phases of anticipation, on-site experiences and recollection). According to the suggested model, travel is planned based on previous travel experiences. After someone starts to travel, he/she would plan the next trip in a way perceived as better than the last trip, based on past experiences. According to this theoretical analysis, the motivations for novelty and change both affect the happiness loop *and* increase the distance component. Figure 3.2 presents this concept with a metaphor of a three-gear wheel system in which one wheel (the novelty and change motivations) is activating the other two, the happiness loop and travel distance.

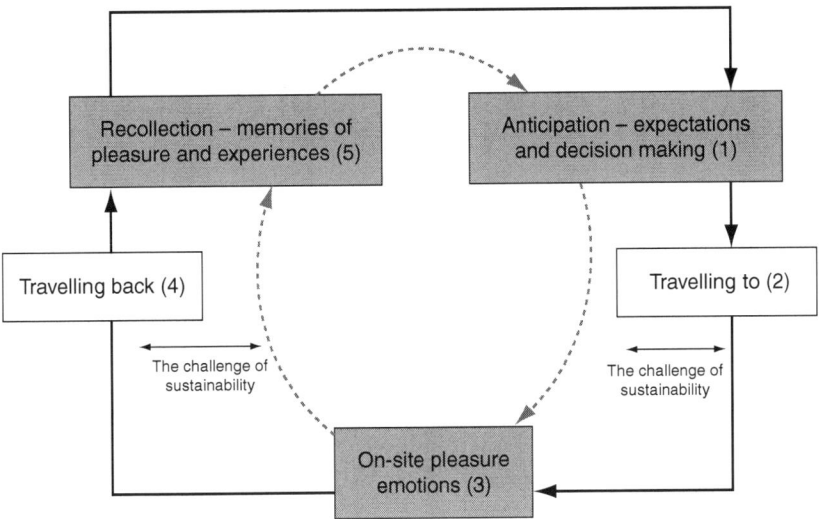

Figure 3.1 The tourism "happiness loop".

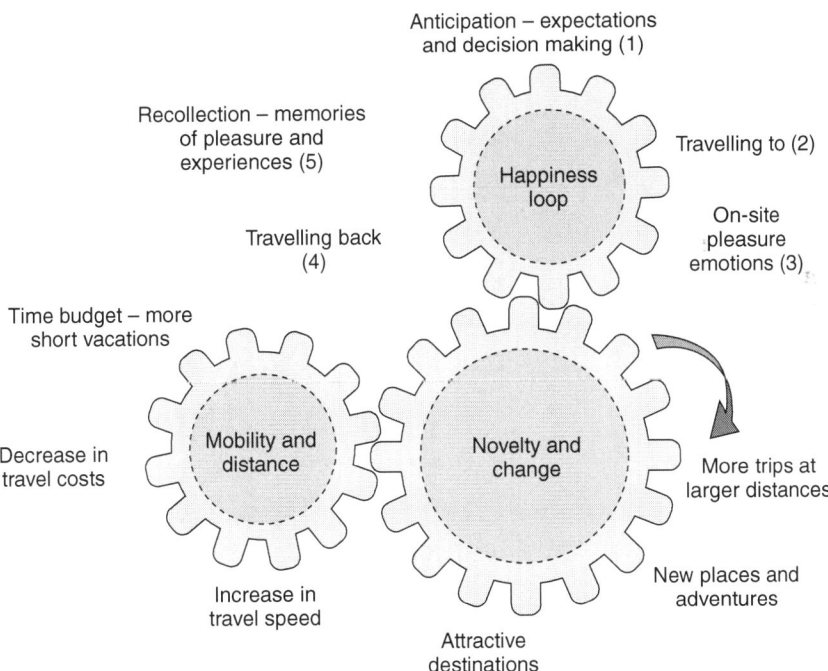

Figure 3.2 The "three-gear" model of unsustainable tourist behaviour.

Table 3.1 Overview of the main constructs in the "three-gear model"

The construct	References	Possible relations of the construct with the "happiness loop"	Possible relations of the construct with other components of the "three-gear model"
Tourists' motivations – specifically novelty and change	Iso-Ahola (1983), Pearce and Lee (2005)	Satisfying tourists' motivations driving the happiness loop	Mobility and (increasing) distance are essential elements in satisfying tourist motivations
Perception of distance	Buhalis (2000), Burns and Bibbings (2009), Dickinson and Peeters (in press), Goffman (1963), Selwood et al. (2004), Uriely et al. (2011)	The perception that novelty and change are better found at greater travel distance satisfies tourist motivations, which drives the happiness loop (indirect relation to the happiness loop)	The perception of the "value" of distance requires mobility
Mobility	Arentze and Timmermans (2011), Banister (2011), Dickinson et al. (2010), Lumsdon and Page (2004), Metz (2008), Peeters (2010), Schäfer et al. (2009)	Mobility becomes cheaper and faster and thus enables longer distances, which drives the happiness loop (indirect relations to the happiness loop)	Mobility is a fundamental condition for consuming distance; cheaper/faster mobility incentivises longer distances

The three-gear model does not suggest that the motivations for novelty and change increase over time. Instead, the motivations are the driver for gradual increases in travelled distance over time because the reference point changes with experience. This outcome was also supported by Pearce and Lee's (2005) study, which demonstrated that, on average, older people have more experience in travel than younger people, although both groups share the same motivations for novelty and change.

In general, a transport policy–infrastructure–demand system favours faster and cheaper transport increasing the demand for distances travelled. For tourism travel, this tendency toward longer travel distances is amplified by the happiness loop. This makes the quest for sustainable tourism transportation even more challenging than other transport motives. Links between distance, mobility and happiness directly hinder two of the most desirable behavioural changes with regards to sustainable tourist mobility, namely, shifting from long-haul travel to medium/short-haul travel, and the call for renouncing aircraft and private cars for trains and coaches.

A critical review of the model

Xin *et al.* (2013) highlighted the need for concept scepticism as an integral part of conceptual analysis. According to this principle, issues that might contradict the model should always be discussed. Three subjects were identified as possible critiques for the three-gear model: the habituation tendency, the overemphasising of good experiences compared to bad experiences and the causation issue between happiness and travel frequency. According to the habituation theory (see Groves and Thompson, 1970), there is an erosion of responses to repetitive stimuli. Thus, the three-gear model will eventually stop working, not because of external forces, but because of habituation. Alternatively, the model will require more energy to continue working. The latter view is supported by the hedonic treadmill theory and metaphor (Brickman and Campbell, 1971). This believes that individuals compare purchases and events over time; pleasure derived from an initial event or purchase is always greater than that from the consecutive purchase or event. In other words, individuals adapt to an event and habitation takes place, establishing a new norm (Kahneman and Miller, 1986). This is most likely the case for vacations. Happiness derived from the same product (a vacation) means that the next vacation must be better (longer, farther, more exotic) to derive the same happiness. Experimental research (Raghunathan and Irwin, 2001) determined that when individuals are asked to report the happiness they expect to derive from visits to certain vacation spots, they compare destinations and their expected predicted effect on happiness. Raghunathan and Irwin's (2001) study supports the hedonic treadmill theory in a vacation context. Not all tourists seek novelty in their vacations, but both McKercher and Du Cros (2003) and Burns and Bibbings (2009) argued that social pressure influences many to look for "better" vacations.

The second subject, the importance of positive emotions and experiences over negative emotions and experiences, was supported by studies from different disciplines. First, people are more likely to remember positive events (compared to bad events) to enhance their self-esteem and well-being (Walker *et al.*, 1996). Thus, their whole information process is based on biased practices (Frey, 1986), including selective attention that blocks unpleasant information (Hart *et al.*, 2009) and selective recall of positive memories (Walker *et al.*, 1996). Second, given that people feel much better on vacations compared to their everyday lives (Nawijn, 2011a), their happy memories are reference points, triggering new trips. Bad experiences may reverse the three-gear model's direction, but as long as there is a social acceptance of travel as pleasure sources (Burns and Bibbings, 2009), and good experiences tend to be remembered (Hart *et al.*, 2009; Tung and Ritchie, 2011a; Walker *et al.*, 1996), then the reverse process would not occur frequently. However, one may consider inserting an additional loop from "bad experience" to "congenial ignorance" and back to the positive loop in the current model.

The third issue addresses the causation problem: Are happy people going on vacations, or the opposite – do vacations make people happy? Although important, this question is beyond the scope of this study, with its main argument that happiness plays an important role in the tourist experience and hinders attempts for behavioural change in leisure mobility.

Discussion and conclusion

Happiness is an integral part of the tourist experience. It is part of the decision making (Hart *et al.*, 2009), the experience (Nawijn, 2011a) and the memories (Tung and Ritchie, 2011a). Understanding happiness enhances the understanding of tourist behaviour in general. Including the concept of happiness in our three-gear conceptual model gives a better understanding of both the awareness-attitude gap and the contextual gap. The awareness-attitude gap reflects the contradiction between tourists' understanding of the negative impacts of their mobility patterns and their attitudes favouring such behaviours. The contextual gap refers to people's resistance to extend environmentally friendly daily mobility practices to the tourism context. According to the three-gear model, these two gaps result from tourists' search for happiness, which for many depends on perceived distance and mobility. Furthermore, the tourists' demand to travel farther and faster increases over time through a continuous process that involves elements of the tourism happiness loop together with the need for change and novelty. Happiness has been shown to play a role in the unsustainable development of tourism. Specifically, the role of seeking happiness in the continued increase of distances tourists travel causes further unsustainable development. The happiness loop also hinders change towards less distance-intense forms of tourism behaviour.

Using the three-gear model, there are three options for policies promoting behavioural changes in tourist mobility. First, policies may attach to the wheel of distance by changing current transport infrastructure policies, thus disrupting the speed–distance–demand feedback loop. Second, policies may be aimed at altering the motivations of novelty and change (e.g. the wheel of motivation). Third, policies may aim to implement elements of the happiness loop wheel for a tourism mobility context.

Policies proposed to break the speed–distance–demand loop include increases in the cost of less environmentally benign transport modes through taxes on carbon emissions (Tol, 2007) and airline tickets (Mayor and Tol, 2007) or by including aviation in carbon trading (Mayor and Tol, 2010). The advantage of taxing is that rebounds are generally small or non-existent, whereas they are common in efficiency measures that increase demand by reducing fuel consumption and cost, for example. However, all past studies reveal that the impact of financial measures in aviation is small and sometimes the changes cause undesired side-effects, such as in the UK ticket tax (Mayor and Tol, 2007). Still, for car and petrol taxes, for example, the historical effect has been shown to be substantial (Sterner, 2007). Another potentially effective approach would be to increase infrastructure capacity selectively such as by prioritising rail infrastructure over airports (Åkerman, 2011). Interestingly, scenario analysis indicates that a shortage of airport capacity will significantly affect future air travel demand (Evans and Schäfer, 2011). This means that the standard policy of demand following infrastructure should be reversed so that environmental limits induce constraints on capacity, which could potentially be more effective than most pricing policies. Of course, such policy changes are not easy because it requires strong behavioural change from both policy makers and the public. Behavioural change by policy makers is probably the main issue: Dickinson *et al.* (2010) found that holiday travel is constrained by existing structures within the travel and tourism industries.

Most previous research that focused on the motivations of tourists to engage in behavioural change (the wheel of motivation in the three-gear model) (Antimova *et al.*, 2012; Hares *et al.*, 2010; Miller *et al.*, 2010) found that tourists would not voluntarily change their motives. Interestingly, they pointed back to the wheel of policies, and claimed that interventions in mobility policies are required. Becken (2007, p. 365) noted that "everyone waits for someone else to do something". However, Antimova *et al.* (2012) and Miller *et al.* (2010) argued that societal change would be more efficient than policy change because social norms and role models have more influence on personal motivations and would more effectively bridge the gap between awareness and attitudes.

Focusing on the wheels of motivations and policies, as in the studies above, excluded two factors from the discussion: the "happiness loop" and the gap between everyday practices and tourism mobility. Therefore, one negative theoretical outcome in implementing policies based on motivations alone can be that people will act in sustainable ways at home but use that behaviour to ignore the negative environmental impacts of their vacations, which are generally more severe.

Our conceptual model supports shifting policies for behavioural change to the third wheel, the wheel of the "happiness loop". The "happiness loop" is composed of three distinct parts: the anticipation stage, the on-site experience and the recollection stage. We observe that the environmental impacts are located outside the "happiness loop", in the stages of travelling. This means that strategies to further improve the anticipation, the onsite experience and recollection stages could theoretically improve happiness. However, this can only be achieved if the distance-generating transport policy related loop would be disrupted by policies that limit capacity rather than price. This would be an important departure from current transport policies that follow demand, disregarding transport mode and volume.

Theoretical contribution

In his review of sustainable tourism, Buckley (2012, p. 537) pointed to immediate priorities for future research. He indicated the issue of individuals' reaction to "responsibility in light of global change" as one of the two most important and promising fields in this context (the other is land-use change). This study corresponds well with this call and suggests a model that focuses on tourists' reactions. Our model has three theoretical contributions. First, it presents a synthesis of theories to interrogate two behavioural gaps known from previous studies: the gap between awareness and attitudes and the gap in sustainability practices between everyday life and vacation circumstances. Second, the model indicates the desire for happiness as a potential cause for both gaps, influenced by motivations of novelty and change as well as the transport-infrastructure–policy–demand feedback loop causing distances travelled to increase over time. Third, it provides a theoretical basis for future empirical studies: the conceptual research described here is a necessary preliminary stage in the empirical and objective method (Xin *et al.*, 2013). Further, the chapter has three contributions to the discussion about the links between psychology and tourism. First, by addressing the issue of behavioural change in general; second, by providing an important bridge between psychology and sustainable tourism; and third, by focusing on emotions, whereas most previous studies in the field of tourism concentrated on more cognitive aspects, such as decision making processes, memory and awareness.

Practical contribution

Previous studies of behavioural change and tourism mobility highlighted the need for policies that relate to social change, because the industry will not improve itself (Antimova *et al.*, 2012; Miller *et al.*, 2010; Verbeek and Mommaas, 2008). These studies called for making the tourists the change agents, through their sense of responsibility (Miller *et al.*, 2010), their guilty feelings (Cohen *et al.*, 2011) and their patterns of political consumerism (Verbeek and Mommaas, 2008). This study questions these insights by placing the happiness of tourists, rather than their consciousness or rationality, at the centre of its analysis. It shows that future solutions for the current unsustainable tourism mobility problem should address the happiness of tourists. This sheds new light on Buckley's (2012, p. 535) remark:

> People want holidays, and on holiday they act hedonistically. The most populous nations are richer, so more people travel.... Amidst these pressures, large-scale voluntary improvements in sustainability are improbable, especially given low public pressure for sustainability and the particular ambivalence to tourism.

In practical terms, tourism and transport policy makers may consider developing a "happiness scale" to test the implications of regulating leisure mobility. In this context, it is worth mentioning that soft measures, such as a "happiness scale", are becoming more common. For example, the Human Development Index, which follows the traditional gross domestic product rankings and well-being scales, is utilised as an integral part of reports of the Organisation of Economic Co-operation and Development.

In addition, future policies for breaking the speed–distance–demand loop may refer to tourists' emotional state, particularly during the stages of on-site experience and recollection. Policy makers should keep in mind that the happiness of tourists is a fundamental condition for successful implementation of policies about leisure mobility.

Study limitations

Future studies based on the models presented here face two major limitations. The first limitation relates to the wide range of models and theories describing human behaviour. The motivation theory that was taken as the base for our model is only one of many. However, the entire range of human behaviour is beyond the scope of the current study, which focused on behavioural change and aimed to explain the two gaps in the leisure mobility literature: the awareness-attitude gap and the contextual gap. The second limitation focuses on individual differences. The current model overlooked individual differences and suggested a homogenous model of happiness and motivations. Future studies may refer to this issue, further elaborating on the theory, and include other motivations, such as seeking personal meaning or relaxation. Uriely (2005) noted that conceptual theories about the "tourist experience" tend to be developed from an initial homogenous concept and expand later to portray interpersonal differences.

References

Ajzen, I. (1991). The theory of planned behavior. *Organizational Behavior and Human Decision Process, 50*(2), 179–211.

Åkerman, J. (2011). The role of high-speed rail in mitigating climate change – the Swedish case Europabanan from a life cycle perspective. *Transportation Research Part D: Transport and Environment, 16*(3), 208–217.

Antimova, R., Nawijn, J. and Peeters, P. (2012). The awareness/attitude-gap in sustainable tourism: A theoretical perspective. *Tourism Review*, *67*(3), 7–16.

Arentze, T.A. and Timmermans, H.J.P. (2011). A dynamic model of time-budget and activity generation: Development and empirical derivation. *Transportation Research Part C: Emerging Technologies*, *19*(2), 242–253.

ATAG. (2010). *The right flight path to reduce aviation emissions*. Geneva: Air Transport Action Group.

Banister, D. (2011). The trilogy of distance, speed and time. *Journal of Transport Geography*, *19*(4), 950–959.

Barr, S., Gilg, A. and Shaw, G. (2011). "Helping people make better choices": Exploring the behaviour change agenda for environmental sustainability. *Applied Geography*, *31*(2), 712–720.

Barr, S., Shaw, G., Coles, T. and Prillwitz, J. (2010). "A holiday is a holiday": Practicing sustainability, home and away. *Journal of Transport Geography*, *18*(3), 474–481.

Becken, S. (2007). Tourists' perception of international air travel's impact on the global climate and potential climate change policies. *Journal of Sustainable Tourism*, *15*(4), 351–368.

Beedie, C.J., Terry, P.C. and Lane, A.M. (2005). Distinctions between emotions and mood. *Cognition & Emotion*, *19*(6), 847–878.

Berger, I.E. (1997). The demographics of recycling and the structure of environmental behavior. *Environment and Behavior*, *29*(4), 515–531.

Brickman, P. and Campbell, D.T. (1971). Hedonic relativism and planning the good society. In M.H. Aplley (ed.), *Adaptation level theory: A symposium* (pp. 287–302). New York: Academic Press.

Buckley, R. (2012). Sustainability tourism: Research and reality. *Annals of Tourism Research*, *59*(2), 528–546.

Buehler, R. and McFarland, C. (2001). Intensity bias in affective forecasting: The role of temporal focus. *Personality and Social Psychology Bulletin*, *27*(11), 1480–1493.

Buhalis, D. (2000).Marketing the competitive destination of the future. *Tourism Management*, *21*(1), 97–116.

Burns, P. and Bibbings, L. (2009). The end of tourism? Climate change and societal challenges. *Twenty-First Century Society: Journal of the Academy of Social Sciences*, *4*(1), 31–51.

Cabrini, L., Simpson, M. and Scott, D. (2009). *From Davos to Copenhagen and beyond: Advancing tourism's response to climate change*. UNWTO Background Paper. Madrid: UNWTO.

Catlin, J.R. and Wang, Y. (2013). Recycling gone bad: When the option to recycle increases resource consumption. *Journal of Consumer Psychology*, *23*(1), 122–127.

Clawson, M. and Knetsch, J.L. (1966). *Economics of outdoor recreation*. Baltimore, MD: Johns Hopkins.

Cohen, S.A., Higham, J.E.S. and Cavaliere, C.T. (2011). Binge flying: Behavioural addiction and climate change. *Annals of Tourism Research*, *38*(3), 1070–1089.

de Bruijn, K., Dirven, R., Eijgelaar, E. and Peeters, P. (2012). *Travelling large in 2010: The carbon footprint of Dutch holidaymakers in 2010 and the development since 2002*. Breda: NHTV University for Applied Sciences in collaboration with NBTC-NIPO Research.

Dickinson, J. and Peeters, P. (in press). Time, tourism consumption and sustainable development. *International Journal of Tourism Research*. doi: 10.1002/jtr.1893.

Dickinson, J.E., Robbins, D. and Lumsdon, L. (2010). Holiday travel discourses and climate change. *Journal of Transport Geography*, *18*(3), 482–489.

Diekmann, A. and Preisendörfer, P. (2003). Green and greenback: The behavioral effects of environmental attitudes in low-cost and high-cost situations. *Rationality and Society*, *15*(4), 441–472.

Diener, E. (1984). Subjective well-being. *Psychological Bulletin*, *95*(3), 542–575.

Dubois, G., Ceron, J.P., Peeters, P. and Gössling, S. (2011). The future tourism mobility of the world population: Emission growth versus climate policy. *Transportation Research – A*, *45*(10), 1031–1042.

Evans, A. and Schäfer, A. (2011). The impact of airport capacity constraints on future growth in the US air transportation system. *Journal of Air Transport Management*, *17*(5), 288–295.

Festinger, L. (1957). *A theory of cognitive dissonance*. Evanston, IL: Row Peterson.

Frey, D. (1986). Recent research on selective exposure to information. In L. Berkowitz (ed.), *Advances in experimental social psychology* (Vol. 19, pp. 41–80). Orlando, FL: Academic Press.

Fridgen, J.D. (1984). Environmental psychology and tourism. *Annals of Tourism Research*, *11*(1), 19–39.

Goffman, E. (1963). *Stigma: Notes on the management of spoiled identity*. Englewood Cliffs, NJ: Prentice-Hall.

Groves, P.M. and Thompson, R.F. (1970). Habituation: A dual-process theory. *Psychological Review*, *77*(5), 419.

Haggett, P. and Chorley, R.J. (1967). Models, paradigms and the new geography. In R.J. Chorley and P. Haggett (eds), *Models in geography* (pp. 20–41). London: Methuen.

Hansen, J. (2008). *Global warming twenty years later: Tipping points near*. Retrieved from www.tipping-points.com/?cat=10.

Hansen, J., Sato, M., Kharecha, P., Beerling, D., Berner, R., Masson-Delmotte, V., Pagani, M., Raymo, M., Royer, D. L. & Zachos, J. C. (2008) Target Atmostpheric CO2: Where Should Humanity Aim? The Open Atmospheric Science Journal, 2 (15), 217–231.

Hares, A., Dickinson, J. and Wilkes, K. (2010). Climate change and the air travel decisions of UK tourists. *Journal of Transport Geography*, *18*(3), 466–473.

Hart, W., Albarracín, D., Eagly, A.H., Brechan, I., Lindberg, M.J. and Merrill, L. (2009). Feeling validated versus being correct: A meta-analysis of selective exposure to information. *Psychological Bulletin*, *135*(4), 555–588.

Higham, J.E.S. and Cohen, S.A. (2011). Canary in the coalmine: Norwegian attitudes towards climate change and extreme long-haul air travel to Aotearoa/New Zealand. *Tourism Management*, *32*(1), 98–105.

Hupkes, G. (1982). The law of constant travel time and trip-rates *Futures*, *14*(1), 38–46.

Iso-Ahola, S.E. (1983). Towards a social psychology of recreational travel. *Leisure Studies*, *2*(1), 45–56.

Jang, S. and Feng, R. (2007). Temporal destination revisit intention: The effects of novelty seeking and satisfaction. *Tourism Management*, *28*(2), 580–590.

Kahneman, D., Krueger, A.B., Schkade, D., Schwarz, N. and Stone, A.A. (2006).Would you be happier if you were richer? A focusing illusion. *Science*, *312*(5782), 1908–1910.

Kahneman, D. and Miller, D.T. (1986). Norm theory: Comparing reality to its alternatives. *Psychological Review*, *93*(2), 136–153.

Lee, D.S., Lim, L.L. and Owen, B. (2013). *Bridging the aviation CO$_2$ emissions gap: Why emissions trading is needed*. Manchester: Manchester Metropolitan University. Retrieved from www.cate.mmu.ac.uk/wp-content/uploads/Bridging_the_aviation_emissions_gap_010313.pdf.

Lounsbury, J.W. and Hoopes, L.L. (1986). A vacation from work: Changes in work and nonwork outcomes. *Journal of Applied Psychology, 71*(3), 392–401.

Lumsdon, L. and Page, S.J. (2004). Progress in transport and tourism research: Reformulating the transport-tourism interface and future research agendas. In L. Lumsdon and S.J. Page (eds), *Tourism and transport. Issues and agenda for the new millennium* (1st edn, pp. 1–27). Amsterdam: Elsevier.

Madrigal, R. (1995). Personal values, traveler type, and leisure travel style. *Journal of Leisure Research, 27*(2), 125–142.

Maslow, A.H. (1968). *Toward a psychology of being.* New York: Van Nostrand.

Maslow, A.H. (1970). *Motivation and personality* (2nd edn). New York: Harper & Row.

Mayor, K. and Tol, R.S.J. (2007). The impact of the UK aviation tax on carbon dioxide emissions and visitor numbers. *Transport Policy, 14*(6), 507–513.

Mayor, K. and Tol, R.S.J. (2010). The impact of European climate change regulations on international tourist markets. *Transportation Research Part D: Transport and Environment, 15*(1), 26–36.

McGuiggan, R. and Foo, J.A. (2004). Who plays which tourist rules? – An Australian perspective. *Journal of Travel & Tourism Marketing, 17*(1), 41–54.

McKercher, B. and Du Cros, H. (2003). Testing a cultural tourism typology. *International Journal of Tourism Research, 5*(1), 45–58.

McKercher, B., Prideaux, B., Cheung, C. and Law, R. (2010). Achieving voluntary reductions in the carbon footprint of tourism and climate change. *Journal of Sustainable Tourism, 18*(3), 297–317.

Metz, D. (2008). *The limits to travel: How far will we go?* London: Earthscan.

Miller, G., Rathouse, K., Scarles, C., Holmes, K. and Tribe, J. (2010). Public understanding of sustainable tourism. *Annals of Tourism Research, 37*(3), 627–645.

Mitas, O., Yarnal, C., Adams, R. and Ram, N. (2012). Taking a "peak" at leisure travelers' positive emotions. *Leisure Sciences, 34*(2), 115–135.

Molz, J.G. (2009). Representing pace in tourism mobilities: Staycations, slow travel and the amazing race. *Journal of Tourism and Cultural Change, 7*(4), 270–286.

Moscardo, G. and Pearce, P.L. (1986). Historical theme parks: An Australian experience in authenticity. *Annals of Tourism Research, 13*(3), 467–479.

Nawijn, J. (2010). The holiday happiness curve: A preliminary investigation into mood during a holiday abroad. *International Journal of Tourism Research, 12*(3), 281–290.

Nawijn, J. (2011a). Determinants of daily happiness on vacation. *Journal of Travel Research, 50*(5), 559–566.

Nawijn, J. (2011b). Happiness through vacationing: Just a temporary boost or long-term benefits? *Journal of Happiness Studies, 12*(4), 651–665.

Nawijn, J., De Bloom, J. and Geurts, S. (2013). Pre-vacation time: Blessing or burden? *Leisure Sciences, 35*(1), 33 44.

Nawijn, J., Marchand, M., Veenhoven, R. and Vingerhoets, A. (2010). Vacationers happier, but most not happier after a holiday. *Applied Research in Quality of Life, 5*(1), 35–47.

Nawijn, J., Mitas, O., Lin, Y. and Kerstetter, D. (2013). How do we feel on vacation? A closer look at how emotions change over the course of a trip. *Journal of Travel Research, 52*(2), 265–274.

Nicolau, J.L. (2008). Characterizing tourist sensitivity to distance. *Journal of Travel Research, 47*(1), 43–52.

Ölander, F. and Thøgersen, J. (1995). Understanding of consumer behaviour as a prerequisite for environmental protection. *Journal of Consumer Policy, 18*(4), 345–385.

Owen, B., Lee, D.S. and Lim, L. (2010). Flying into the future: Aviation emissions scenarios to 2050. *Environmental Science & Technology*, *44*(7), 2255–2260.

Parry, M., Lowe, J. and Hanson, C. (2009). Overshoot, adapt and recover. *Nature*, *458*(7242), 1102–1103.

Pearce, D.G. (2008). A needs-functions model of tourism distribution. *Annals of Tourism Research*, *35*(1), 148–168.

Pearce, P.L. (1988). *The Ulysses factor: Evaluating visitors in tourist settings*. New York: Springer-Verlag.

Pearce, P.L. (1993). Fundamentals of tourist motivation. In D.G. Pearce and R.W. Butler (eds), *Tourism research: Critiques and challenges* (pp. 113–134). London and New York: Routledge.

Pearce, P.L. and Lee, U.-I. (2005). Developing the travel career approach to tourist motivation. *Journal of Travel Research*, *43*(3), 226–237.

Pearce, P.L. and Packer, J. (2013). Minds on the move: New links from psychology to tourism. *Annals of Tourism Research*, *40*(1), 386–411.

Peeters, P. (2010). Tourism transport, technology, and carbon dioxide emissions. In C. Schott (ed.), *Tourism and the implications of climate change: Issues and actions* (Vol. 3, pp. 67–90). Bingley: Emerald.

Peeters, P.M. and Dubois, G. (2010). Tourism travel under climate change mitigation constraints. *Journal of Transport Geography*, *18*, 447–457.

Peeters, P., Gössling, S. and Lane, B. (2009). Moving towards low-carbon tourism: New opportunities for destinations and tour operators. In S. Gössling, C.M. Hall and D.B. Weaver (eds), *Sustainable tourism futures: Perspectives on systems, restructuring and innovations* (Vol. 15, pp. 240–257). New York: Routledge.

Peters, P. (2006). *Time, innovation and mobilities: Travel in technological cultures*. London: Taylor & Francis.

Pine II, B.J. and Gilmore, J.H. (1998). Welcome to the experience economy. *Harvard Business Review*, *July–August*, 97–105.

Plog, S. (1974). Why destination areas rise and fall in popularity. *Cornell Hotel and Restaurant Administration Quarterly*, *14*(4), 55–58.

Plog, S. (2002). The power of psychographics and the concept of venturesomeness. *Journal of Travel Research*, *40*(3), 244–251.

Raghunathan, R. and Irwin, J.R. (2001). Walking the hedonic product treadmill: Default contrast and mood-based assimilation in judgments of predicted happiness with a target product. *Journal of Consumer Research*, *28*(3), 355–368.

Rashid, N.R.N.A. and Mohammad, N. (2012). A discussion of underlying theories explaining the spillover of environmentally friendly behavior phenomenon. *Procedia – Social and Behavioral Sciences*, *50*, 1061–1072.

Rockstrom, J., Steffen, W., Noone, K., Persson, A., Chapin, F.S., Lambin, E.F., … and Foley, J.A. (2009). A safe operating space for humanity. *Nature*, *461*(7263), 472–475.

Ryan, R.M. and Deci, E.L. (2000). Self-determination theory and the facilitation of intrinsic motivation, social development, & well-being. *American Psychologist*, *55*(1), 68–78.

Schäfer, A. (2000). Regularities in travel demand: An international perspective. *Journal of Transportation and Statistics*, *3*(3), 1–31.

Schäfer, A., Heywood, J.B., Jacoby, H.D. and Waitz, I.A. (2009). *Transportation in a climate constrained world*. Cambridge, MA: MIT Press.

Scott, D., Peeters, P. and Gössling, S. (2010). Can tourism deliver its "aspirational" greenhouse gas emission reduction targets? *Journal of Sustainable Tourism*, *18*(3), 393–408.

Selwood, J., Tonts, M., Hall, C. and Müller, D. (2004). Recreational second homes in the south west of Western Australia. In C.M. Hall and D.K. Müller (eds), *Tourism, mobility and second homes: Between elite landscape and common ground* (pp. 149–161). Clevedon: Channel View.

Sirgy, M.J., Kruger, P.S., Lee, D.-J. and Yu, G.B. (2011). How does a travel trip affect tourists' life satisfaction? *Journal of Travel Research, 50*(3), 261–275.

Smith, S.L.J. (1990). A test of Plog's allocentric/psychocentric model: Evidence from seven nations. *Journal of Travel Research, 28*(4), 40–43.

Stern, P.C. and Oskamp, S. (1987). Managing scarce environmental resources. In D. Stokols and I. Altman (eds), *Handbook of environmental psychology* (Vol. 2, pp. 1043–1088). New York: Wiley.

Sterner, T. (2007). Fuel taxes: An important instrument for climate policy. *Energy Policy, 35*(6), 3194–3202.

Thøgersen, J. (2004). A cognitive dissonance interpretation of consistencies and inconsistencies in environmentally responsible behavior. *Journal of Environmental Psychology, 24*(1), 93–103.

Tol, R.S.J. (2007). The impact of a carbon tax on international tourism. *Transportation Research Part D: Transport and Environment, 12*(2), 129–142.

Tung, V.W.S. and Ritchie, J.R.B. (2011a). Exploring the essence of memorable tourism experiences. *Annals of Tourism Research, 38*(4), 1367–1386.

Tung, V.W.S. and Ritchie, J.R.B. (2011b). Investigating the memorable experiences of the senior travel market: An examination of the reminiscence bump. *Journal of Travel & Tourism Marketing, 28*(3), 331–343.

UNWTO. (2012). *Tourism highlights: 2012 edition.* Madrid: UNTO.

UNWTO-UNEP-WMO. (2008). *Climate change and tourism: Responding to global challenges.* Madrid: UNWTO.

Uriely, N. (2005). The tourist experience: Conceptual developments. *Annals of Tourism Research, 32*(1), 199–216.

Uriely, N., Ram, Y. and Malach-Pines, A. (2011). Psychoanalytic sociology of deviant tourist behavior. *Annals of Tourism Research, 38*(3), 1051–1069.

Veenhoven, R. (1984). *Conditions of happiness.* Dordrecht: Kluwer Academic.

Veenhoven, R. (2009). How do we assess how happy we are? Tenets, implications and tenability of three theories. In A.K. Dutt and B. Radcliff (eds), *Happiness, economics and politics: Towards a multidisciplinary approach* (pp. 45–69). Cheltenham: Edward Elgar.

Verbeek, D. and Mommaas, H. (2008). Transitions to sustainable tourism mobility: The social practices approach. *Journal of Sustainable Tourism, 16*(6), 629–644.

Walker, W.R., Skowronski, J.J. and Thomson, C.P. (1996). Life is pleasant – and memory helps to keep it that way! *Review of General Psychology, 7*(2), 203–210.

Wilson, T.D. and Gilbert, D.T. (2005). Affective forecasting: Knowing what to want. *Current Directions in Psychological Science, 14*(3), 131–134.

Wolf, J. and Moser, S.C. (2011). Individual understandings, perceptions, and engagement with climate change: Insights from in-depth studies across the world. *Wiley Interdisciplinary Reviews: Climate Change, 2*(4), 547–569.

Xin, S., Tribe, J. and Chambers, D. (2013). Conceptual research in tourism. *Annals of Tourism Research, 41*(1), 66–88.

4 Air travellers' willingness to donate frequent flyer points for charitable purposes

A Scandinavian case study

Eljas Johansson and Stefan Gössling

Introduction

There exists a considerable body of research showing that air travellers and tourists are generally more unwilling to change behaviour in favour of more environmentally friendly transport modes or fewer journeys, even among environmentally aware travellers (Barr *et al.*, 2010; McKercher *et al.*, 2010). As a result, offsetting of emissions has been discussed as an alternative way of dealing with rising emissions from in particular air travel, for which specific consumer interest has been detected (e.g. Gössling *et al.*, 2009; Lu and Shon, 2012; Mair, 2011; McKercher *et al.*, 2010). However, setting aside various systemic and practical problems associated with carbon offsets (e.g. Gössling *et al.*, 2007), voluntary uptake of offsetting purchases has been slow, and generally on the order of a few percent of overall emissions (Gössling *et al.*, 2009; McKercher *et al.*, 2010). Various reasons for this have been presented, including a disbelief in climate change, a belief that reducing emissions is not one's own responsibility, or that carbon offsetting is not a valid and credible approach to dealing with emissions from aviation (Gössling *et al.*, 2009). In light of this, yet another approach would seek to understand travellers' willingness to contribute to carbon offsetting or pro-environmental initiatives more generally, i.e. based on measures not involving a direct cost. This chapter presents the results of a study investigating willingness to donate frequent flyer points for charitable purposes.

Frequent flyer programmes (FFPs) are widely used by airline companies to build loyalty. These programmes issue frequent flyer points to their members depending on service class, flight length or sector flown. At designated levels earned "award points" can be redeemed for various consumption items, though they will expire if not redeemed within certain time limits. According to anecdotal evidence (Fleetwood, 2011), billions of points expire annually, with an estimated monetary value attached to these points on the order of hundreds of millions of US dollars (Winship, 2011). Frequent flyer programmes have been studied from various perspectives. For instance, Long *et al.* (2006) investigated the effects of such programmes on building customer relationships and loyalty, finding differences between leisure and business travellers; the latter being more interested in the symbolic dimensions of these programmes ("keeping score"). The importance of such non-material aspects of consumption was confirmed in other studies investigating the social status embedded in frequent flyer structures, also showing that various strategies are employed by air travellers to reach higher frequent flyer status (Gössling and Nilsson, 2010). The study concluded that both award and status points represent tangible benefits to travellers, and that frequent flyer programmes need to be seen as institutionalised frameworks to reproduce and foster high mobility patterns.

An aspect that has so far received little attention is the use of frequent flyer points for charitable purposes. Most major airlines offer the opportunity to donate points to charity, and many air travellers have collected more award points than they can consume. However, it remains unclear to which extent travellers do actually donate points, or how they value and understand opportunities to donate points for charitable causes. The following sections present the results of a first study on these interrelationships.

Methodology

To conduct a quantitative survey of members of a frequent flyer programme, it was necessary to collaborate with an airline in order to reach out to their customer base. Co-operation was agreed upon with an airline in Europe that does wish not to be disclosed. The airline offers an FFP, and its members can collect award and status points that can be used for different purposes, such as free flights, upgrades or other consumption (award points) or the use of lounges, speedy check-in, etc. (depending on status gained through status points). This study focuses on the programme's award points, which can be earned through flying or through other means of consumption, for instance if purchases are made through the airline's branded credit cards. Points are valid for three years from the date they are earned. Alongside free flights, upgrading and other rewards, the programme allows its members to redeem award points to the benefit of various non-profit organisations. The minimum donation is 1,000 points, which will translate into an estimated €10 to the organisation receiving the donation.

An online survey was created to collect data. Questions addressed motivations to collect award points, travel patterns and opinions regarding point donation, the latter including the categories Animals (animal rights, wildlife conservation); Art, Culture, Humanities (libraries, landmark preservation, museums, performing arts); Education (public schooling, private schooling); Environment (carbon offset programmes, environmental conservation); Health (diseases and disorders, patient and family support, treatment, medical research); Human services (children and family services, youth development, food banks, homeless services); International aid (development and relief services, international peace and security); Public benefits (advocacy and civil rights, research and policy institutions, community development). Where opinions were evaluated, these were measured using a four-point Likert scale. Respondents were also given the opportunity to answer a number of open questions, in order to better understand their reasoning. The questionnaire was first designed in English, and then translated into Finnish. Both versions were tested, until their clarity was ascertained. In order to find respondents for the main survey, a link to the survey was attached to the FFP's electronic newsletter that has about 120,000 subscribers. Additionally, the link was posted on the company's Twitter account that has over 11,000 followers. Overall, 307 responses were submitted between April and May 2013. Respondent rates are extremely low, and there is a considerable non-response bias and low representativeness of the results, which consequently need to be seen as indicative.

The FFP has four membership levels, depending on mobility patterns. Obtaining responses from all membership levels was considered important because these are likely to have distinct characteristics. Therefore, responses were stratified on the basis of the share of members in each membership level, as obtained from the airline. A random sample was then drawn from the responses collected, proportionally to membership in the overall FFP membership population (i.e. basic level=96.3%; tier 1=2.7%; tier 2=0.7% and tier 3=0.3%). On the basis of this strategy, 23 responses were excluded from the sample, together with another 13 where more than 20% of questions remained unanswered. The final sample comprised 271 responses. Results were analysed with SPSS (v. 21.0) (Johansson, 2013).

Results

Respondent characteristics

A clear majority (96.7%) of respondents stated that they lived in Finland, the remainder (3.3%) in other countries. A majority of respondents (58%) was female, which is unlikely to represent the true gender distribution of the FFP studied. Nearly half of the sample (48%) described themselves as employees, followed by pensioners (21.8%), managers/entrepreneurs (13.7%), students (6.8%) and unemployed (3.0%; other/no answer: 5.3%). With regard to age, travellers 61+ were possibly overrepresented (28.4%), followed by age groups 31–40 (24.7%), 51–60 (18.5%), 41–50 (15.9%) and those under the age of 31 (12.5%).

Travel behaviour

Respondents were asked to state the number of individual flights taken per year. Given the wide spectrum of answers, travellers were divided into three mobility groups, "low mobility" (0–5 flights), "medium mobility" (6–10 flights) and "high mobility" (>10 flights). Over half (55%) of the respondents stated that they took more than five individual flights per year, and 20.3% participated in more than ten individual flights per year. This indicates considerably lower flight intensities than found in other studies in Nordic contexts. For example, over half (51%) of respondents, in a survey conducted at Gothenburg Landvetter airport,

Sweden, reported participating in up to five round trips per year (Gössling *et al.*, 2009). In another survey, 89% of FFP cardholders at Copenhagen Airport, Denmark, had taken at least four return flights per year (Gössling and Nilsson, 2010). It is worth noting that the data collection in the previous surveys took place at airports, whereas the present study's results are derived from an online survey, which may have affected the representativeness of the sample.

A strong relationship was also found with regard to the number of flights taken and the share of these flights made for business. Table 4.1 also shows a trend for frequent flyers to use business class more often than economy class.

Award points

Respondents stated to have accumulated between 0 and 180,000 award points, with a mean of 23,387. Aggregated, the respondents' award points totalled almost five million, with a monetary value of about €50,000 at a value of 1 cent/ point (e.g. McGee, 2007; Winship, 2011). Extrapolated to all frequent flyer pro- gramme members of the airline, as many as 20 billion award points for the airline studied might be in circulation, with a corresponding value of €200 million. Results also indicated significant differences between mobility groups in terms of frequent flyer points accumulated, with low mobility members owning an average of 12,920 points, medium mobility group members 24,756 points and high mobility group members 42,500 points. This raises a question about the use of points, as it could be assumed that those with high quantities of collected points would also redeem these, in which case the number of accumu- lated points should be more evened out across the mobility groups.

Preferred ways to use points

Respondents were asked to rank alternative redemption options in preferred order. As shown in Table 4.2, a broad majority (61%) of the respondents indi- cated that free flights were their most preferred redemption choice, followed by upgrading (36%), frequent flyer status upgrades (33%), hotel nights/car rental (37%), shopping (41%) and donations to charity, the least popular way of spend- ing points (66%). Results are thus in line with previous research with regard to the importance of free flights and upgrades (Gössling and Nilsson, 2010), though this survey reveals that there may also be a small minority (4%) for whom donat- ing points to charity is a priority (first or second choice).

As donating points to charity can involve expiring points or accumulated point numbers too low to be used for free flights or other consumption/services, this option was tested among respondents. A majority (78.8%) in each mobility group (low mobility=72.1%; medium mobility=85.9%; high mobility=81.8%) stated that they would rather donate their points than see them expire. Only 3% stated that they would not donate their points, even though there was also a comparably large share (18.2%) of respondents who were undecided or had no opinion. A possible explanation for this is uncertainty about charities, which will be discussed later.

Table 4.1 Flight frequency/purpose of trip/service class (n=271)

	Low mobility (0–5 flights)		Medium mobility (6–10 flights)	High mobility (>10 flights)				Total
No. of flights	0–2	3–5	6–10	11–20	21–30	31–50	51+	Total
%	15.5	29.5	34.7	12.2	4.8	1.8	1.5	100
Leisure %	75	72	62	33	26	27	33	
Business %	6	11	16	42	67	70	55	
VFR %	9	14	19	18	4	2	11	
Other %	10	3	2	6	3	0	0	
Flights in economy class %	97.18	95.63	95.27	93.44	84.62	90	82.5	94.7

Table 4.2 Preferred redemption choices (*n*=269)

%	1st choice	2nd choice	3rd choice	4th choice	5th choice	6th choice
Free flights	**61**	22	6	3	2	4
Upgrades	22	**36**	20	11	9	2
Buy tier-points*	3	7	**33**	21	23	13
Hotel nights and car rental	4	18	22	**37**	15	4
Shopping	8	14	12	15	**41**	10
Donating to charity	1	3	7	13	10	**66**

Note

* Award points can be used to buy tier points to move up tier levels; higher frequent flyer status can then be used for other benefits such as accesses to airport lounges, priority check-in and priority boarding.

Expired award points

As outlined, points earned in the FFP studied in the context of this chapter are valid for three years. In order to further explore aspects of point expiration, respondents were asked if they had been part of the programme for three years or more. In total, 166 respondents fell into this category, constituting the sub-sample for the question whether any of their points had expired in the past 12 months. In total, 15.7% reported expired points, in the range between 4 and 20,000, with a mean of 4,227 points. Extrapolated, this may correspond to 342 million points or €3.4 million in monetary value that could potentially be donated. No significant differences in the relative share of expired points were found between low, medium or high mobility groups (Johansson, 2013). While infrequent travellers' points are likely to expire due to insufficient accumulation of points (DeKay *et al.*, 2009), it is more difficult to explain why frequent flyers would let their points expire. Possible reasons could be disregard of benefits perceived to be intangible, but also poor availability of free flights on desired dates (Whyte, 2004). Points about to expire may as well be donated. The following section investigates why frequent flyers would choose to make donations.

Awareness of point donation

Respondents were asked if they had been aware of the opportunity to donate award points to charity prior to seeing the questionnaire. About one-third answered negatively, while the majority was aware of this option. Results indicate that travellers who had been members of the FFP for three years or more were more aware of donation options compared to those with shorter memberships (69%, compared to 52%, respectively). In order to find out if point donation could be stimulated by providing better information, respondents were asked to indicate the extent to which they agree with the following statement: *If airlines provided more information about donating unused points to charity, this* would increase the likelihood of me donating my unused points to charity. Response options were presented on a 4-point Likert scale with bipolar choices "to a great extent (1)" and "not at all (4)". A broad majority (66.8%) indicated agreement with the statement. However, whether travellers are willing to donate points also depends on the perceived credibility of the charities. For instance, as one respondent pointed out, "donators do not necessarily feel convinced that the help goes toward those in need … which is an obstacle to making donations". Another respondent emphasised the importance of knowing "to which individual causes the points are given: whether they go toward children's food, health, or down the drain". Yet another respondent felt that "the only people benefiting from charitable programmes are those running them".

The marketing literature suggests that the likelihood of making donations declines if the perceived credibility of the linkage between the charitable cause and the organisation processing the donation is low (Robinson *et al.*, 2012; Youn and Kim, 2008). It might be argued that the linkage between a charitable cause and the airline industry is often low. Robinson *et al.* (2012) argue that the likelihood of donations increases if consumers are given the opportunity to choose the beneficiary. Consequently, it was hypothesised that point donation could be stimulated by expanding the selection of charitable causes in the FFP donation programme. In order to test this, the respondents were asked to indicate the extent to which they agree with the following statement: *If airlines provided a greater range of charitable causes to which I can donate my award points, this would increase the likelihood of me donating my unused points to charity.* Response options were presented on a four-point scale with bipolar items "to a great extent" and "not at all". More than half of the respondents (56.4%) supported the statement, with 8.1% being undecided or having no opinion. With regard to preferred charity causes, more than a quarter (28.3%) supported Health related, and another 24.9% Human services, followed by International Aid (12.3%) and Environment (11.9%). Other causes were less often mentioned, with 5.2% being generally unwilling to donate any points (Figure 4.1). With

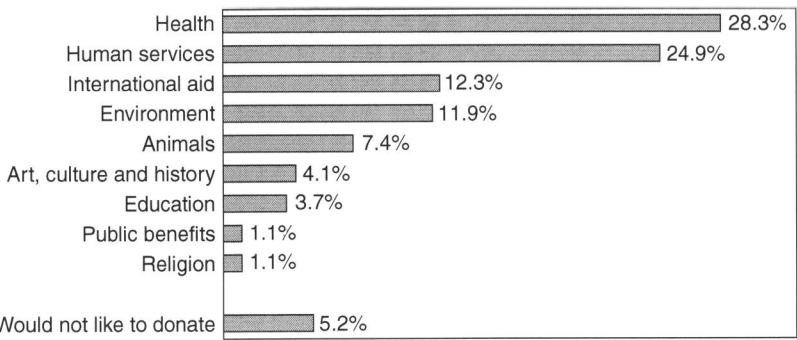

Figure 4.1 Most preferred charitable causes (*n* = 269).

regard to "Environment", it is interesting to note that a range of written comments were added by respondents, perhaps also because this category has a "clear link to flying", as noted by one traveller. Other comments regarding "Environment" included "because flying results in emissions", "carbon offsetting", "donating points to an environmental cause is just logical because of the environmental impact of flying" and "because flying is a polluting business, and because I am a recreational user and I could also take the train or other transportation mode that pollutes less when taking my holiday". This points at a specific, and potentially more attractive, role for offsetting in point donations.

Conclusions

The survey of FFP members had the purpose to better understand perceptions of donations of frequent flyer points to charitable causes, and in particular environmental projects. Though not representative, results indicate that air travellers are generally reluctant to donate their award points, and that personal use is clearly preferred. However, adding first and second choices for spending points, about 4% in this sample indicated that giving points to charity may be more relevant than personal consumption, and a vast majority of respondents agreed that donating points is favourable over having these expire. Throughout traveller classes, including low, medium and highly mobile travellers, there is evidence that a share of points remains unused and expires, though most travellers apparently seek to spend their points in time. Expiring points might be given to charity, but travellers appear keen to know details about the charitable causes supported. Environmental causes, including offsetting, are not spending priorities for air travellers. For those interested in environmental charity, many stated concerns about the environmental impacts of air travel, and indicated an interest to donate points to address this situation. Overall, however, the study does not indicate a considerable potential of donations for charitable causes and specifically offsetting. Even though the potential for donations is considerable in absolute terms, with billions of points expiring every year, these constitute only a fraction of the points used for personal consumption. To address climate change through point donations thus appears to have a very limited potential.

References

Barr, S., Shaw, G., Coles, T. and Prillwitz, J. (2010). "A holiday is a holiday": Practicing sustainability, home and away. *Journal of Transport Geography*, *18*(3), 474–481.

DeKay, F., Toh, R. S. and Raven, P. (2009). Loyalty programs: Airlines outdo hotels. *Cornell Hospitality Quarterly*, *50*(3), 371–382.

Fleetwood, B. (2011, 5 April). Frequent-flyer programs are convoluted, mysterious, and a maddening fraud. *Huffington Post* [online]. Retrieved from www.huffingtonpost.com/blake-fleetwood/frequentflier-programs-ar_b_856623.html.

Gössling, S. and Nilsson, J. H. (2010). Frequent flyer programmes and the reproduction of aeromobility. *Environment and Planning*, *42*, 241–252.

Gössling, S., Broderick, J., Upham, P., Peeters, P., Strasdas, W., Ceron, J.-P. and Dubois, G. (2007). Voluntary carbon offsetting schemes for aviation: Efficiency and credibility. *Journal of Sustainable Tourism, 15*(3), 223–248.

Gössling, S., Haglund, L., Kallgren, H., Revahl, M. and Hultman, H. (2009). Swedish air travellers and voluntary carbon offsets: towards the co-creation of environmental value? *Current Issues in Tourism, 12*(1), 1–19.

Johansson, E. (2013). *Exploring the Prospects for Using Frequent Flyer Points for Charity* (Unpublished master's thesis). Department of Service Management, Lund University, Sweden.

Long, M. M., McMellon, C., Clark, S. D. and Schiffman, L. G. (2006). Building relationships with business and leisure flyers: Perceived loyalty and frequent flyer programs. *Services Marketing Quarterly, 28*(1), 1–15.

Lu, J.-L. and Shon, Z. Y. (2012). Exploring airline passengers' willingness to pay for carbon offsets. *Transportation Research Part D – Transport and Environment, 17*(2), 124–128.

Mair, J. (2011). Exploring air travellers' voluntary carbon-offsetting behaviour. *Journal of Sustainable Tourism, 19*(2), 215–230.

McGee, W. J. (2007, June). Mileage mania. *Condé Nast Traveler* [online]. Retrieved from www.concierge.com/cntraveler/traveltips/10839?all=yes.

McKercher, B., Prideaux, B., Cheung, C. and Law, R. (2010). Achieving voluntary reductions in the carbon footprint of tourism and climate change. *Journal of Sustainable Tourism, 18*(3), 297–317.

Robinson, S. R., Irmak, C. and Jayachandram, S. (2012). Choice of cause in cause-related marketing. *Journal of Marketing, 76*(4), 126–139.

Whyte, R. (2004). Frequent flyer programmes: Is it a relationship, or do the schemes create spurious loyalty? *Journal of Targeting, Measurement and Analysis for Marketing, 12*(3), 269–280.

Winship, T. (2011, 15 March). The value of a frequent flyer mile revisited. *Smart Travel* [online]. Retrieved from www.smartertravel.com/travel-advice/the-value-of-frequent-flyer-mile-revisited.html?id=7050813.

Youn S. and Kim, K. (2008) Antecedents of consumer attitudes toward cause-related marketing. *Journal of Advertising Research, 48*(1), 123–137.

5 Sociological barriers to developing sustainable discretionary air travel behaviour

Scott A. Cohen, James E.S. Higham and Arianne C. Reis

Introduction

Mitigating the greenhouse gas emissions from air travel is one of the most challenging aspects of society's response to climate change (Monbiot, 2007). Whereas research from the transport and tourism sectors agrees that air travel emissions are a key environmental challenge (Barr *et al.*, 2010; Becken, 2007; Scott, 2011; Scott *et al.*, 2012), how to best address the climate impacts of discretionary air travel remains an elusive problem (Cohen *et al.*, 2011). Scope for further efficiency gains in aircraft emissions is declining (Scott *et al.*, 2010), and there is not yet a global climate policy for international commercial aviation (Duval, 2013). Signatories of airlines in the United Kingdom, for instance, instead of choosing to transform supply or raise consumer awareness of air travel's climate impacts, presently pin their hopes for a sustainable aviation future on technology, alternative fuels and operational innovations (Sustainable Aviation, 2011). In the context of industry resistance to wholesale supply changes and in the absence to date of a global market-based mechanism for aviation, such as carbon trading, the concept of encouraging voluntary public behaviour change has been presented as a mechanism for moving discretionary air travel consumption towards a more sustainable pathway (Barr *et al.*, 2011a; Miller *et al.*, 2010).

Encouraging pro-environmental behaviour change has been discussed in environmental psychology more generally (e.g. Steg and Vlek, 2009) and with regards to climate impacts specifically (e.g. Barr *et al.*, 2011a; Semenza *et al.*, 2008). The prospects for positive behaviour change in the context of the public's air travel behaviour have been the focus of recent empirical attention, with studies finding gaps between awareness and attitudes (Hares *et al.*, 2010) and attitudes and behaviour (Kroesen, 2013; Miller *et al.*, 2010). These works point to a breakdown in consumer decision-making processes in the context of discretionary air travel behaviour, in which environmental awareness and pro-environmental attitudes are not in practice translating to voluntary sustained behavioural changes (McKercher *et al.*, 2010). Barr *et al.* (2010), who sought to understand tourists' environmental concern in relation to a wider scope of everyday lives and their processes of decision-making, found that participants who were committed to environmental practices at home were unwilling to reduce holiday air travel. This suggests that there are fundamental barriers to changing discretionary air travel behaviour, even amongst pro-environmental consumers.

This chapter takes Barr *et al.*'s (2010) finding that environmental concern may not transcend "home" to the context of "away" as its departure point. Based on 50 in-depth semi-structured interviews carried out in Australia, Norway and the United Kingdom, we illustrate behavioural consistencies and inconsistencies with respect to climate change amongst consumers in both everyday domestic (home) and tourism (away) practices. In contrast to Barr *et al.* (2010), however, we use both modern and postmodern sociological theory to explain why these seemingly contradictory consumption decisions occur. Specifically, we engage modern theory on tourism as liminoid space (Turner, 1982) and postmodern theory that suggests that personal identity (and consequently behaviour) is inconsistent and performed differently across varying contexts (Bell, 2008; Edensor, 2001). The findings of our research, framed within these theoretical perspectives, hold important implications for the viability of climate change mitigation strategies that rely, at least in part, on encouraging voluntary behaviour change amongst consumers.

Climate concern and discretionary air travel

Tourism, as an oil-intensive industry (Becken, 2010), has come under increasing pressure to move to a sustainable emissions path (Gössling, 2009). It is widely acknowledged that the tourism industry is implicated in climate change in terms of both cause and effect (Pang *et al.*, 2013). Of the 4.4 per cent of global carbon emissions for which tourism is directly accountable (Peeters and Dubois, 2010), 40 per cent can be conservatively attributed to tourist air travel (Gössling, 2009). To disentangle tourist air travel, however, from a wider range of discretionary mobilities, such as visiting friends and relatives, or business travel coupled with tourism activities, is a challenging proposition. Regardless, in comparison with alternatives such as rail, road and sea-based passenger modes, air travel is the most harmful for the climate system (Gössling and Peeters, 2007) and presents one of the tourism industry's largest challenges if it seeks to sustain contemporary aeromobility-dependent tourism practices (Burns and Bibbings, 2009).

Correspondingly, much of the recent academic concern over tourism's climate change impacts has centred upon issues surrounding tourist air travel (Gössling and Upham, 2009), including how these issues intersect with tourism demand and behaviour (e.g. Gössling *et al.*, 2012; Mair, 2011). In conjunction with this research line there has been growing interest in whether consumer awareness about climate change manifests itself in perceptual (Becken, 2007; Cohen and Higham, 2011; Huebner, 2012), attitudinal (Higham and Cohen, 2011) and/or intended (or actual) behavioural changes in tourism practices (Kroesen, 2013; McKercher and Prideaux, 2011; Miller *et al.*, 2010). These studies largely suggest a dissonance between awareness or attitudes and actual behavioural change. For instance, on one hand Cohen and Higham (2011, p. 331) report a growing movement of UK consumers who reflect a "carbon conscience" in approaching air travel decisions, and Gössling *et al.* (2008, p. 875) observe that "pro-environmental concerns are clearly emerging among consumers, and may play a significant role in travel decisions in the future". On the other hand, Miller *et al.* (2010) find public reluctance in the UK to actually take fewer holidays in order to reduce personal carbon impacts, and Kroesen (2013) measured in the Netherlands an inconsistency between pro-environmental awareness and air travel behaviour. These latter studies support McKercher and Prideaux's (2011) observation that tourism and air transport are low on personal environmental agendas, at least in terms of manifest behavioural changes.

This attitude–behaviour gap is not unique to discretionary air travel practices: it has also been identified as a problem for work-related air travel (Lassen, 2010), and sustainable tourism (Antimova *et al.*, 2012) and ethical consumption more widely (Bray *et al.*, 2011). The gap is reported in the context of other modes of transport, such as automobiles (Anable, 2005), and has been described as a barrier to the ability of individuals to reduce emissions, as part of what Whitmarsh *et al.* (2011, p. 58) term the public's "carbon capability". In the context of discretionary air travel, however, the issue seems to be compounded by consumer perceptions that tourism spaces are not "appropriate sites in which to be environmentally conscious", with behavioural differences between home and tourism settings reported (Barr *et al.*, 2011b, p. 1243). Barr *et al.* (2011b) go on to argue that tourism research needs to be better connected to wider knowledge on pro-environmental behaviour, so that this additional gap between "home" and "away", which further complicates the attitude–behaviour gap, can be understood. We concur and consequently suggest that by turning to modern and postmodern sociological theory in tourism, complementary explanatory concepts are available that may help us to better understand these behavioural inconsistencies.

Liminoid space and contextualised performances of identity

Despite suggestions that tourism practices are increasingly blended into the fabric of everyday life (Edensor, 2007; Franklin and Crang, 2001; Larsen, 2008), tourism still largely occurs as a bounded experience outside the rhythms of the day-to-day, is often experienced as extraordinary (Tung and Ritchie, 2011) and frequently involves conspicuous consumption (Carr, 2005). With tourism typically experienced as an event set apart from the day-to-day, it is unsurprising that few studies, with the exception of Barr *et al.* (2010), have sought to understand tourist environmental concern in relation to a wider scope of everyday lives and daily decision-making.

Both a modern sociological perspective that positions tourism experiences as an escape from one's everyday self (Cohen and Taylor, 1992), and a postmodern perspective that views selves, and in turn behaviours, as performed and contextually dependent (Bell, 2008), suggest that just because individuals act or perform one way in a situation, does not mean that behaviours transfer consistently across contexts. This has implications for understanding the transferability of pro-environmental attitudes and behaviours across differing life contexts, as each of these theoretical perspectives suggests that behaviour is situationally dependent.

Within modern motivational literature on tourism, the need to escape has long been recognised as a key motivator for travel (Crompton, 1979; Dann, 1977). Crompton (1979, p. 416) observes that the desire to "escape from a perceived mundane environment", or alternatively, the tedium of routine, forms one of the major motives driving tourist behaviour. Under this view tourism is "essentially a temporary reversal of everyday activities – it is a no-work, no-care … situation" (Cohen, 1979, p. 181). Breaking from everyday routine is linked to Turner's (1982) description of the "liminoid", a secularised term conveyed from ritual studies, which is characterised as a departure from the structure of everyday life (Lett, 1983). Liminality is associated with three phases: separation, "limen" and re-aggregation, with limen signifying a metaphorical threshold that one may pass through as a departure from the structure of one's everyday life in society (Turner, 1982). Sharpley (2003, pp. 5–21) applies the liminoid to tourism experiences, noting that whilst away on holiday, tourists may feel "temporarily freed from … household chores, social commitments and, generally, the behavioural norms and values of their society". Furthermore, Kim and Jamal (2007, p. 184) suggest that within "liminal touristic space, conventional social norms and regulations are often temporarily suspended as tourists take advantage of the relative anonymity and freedom from community scrutiny".

Indeed, the notion of tourism occurring in liminoid space melds well with more recent academic literature that holds that tourism is a furtive ground for extraordinary experiences (see Morgan *et al.*, 2010; Tung and Ritchie, 2011). These two-fold discourses, of escape and, in turn, extraordinary experience, attempt to map out tourism space as fundamentally different from the everyday, contributing to a dichotomised framing of "home" and "away". This dualism, between tourism and, in contrast, everyday life, views tourism as liminal, exotic and pleasurable, whereas the day-to-day is represented as bound by rules, and as ordinary and boring (see Larsen, 2008 for a critique of this dualism). An important implication of this modernist perspective on tourism is that behaviour in (liminoid) tourism spaces will be markedly different to behaviour at "home", due to the lack of rules, sense of escape and suspension of behavioural norms associated with the former.

Albeit based on quite different premises to those of modernity, a postmodernist sociological approach, like the above modernist position, also draws into question the degree to which performances of identity, and hence behaviour, are likely to consistently transcend contexts. A performance perspective (Bell, 2008; Edensor, 2001), arising from Goffman's (1959) work on selves, in which individuals perform different "faces" depending on the social situation, suggests that personal identities are too fragmented, contextually dependent and relational (Finnegan, 1997; McAdams, 1997; Vaughan and Hogg, 2002) to expect behavioural consistency. This means that individuals have multiple selves that are "often demonstrated in different interpersonal roles or relationships", and which may "contradict or conflict with each other depending on situation or context" (Bond and Falk, 2013). Tourism, specifically, has been suggested as a space to "display and create new identities, even if only for the duration of the holiday" (Hibbert *et al.*, 2013, p. 32). Thus tourism practices may be characterised by fleeting performances of identity, in which new identities may be tried out, played with and discarded, whilst other identity aspects can be emphasised, or hidden, all depending on social context. Within this postmodernist perspective, dissonance between attitudes and behaviours, and inconsistencies in patterns of behaviour across contexts are both easily reconciled because consistency is not presumed from the start.

These modernist and postmodernist worldviews, with behaviour dichotomised between "here" and "there" in the former, and performances of identity fragmented and unstable across contexts in the latter, may seem distant to our focus on consumer climate concern as it relates to discretionary air travel. However, these issues are paramount if seeking to mitigate tourism's climate change impacts through strategies that attempt to nudge consumers towards pro-environmental behaviour. With identities, and in turn, attitudes and behaviours, largely contingent on social context, there can be no certain expectation that consumer climate concern in daily life practices will necessarily transfer across to, or be sustained in, tourism settings.

Study methodology

The empirical material that follows draws from a wider multi-national research project on consumer attitudinal and behavioural responses to climate change in the contexts of discretionary air travel practices and day-to-day domestic living. Other parts of the broader project have examined consumer attitudes towards, and perceptions of, the climate impacts of long haul air travel from the United Kingdom (Cohen and Higham, 2011) and Norway (Higham and Cohen, 2011) to New Zealand, and the phenomenon of "binge flying" (Cohen *et al.*, 2011, p. 1071). We adopted a critical interpretive research paradigm with a subjectivist epistemological position (Denzin and Lincoln, 2005). Our philosophical stance was influenced by our shared position that aviation greenhouse gas emissions are a significant contributor to anthropogenic climate change and need to be mitigated through various social, political and technical avenues. Our joint view is that significant reductions in levels of discretionary air travel amongst consumers represent a key part of the societal response to climate change.

We considered our study participants to be "individuals whose opinions are valued, and valid" (Sedgley *et al.*, 2012, p. 954) and as such followed the advice of Fontana and Frey (2005) and did not superimpose our worldviews on the study participants. Although we were non-activist in our approach, our research was transformative (Pernecky and Jamal, 2010) in that asking the questions we did was an act of raising self-awareness on the part of the study participants, stimulating reflection upon the potential consequences of their discretionary air travel behaviours. These decisions were aligned with our aim to elicit deeply subjective personal perspectives on air travel behaviour and climate change.

The empirical material is drawn from a cross-section of consumers in three nations: Australia, Norway and the United Kingdom, where tensions exist between global climate change and the conspicuous consumption of aeromobility (Burns and Bibbings, 2009; Randles and Mander, 2009).Whilst we recognise that the Australian public does not have the same options for more sustainable travel as Europeans, owing to the nation's distance from other countries and a less developed and less significant domestic rail network, the governments of these three nations have all been actively engaged in discourses addressing the urgency of climate change mitigation (Gössling, 2009; Hares *et al.*, 2010; Høyer, 2000; Zeppel, 2012), with Australia recently headlining in the media for its new carbon tax (BBC News, 2012). Specifically, we set out to achieve in-depth insights into awareness of, attitudes towards and personal behavioural responses to global climate change, in both domestic living and in tourism contexts. Extensive qualitative materials were generated through one-to-one open-ended personal interviews (Fontana and Frey, 2005), an approach selected for the flexibility it offers in identifying and exploring issues in detail (Jennings, 2001).

The qualitative materials are derived from 50 semi-structured open-ended interviews conducted in Coffs Harbour, Australia (April–July 2011), Stavanger, Norway (June–July 2009) and Bournemouth, United Kingdom (July 2009). The locations where interviews were conducted were based on convenience, as members of the research team were based at Southern Cross University (Australia), the University of Stavanger (Norway) and Bournemouth University (UK) during the fieldwork when the respective interview programmes were conducted. Participants were recruited using convenience and snowball sampling techniques. Access to participants initially relied on key informants in each study site, both from within and outside the university contexts. Selection criteria were that participants self-identify as Australian, Norwegian or British nationals and be willing to be interviewed face-to-face in English. We aimed to access a relatively equal gender distribution across a broad age range, with a minimum age for participation of 18 years. The interviews were conducted at neutral sites, lasted 30–60 minutes, and were digitally recorded. The main themes addressed in the interviews were awareness of and attitudes towards anthropogenic climate change, domestic behavioural responses to climate change in day-to-day living and changes in travel decision-making and behaviour in relation to climate change, including a focus on discretionary air travel practices. A copy of the programme used to semi-structure the interviews can be found as a supplementary file in the *Journal of Sustainable Tourism* article version of this chapter; however, this was employed as a flexible guide that allowed each interview to move in different directions as it developed. Interviewing concluded in each national study site when evidence of data saturation emerged.

The 50 interview participants included 25 females and 25 males (Australia ten females: ten males; Norway 8:7; UK 7:8) with ages from 18 to 67 (Table 5.1).

Table 5.1 Summary profile of Australian, Norwegian and British interview programme participants

Pseudonym	Gender	Age	Nationality	Occupation	Highest qualification
Alex	M	49	Australian	Unemployed	Undergraduate
Danielle	F	31	Australian	Industry work	Undergraduate
Jessamin	F	18	Australian	Undergraduate student	High school
Tina	F	36	Australian	Industry work	Undergraduate
Lauren	F	47	Australian	University administrator	Undergraduate
Josi	F	29	Australian	Industry work	Technical diploma
Martin	M	57	Australian	Industry work	Technical diploma
Grant	M	56	Australian	Unemployed	High school
Justin	M	24	Australian	Postgraduate student	Masters
Camilla	F	24	Australian	Industry work	Undergraduate
Kevin	M	57	Australian	Postgraduate student	Masters
Brian	M	29	Australian	Industry work	Technical diploma
Kay	F	46	Australian	University administrator	Masters
Bruce	M	58	Australian	University administrator	Undergraduate
Tom	M	47	Australian	Industry work	Technical diploma
Lili	F	43	Australian	Unemployed	Undergraduate
Ian	M	43	Australian	Academic	PhD
Eric	M	38	Australian	Unemployed	Technical diploma
Amy	F	43	Australian	Industry work	High school
Jen	F	30	Australian	Teacher	Undergraduate
Frode	M	37	Norwegian	Industry work	Masters
Rita	F	34	Norwegian	Industry work	Masters
Bjørn	M	41	Norwegian	Industry work	PhD
Silje	F	45	Norwegian	Industry work	Masters
Svein	M	35	Norwegian	Industry work	High school
Tone	F	58	Norwegian	Postgraduate student	Masters
Ida	F	52	Norwegian	University administrator	Masters
Grete	F	27	Norwegian	Postgraduate student	Undergraduate
Lars	M	53	Norwegian	Academic	PhD
Pål	M	34	Norwegian	Industry work	Masters
Hilda	F	67	Norwegian	Retiree	Masters
Håkon	M	48	Norwegian	Industry work	Undergraduate
Johannes	M	57	Norwegian	Academic	PhD
Anette	F	35	Norwegian	Industry work	Masters
Grethe	F	27	Norwegian	Postgraduate student	Masters
Cindy	F	42	British	University administrator	High school
Jack	M	35	British	Industry work	Undergraduate
Grace	F	36	British	University administrator	Masters
Jessica	F	48	British	University administrator	High school
Ruby	F	41	British	Industry work	High school
Amy	F	30	British	Academic	PhD
Hannah	F	48	British	Postgraduate student	Masters
Oliver	M	30	British	Academic	Masters
Thomas	M	38	British	Academic	Masters
Harry	M	40	British	Industry work	Undergraduate
Daniel	M	18	British	Undergraduate student	High school
Mia	F	21	British	Undergraduate student	High school
James	M	63	British	Academic	PhD
William	M	42	British	Industry work	Undergraduate
Lewis	M	39	British	Industry work	Undergraduate

Their occupational status reflected 21 industry workers from a variety of professional and non-professionalised fields, nine students, seven university academics, seven university administrators, four unemployed persons, one teacher and one retiree. The participants spanned a range of education levels, however, the majority were highly educated and moderately affluent (albeit less so in the Australian context), reflecting the sampling being driven out of a university context. We recognise that educational, financial and social backgrounds may have significant bearing on the participants' attitudes and behaviours. Our collective participant profile mostly reflects the attitudes and behaviours of individuals with the resources to make frequent flying personally relevant. The majority of the participants, particularly in the European contexts, were in fact highly aeromobile, with air travel at least once annually routine, with several flights per year (and sometimes per month) not uncommon, for a mix of reasons including leisure, business and/or visiting friends and relatives. The participant profile is one of relative privilege such that we do not claim representation of Australian, Norwegian or British society.

The interviews were transcribed and following repeated independent readings and annotation, we applied a triple blind thematic analysis approach in manually interpreting the data (Patton, 2002). This approach involved reducing the data into categories guided by the study participants' narratives, but without losing sight of the research aims, a process that allowed for the identification of emergent themes (Miles and Huberman, 1994; O'Reilly, 2005). The analytical perspectives applied in this chapter, of liminoid space and contextualised identities, were not part of the initial research design but rather emerged as relevant during analysis of the data. During the immersive blinded process, we acted as three independent critical analysts and then engaged in collective "analyst triangulation" (Patton, 2002). This aimed to ensure trustworthiness by checking for congruity of interpretations, blind spots and multiple ways of interpreting the data (Lincoln and Guba, 1985). Our respective blinded interpretations were largely in accordance; however, some individual interpretations did highlight specific data blind spots that were revealed in our collaborative discussions. Through triangulation we set out to promote dependability (through interpreter triangulation), credibility (through theoretical triangulation between our empirical material and existing theories) and transferability (through rich description of the context to facilitate analytical transfer) (Decrop, 2004).

Consistencies between home and away

Our study revealed significant inconsistencies in the participants' climate sensitivities and related behaviours between domestic day-to-day and tourism contexts. The findings were remarkably similar across the study participants from each of the three nations, in that the majority of the participants reduced, suppressed or abandoned their climate concern when in tourism spaces. A minority of the participants, however, held there was no difference between the environmental sustainability of their practices in domestic decisions versus those made whilst away on holiday. For instance, Tom (Australian, 47) explained: "I think the same as I think about the impacts in everyday life, no different in holidays … they are the same decisions that I'd make if I was at home." Oliver (British, 30) maintained a similar view:

> Exactly the same principles would apply. If I'm staying in a hotel, I wouldn't dream of leaving the room with the lights on, for example. If I'm in a hotel, I'm not going to boil more water than I need. I'm not going to stand under the shower for ten minutes longer than is necessary.

Such statements were typically used to discount the notion that economic motives underpinned some pro-environmental behaviour. William (British, 42) placed this issue in a stark light – "When we go to Florida, I wouldn't just leave the air-conditioning on all day and all night because I'm not paying for it. I would be responsible about it." Yet this statement avoids the issue that the energy use of air conditioning is insignificant alongside the decision to fly to Florida. Svein (Norwegian, 35), when asked how important money was in his attitudes towards the environment whilst on holiday, responded:

> For me, economics is not a big issue. I'm not above average in Norway. We're so rich and comfortable here and what I want more of in my life is other qualities than monies and luxury and that kind of wealth. So it's not motivated by money.

Svein prioritised consuming ethically across the different facets of his life. As he recognises, however, this is a position of privilege largely made available through his citizenship in an affluent nation.

For each of these study participants, tourism practices were viewed holistically as part of a broader lifestyle in which consistency was sought in values, attitudes and behaviours across different facets of life. Barr *et al.* (2010, p. 475) describe this notion of a "sustainable lifestyle" as implying that "individuals would demonstrate a series of commitments across lifestyle practices, not merely as part of their routine, but also in tourism contexts". Such a perspective counters the notion of tourism as liminoid space (Sharpley, 2003) and, to a degree, identities as contextually contingent (Finnegan, 1997; Vaughan and Hogg, 2002), by displaying behavioural consistency across domestic day-to-day and tourism spaces. It furthermore illustrates an entanglement of tourism in daily life (Larsen, 2008), whereby everyday environmental concerns and those associated with tourism are enmeshed. Svein further elaborated a view of the everyday and holidays as inextricably interlinked, with the carbon savings accumulated through practices such as cycling to work seen as nonsensical when positioned alongside the prospect of flippant discretionary air travel: "So you can't ride your bike to your job and use a plane everywhere without thinking about it – it would be stupid." This type of consistent rational actor approach, however, was relatively rare amongst the study participants.

Inconsistencies between home and away

As opposed to achieving alignment between approaches to environmental sustainability in everyday practices and those whilst on holiday, participants evidenced that tourism spaces are often the stage for performances of less stringent, suppressed or non-existent climate concern and more environmentally destructive consumption practices. This supports the work of Barr *et al.* (2010, 2011b), and others' observations (e.g. Kroesen, 2013; Lassen, 2010) that air travel behaviour fails to correlate significantly with broader environmental awareness. These study participants, who may be committed to environmental practices in and around the home (e.g. reducing waste and energy use, buying organic, "ethical" purchasing, cycling instead of driving), are indeed often unwilling to reduce holiday air travel. For instance, Harry (British, 40), whose undergraduate degree was paradoxically in environmental management, undertook a range of practices in everyday life to mitigate his climate impact, but was unwilling to transfer sustainable practices to the realm of discretionary air travel, where he privileged speed and convenience over environmental sustainability:

> I have a small car with a small engine and that is purely from a global warming point of view, from a pollution point of view. I do see the impact [of air travel] and I would get on an airplane and go on a long-haul flight because I want to travel, I want to get to this place, and I can't think of another way to do it reasonably quickly, reasonably safely, minimum of fuss. It's the convenience, it boils down to that.

Harry thus illustrates how the two gaps, one in attitude/behaviour and the other in home/away are interlinked, whereby his pro-environmental attitudes lead to positive environmental behaviour in domestic life, yet a gap remains between those attitudes and how his behaviour manifests in the context of tourism travel. In order to get "away", he is willing to suspend the climate sensitivities he performs at "home" which maintain his identification with environmental management. Likewise, Jen (Australian, 30) recognised through the interview process an inconsistency between climate concern in her daily life and tourism practices, and suggested that convenience took priority in holiday flying decisions:

> I never really thought about climate change in relation to travel much and it's interesting to actually start thinking about it, because I think about it in every other area, I mean, I do when I'm driving my car but not when I'm actually going on holiday … it's probably something that a lot of people don't, you don't think about it, because you're just thinking about the convenience of flying somewhere.

Also supporting behavioural inconsistency between home and away, Frode (Norwegian, 37) took great interest in reducing waste in his everyday life, but chose not to buy voluntary carbon offsets or reduce his frequent air travel:

> I'm not buying CO_2 quota on the planes when I'm flying, I'm not buying that. What I'm doing – I'm recycling quite a bit. I think that's the most important thing that I'm doing – I'm quite concerned about how I distribute my garbage. So good with garbage, not that good with travel – travel like always.

Frode's concern about managing his garbage, but lack of concern in the context of air travel, may reflect the deep socially embedded nature of environmental practices such as recycling within his society, which may have become habit. Quite oppositely, Randles and Mander (2009) suggest that tourist air travel itself has become habit, for some sections of society, and that there are only "flickerings" of evidence of consumer environmental concern over aviation. Bjørn (Norwegian, 41), however, argued that decisions, rather than being habit, are often consciously weighed, but typically cannot be attributed to a singular motivation, such as climate concern. In his case, as a father, climate concern needed to be balanced against a range of other personal considerations, such as cost, time and comfort:

> These values are a little bit related to how much does it cost for me also, I must admit. I feel like a bit schizophrenic in terms of climate, because on one end I want to contribute and at the same time I have all these requirements during every day with small kids, going to shopping, all this practical stuff you have to do. There is a set of motivating factors, and environmental is one aspect of many. And the importance of that aspect is partly related to your situation in life at the present moment.

Bjørn's words suggest that the primacy given to environmental values may vary through the lifecycle, as other demands, such as family, compete in consumption decisions. His identity as a father is thus in conflict with his environmental values, reflecting how interpersonal relationships and responsibilities can shape mobility decisions (Hibbert *et al.*, 2013), with particular aspects of identity coming to the fore in different social contexts. Bjørn describes this as "schizophrenic", but a perspective of identities as multiple and performed (Bell, 2008), as discussed earlier, would view such behaviour as commonplace. This explanation supports also Ryan's (1997) observation that motivations are often multidimensional and contextual. Thus, whilst participants may have behaved in particular ways due to some extent to climate concern, such behaviours typically emerged out of a mix of motivational and context-specific factors, with the implication that pro-environmental behaviour was unlikely to remain consistent as situational factors shifted. These findings, in which behavioural inconsistencies between home and away were typical, therefore, further substantiate McKercher *et al.*'s (2010, p. 299) view that "consumer reaction to climate change issues can be described as contradictory at best".

Liminoid tourism spaces

Several of the participants perceived tourism practices as existing largely outside of the social norms that they use, consciously or subconsciously, to structure their behaviour in everyday life. In this sense, tourism space was experienced as liminoid (Kim and Jamal, 2007), and hence perceived as relatively free from the behavioural norms and values of the day-to-day. As such, Pål (Norwegian, 34), when asked if he saw a difference in energy consumption decisions he might take in daily life versus on a holiday, replied: "I think so – because when you're on holiday you're in a different mode. You are somewhere else and you want to get the most out of it and go home and be filled with impressions and experiences." Pål viewed the spaces of tourism as extraordinary, wherein climate change sensitivity took backseat to securing memorable experiences, in theory on offer through tourist activities (Morgan *et al.*, 2010; Tung and Ritchie, 2011). Ida (Norwegian, 52) emphasised making the most of her holidays, which did not include taking time to consider its climate impact: "I'm not stopping and thinking, no. I'm there and I want to see much and do what I want to do." Similarly, Eric (Australian, 38) expressed reluctance to associate holiday spaces with environmental concern of any type, and placed his own level of enjoyment, and desire to relax, as the key factors driving his decision-making:

> [W]hatever I will do on a holiday it will come more out of my personal enjoyment of doing whatever it may be, and yeah, I won't be consciously thinking ok, does this activity impact that? That's not my thought process. If I am on holiday I am there to bloody relax, not feel more responsible and guilty that I am killing the world.

These participants attached too high an importance to their holidays to consider adapting them because of climate change, such as through travelling less, taking a domestic holiday instead of international (Miller *et al.*, 2010) or travelling slowly by more environmentally benign transport modes such as rail or coach (Dickinson *et al.*, 2010). This mirrors the findings of Hares *et al.* (2010), in which there is reluctance to forgo the perceived positive benefits made accessible by tourist air travel in order to reduce personal emissions.

For Rita (Norwegian, 34), both the importance of escaping to an attractive overseas destination to relax and the trip's corresponding economic cost outweighed concern over the climate impacts of her holidays. She attributed this to the relative infrequency of her holidaying:

> Holiday trips are maybe once a year and other issues would be more important – where to go and economic questions – would be more important on my annual travels. The things I can do every day are easier to be conscious about and to make a decision about then what you do once a year. Because then it's more important to me to go to a nice place and relax for two weeks.

Rita's viewpoint contrasts, at least in a corporeal sense, the notion of tourism as part of everyday life (Edensor, 2007; Franklin and Crang, 2001; Larsen, 2008), instead viewing it as extraordinary and helping to perpetuate a dualistic separation between daily domestic and less frequent tourism experiences. For Rita, holidays are liminoid spaces of pleasure where everyday rules and social norms are relaxed (Kim and Jamal, 2007). The infrequency of tourism travel was particularly significant amongst the Australian participants, for whom most holiday flying was long-haul, overseas and justified through its uncommonness. This point on the difference in regularity between domestic and tourism decisions was also cited as an important factor by Tone (Norwegian, 58):

> Daily life is more important. I'm more concerned about daily life because we don't travel all the time. It's [flying] kind of abstract, because you are not doing this every day and it is a little bit away from you when you have landed and then you go home.

Tone illustrates how physical distancing from spaces of daily life through air travel contributes to experiences of tourism spaces as liminoid: her words reflect Turner's (1977) phases of liminality, in which one separates from "normal" life (take-off), passes through the "limen", or metaphorical threshold (via flight), into a transitory state of liminality characterised by a perceived lack of structure (at a new destination), and finally re-aggregates by returning (flying) home. Notably, such a cycle assumes (often wrongly) a lack of familiarity with the social norms of the destination.

Nonetheless, the relative infrequency of tourism practices, combined with their typical occurrence in spaces outside of everyday life, provided justification for sustainability practices to be temporarily suspended. A temporary suspension of environmental norms when on holidays lends support to Barr et al.'s (2010, p. 475) observation that a sustainable lifestyle will only exist once "individuals are able to transfer their behaviours between contexts, as part of an embedded set of lifestyle practices". As discretionary air travel is often employed as a gateway to liminoid experiences, tourism practices when viewed from this conceptual perspective pose a significant barrier to achieving sustainable lifestyles.

Contextualised performances of consumer concern

Rather than consistent performances of identity aligned with an embedded set of lifestyle practices, through a commitment to reducing climate impacts across all life contexts, many of the participants narrated performances of consumer concern that were contingent upon context, reflecting how multiple identities can be performed depending on the social situation at hand (Bell, 2008; Bond and Falk, 2013). In some cases, the contradiction between striving for sustainable practices in everyday life, only for a single long-haul flight taken to exceed annual per capita sustainable emission levels (Gössling et al., 2009), was openly acknowledged:

I think it's a contradiction. I think a lot of people do it. But you kind of, you kind of try to put it back of your mind and try not to worry about it. Well, you think, I'm seeing the world and it's great for the kids to see the world. So you try to put it to one side. Silly really.

(Ruby, British, 41)

A lot of the work I do ... focuses on improving resilience to the effects of climate change ... I recycle, train, live with the least sort of environmental footprint impact as I can ... I am aware [of] the impact that planes flying every which way, all of the time, has on the atmosphere and things like that ... but it probably doesn't influence my travel decision.

(Camilla, Australian, 24)

In another instance, a participant who regularly stayed in the UK and went camping for her holidays, both because of lower costs and pro-environmental attitudes, admitted that if her financial circumstances were to change, that she would probably not be able to resist taking tourism trips via long-haul air travel – "Say I won a load of money tomorrow – I'd probably go [to New Zealand]. It's awful, isn't it? You feel guilty but you justify it to yourself in some respect" (Grace, British, 36). Thus, for Grace, her travel behaviour was contingent on the social and economic context in which she might be positioned, rather than an enduring set of core values or a steadfast environmental identity. Equally, Tom (Australian, 47) further speculated on the contextually contingent climate concern of others:

[A] lot of people got those solar power subsidy deals and, you know, a lot of those people are concerned about climate change and did it for that reason, but those same people would be more than happy to jump in a plane and fly to Europe given the opportunity.

For another participant, different performances were offered between home and away, which whilst inconsistent, were not recognised as conflictive with the participants' environmental values:

I probably don't think about it actually. You know what, I went to Turkey last year, and it was 40 degrees and we had air conditioning and we left it on. We went out and left the air conditioning on. And I don't think that I, for one moment, thought about the effects on the environment. And I've even done an environmental degree.

(Harry, British, 40)

For Harry, who closely monitored his domestic energy consumption at home, both air travel and energy usage once in the destination were subject to a lower level of climate concern than in daily life practices.

The inconsistencies between these different "faces" performed depending on context, which constitute what we may term "multiple environmental identities", were not experienced by the participants as a source of concern that needed any mediation or reconciliation. Multiple malleable and fluid identities (McAdams, 1997), with climate change sensitivities adapted to suit the participant's needs in each situation, were narrated to make sense of and justify what may be externally perceived as behavioural contradictions. Lewis (British, 39), who was gravely concerned over the implications of climate change for the futures for his two young children, reflected this capacity for multiple identity performances in justifying the environmental impact from his last holiday in Florida:

> [T]here were four families with four cars and we drove everywhere every single day to a different a location to do something. And even when you're in those locations chances are you're using amenities that are extremely wasteful on electricity and emissions as well. So you think a lot less about the environmental impacts then. You're in an apartment, you pay for it, it's not yours, whether you go out and leave the lights on – chances are you're a lot less environmentally aware when you're on holiday than when you're not.

As illustrated through these latter examples, not only the decision to fly but also a range of other tourism practices with environmental consequences relating to accommodation, activities and ground transport reflect inconsistent performances of consumer concern between "home" and "away". Such seemingly contradictory consumption decisions, however, are routine within postmodern theory that assumes personal identity (and consequently behaviour) is inconsistent and performed differently across varying contexts.

Conclusion

In this chapter, we have sought to further the understanding of why there is an attitude/behaviour gap, compounded by a dissonance between "home" and "away" (Barr *et al.*, 2010, 2011b), which makes voluntary consumer behaviour change in the context of discretionary air travel an intractable situation. The chapter does have limitations: it is based on Western countries, with a participant profile that is for the most part highly educated and moderately affluent. Nonetheless, empirical evidence from our study participants, who were in the main highly aeromobile, demonstrates that tourism spaces are often subject to lower levels of environmental concern than day-to-day contexts. Although there was some evidence of consistency amongst the study participants between the environmental sustainability of their domestic practices and those made when on holiday, such cases were exceptional. More common was the tendency to subject tourism settings to less stringent, or even to consider them as exempt from, climate concern.

This chapter echoes the findings of other works that suggest that the public is largely unwilling to voluntarily change their holiday flying behaviour for environmental reasons (Barr *et al.*, 2010; Cohen *et al.* 2011; Hares *et al.* 2010; Miller *et al.* 2010), a resistance also found in work-related air travel (Lassen, 2010), and in flying more generally (Kroesen, 2013). This chapter is, however, unique in providing an in-depth sociological explanation of "why" an attitude/behaviour gap exists between climate change sensitivity and discretionary air travel. Drawing heavily from insights from the field of knowledge in tourism studies, the chapter responds to calls to better connect tourism research to knowledge on pro-environmental behaviour (Barr *et al.*, 2011b). In doing so, it provides a nuanced understanding of why behavioural inconsistencies exist between public climate concern and actual holiday flying practices.

Whether these behavioural inconsistencies and contradictions are understood as a modern expression of tourism practices occurring in liminoid space (Turner, 1982), or as a postmodern reflection of multiple, contextually dependent identities (Bell, 2008), the implications are the same. Our consistent findings from study participants not only show how these sociological theories are powerful devices in explaining both attitude/behaviour and home/away gaps within discretionary air travel decisions, but also provide a firmer basis for evaluating the prospects of sustained behavioural change amongst the travelling public. We argue that scope for positive behaviour change in the context of discretionary air travel practices is limited. In this sense we provide theoretical and empirical support for the suggestion of McKercher *et al.* (2010, p. 297) that "government intervention may be required to create meaningful behavioural change in tourism patterns".

These findings are of importance to policy makers who may seek reductions in the climate impacts of discretionary air travel through even a partial reliance on public behavioural change. Given the declining scope for technical gains in aircraft energy efficiency (Scott *et al.*, 2010) and in the absence of a global market-based mechanism (e.g. carbon trading) for the aviation sector (Duval, 2013), governments' ambitions to encourage public behaviour change (see Barr *et al.*, 2011b and Miller *et al.*, 2010 in the UK case) may be misguided in the context of the tourism and travel sectors. Strategies that seek to tackle these issues, whether through education or media, by aiming to nudge individual lifestyles towards less carbon intensive consumption choices, need to be tempered with an awareness that environmental identities, like other aspects of personal identities, cannot be relied upon to lead to consistent behaviour. This is a significant challenge for the governance of climate change and aviation.

Academia, in cooperation with third-sector pressure groups, needs to focus on changing the attitudes of policy makers and key industry stakeholders, at national and supranational levels, to help pave the difficult road towards strong (global) policy interventions aimed at reducing the greenhouse gas emissions from the aviation industry. This chapter demonstrates why scope for positive voluntary public behaviour change in the context of discretionary air travel is limited. When set alongside the wealth of evidence that corroborates this conclusion (e.g. Barr *et al.*, 2011b; Hares *et al.*, 2010; Kroesen, 2013; Lassen, 2010; Miller *et al.*, 2010), it is abundantly clear that urgent policy interventions for stronger climate "governance" of aviation are required. Future research, there-fore, needs to now turn to the issues that may hinder policy development in this area, such as concerns over social equity in the distribution of air travel, cultural nuances across nation states, including non-Western countries, which may impede (or facilitate) collective agreements, structural barriers in alternatives to short-haul air travel, and how a redistribution of receipts from tourism and air travel more generally can be accommodated. Such work will be critical to the international policy interventions that are clearly necessary, to put in place and implement binding mitigation targets for the aviation industry.

References

Anable, J. (2005). "Complacent car addicts" or "aspiring environmentalists"? Identifying travel behaviour segments using attitude theory. *Transport Policy, 12,* 65–78.

Antimova, R., Nawijn, J. and Peeters, P. (2012). The awareness/attitude gap in sustainable tourism: A theoretical perspective. *Tourism Review, 67*(3), 7–16.

Barr, S., Gilg, A. and Shaw, G. (2011a). "Helping people make better choices": Exploring the behaviour change agenda for environmental sustainability. *Applied Geography, 31,* 712–720.

Barr, S., Shaw, G. and Coles, T. (2011b). Times for (un)sustainability? Challenges and opportunities for developing behaviour change policy. A case-study of consumers at home and away. *Global Environmental Change, 21,* 1234–1244.

Barr, S., Shaw, G., Coles, T. and Prillwitz, J. (2010). "A holiday is a holiday": Practicing sustainability, home and away. *Journal of Transport Geography, 18*(3), 474–481.

BBC News. (2012). *Australia introduces controversial carbon tax.* Retrieved 30 April 2013, from www.bbc.co.uk/news/world-asia-18662560.

Becken, S. (2007). Tourists' perception of international air travel's impact on the global climate and potential climate change policies. *Journal of Sustainable Tourism, 15*(4), 351–368.

Becken, S. (2010). A critical review of tourism and oil. *Annals of Tourism Research, 38*(2), 359–379.

Bell, E. (2008). *Theories of performance.* Los Angeles: Sage.

Bond, N. and Falk, J. (2013). Tourism and identity-related motivations: Why am I here (and not there)? *International Journal of Tourism Research, 15*(5), 430–442.

Bray, J., Johns, N. and Kilburn, D. (2011). An exploratory study into the factors impeding ethical consumption. *Journal of Business Ethics, 98*(4), 597–608.

Burns, P. and Bibbings, L. (2009). The end of tourism? Climate change and societal challenges. *21st Century Society, 4*(1), 31–51.

Carr, N. (2005). Poverty, debt, and conspicuous consumption: University students' tourism experiences. *Tourism Management, 26,* 797–806.

Cohen, E. (1979). A phenomenology of tourist experience. *Sociology, 13,* 179–202.

Cohen, S.A. and Higham, J.E.S. (2011). Eyes wide shut? UK consumer perceptions on aviation climate impacts and travel decisions to New Zealand. *Current Issues in Tourism, 14*(4), 323–335.

Cohen, S. and Taylor, L. (1992). *Escape attempts: The theory and practice of resistance to everyday life.* London: Routledge.

Cohen, S.A., Higham, J.E.S. and Cavaliere, C.T. (2011). Binge flying: Behavioural addiction and climate change. *Annals of Tourism Research, 38*(3), 1070–1089.

Crompton, J. (1979). Motivations for pleasure vacation. *Annals of Tourism Research*, 6(4), 408–424.

Dann, G. (1977). Anomie, ego-enhancement and tourism. *Annals of Tourism Research*, 4(4), 184–194.

Decrop, A. (2004). Trustworthiness in qualitative tourism research. In J. Phillimore and L. Goodson (eds), *Qualitative research in tourism: Ontologies, epistemologies and methodologies* (pp. 156–169). London: Routledge.

Denzin, N.K. and Lincoln, Y.S. (2005). Introduction: The discipline and practice of qualitative research. In N.K. Denzin and Y.S. Lincoln (eds), *The Sage handbook of qualitative research* (3rd edn) (pp. 1–32). Thousand Oaks, CA: Sage.

Dickinson, J., Robbins, D. and Lumsdon, L. (2010). Holiday travel discourses and climate change. *Journal of Transport Geography*, 18(3), 482–489.

Duval, D.T. (2013). Critical issues in air transport and tourism. *Tourism Geographies*, 15(3), 494–510.

Edensor, T. (2001). Performing tourism, staging tourism. *Tourist Studies*, 1(1), 59–81.

Edensor, T. (2007). Mundane mobilities, performances and spaces of tourism. *Social & Cultural Geography*, 8(2), 199–215.

Finnegan, R. (1997). "Storying the self": Personal narratives and identity. In H. Mackay (ed.), *Consumption and everyday life* (pp. 66–111). London: Sage.

Fontana, A. and Frey, J.H. (2005). The interview: From neutral space to political involvement. In N.K. Denzin and Y.S. Lincoln (eds), *The Sage handbook of qualitative research* (3rd edn) (pp. 695–728). Thousand Oaks, CA: Sage.

Franklin, A. and Crang, M. (2001). The trouble with tourism and travel theory? *Tourist Studies*, 1(1), 5–22.

Goffman, E. (1959). *The presentation of self in everyday life*. Harmondsworth: Penguin Books.

Gössling, S. (2009). Carbon neutral destinations: A conceptual analysis. *Journal of Sustainable Tourism*, 17(1), 17–37.

Gössling, S. and Peeters, P. (2007). "It does not harm the environment!" An analysis of industry discourses on tourism, air travel and the environment. *Journal of Sustainable Tourism*, 15(4), 402–417.

Gössling, S. and Upham, P. (eds). (2009). *Climate change and aviation: Issues, challenges and solutions*. London: Earthscan.

Gössling, S., Haglund, L., Kallgren, H., Revahl, M. and Hultman, J. (2009). Swedish air travellers and voluntary carbon offsets: Towards the co-creation of environmental value. *Current Issues in Tourism*, 12(1), 1–19.

Gössling, S., Peeters, P. and Scott, D. (2008). Consequences of climate policy for international tourist arrivals in developing countries. *Third World Quarterly*, 29(5), 873–901.

Gössling, S., Scott, D., Hall, C.M., Ceron, J-P. and Dubois, G. (2012). Consumer behaviour and demand responses of tourists to climate change. *Annals of Tourism Research*, 39(1), 36–58.

Hares, A., Dickinson, J. and Wilkes, K. (2010). Climate change and the air travel decisions of UK tourists. *Journal of Transport Geography*, 18(3), 466–473.

Hibbert, J., Dickinson, J.E. and Curtin, S. (2013). Understanding the influence of interpersonal relationships on identity and tourism travel. *Anatolia: An International Journal of Tourism and Hospitality Research*, 24(1), 30–39.

Higham, J.E.S. and Cohen, S.A. (2011). Canary in the coalmine: Norwegian attitudes towards climate change and extreme long-haul air travel to *Aotearoa*/New Zealand. *Tourism Management*, 32(1), 98–105.

Høyer, K. (2000). Sustainable tourism or sustainable mobility? The Norwegian case. *Journal of Sustainable Tourism, 8*(2), 147–160.

Huebner, A. (2012). Public perceptions of destination vulnerability to climate change and implications for long-haul travel decisions to small island states. *Journal of Sustainable Tourism, 20*(7), 939–951.

Jennings, G. (2001). *Tourism research.* Milton: John Wiley & Sons.

Kim, H. and Jamal, T. (2007). Touristic quest for existential authenticity. *Annals of Tourism Research, 34*(1), 181–201.

Kroesen, M. (2013). Exploring people's viewpoints on air travel and climate change: Understanding inconsistencies. *Journal of Sustainable Tourism, 21*(2), 271–290.

Larsen, J. (2008). De-exoticizing tourist travel: Everyday lives and sociality on the move. *Leisure Studies, 27*(1), 21–34.

Lassen, C. (2010). Environmentalist in business class: An analysis of air travel and environmental attitude. *Transport Reviews, 30*(6), 733–751.

Lett, J.W. (1983). Ludic and liminoid aspects of charter yacht tourism in the Caribbean. *Annals of Tourism Research, 10*(1), 35–56.

Lincoln, Y.S. and Guba, E.G. (1985). *Naturalistic inquiry.* Newbury Park, CA: Sage.

Mair, J. (2011). Exploring air travellers' voluntary carbon-offsetting behaviour. *Journal of Sustainable Tourism, 19*(2), 215–230.

McAdams, D.P. (1997). The case for unity in the (post)modern self: A modest proposal. In R.D. Ashmore and L. Jussim (eds), *Self and identity: Fundamental issues* (pp. 46–78). New York: Oxford University Press.

McKercher, B. and Prideaux, B. (2011). Are tourism impacts low on personal environmental agendas? *Journal of Sustainable Tourism, 19*(3), 325–345.

McKercher, B., Prideaux, B., Cheung, C. and Law, R. (2010). Achieving voluntary reductions in the carbon footprint of tourism and climate change. *Journal of Sustainable Tourism, 18*(3), 297–317.

Miles, M.B. and Huberman, A.M. (1994). *Qualitative data analysis: An expanded sourcebook.* Thousand Oaks, CA: Sage.

Miller, G., Rathouse, K., Scarles, C., Holmes, K. and Tribe, J. (2010). Public understanding of sustainable tourism. *Annals of Tourism Research, 37*(3), 627–645.

Monbiot, G. (2007). *Heat: How to stop the planet burning.* London: Penguin Books.

Morgan, M., Lugosi, P. and Ritchie, J.R.B. (eds). (2010). *The tourism and leisure experience: Consumer and managerial perspectives.* Clevedon: Channel View.

O'Reilly, K. (2005). *Ethnographic methods.* London: Routledge.

Pang, S.F.H., McKercher, B. and Prideaux, B. (2013). Climate change and tourism: An overview. *Asia Pacific Journal of Tourism Research, 18*(1–2), 4–20.

Patton, M.Q. (2002). *Qualitative research and evaluation methods.* Thousand Oaks, CA: Sage.

Peeters, P. and Dubois, G. (2010). Tourism travel under climate change mitigation constraints. *Journal of Transport Geography, 18*(3), 447–457.

Pernecky, T. and Jamal, T. (2010). (Hermeneutic) phenomenology in tourism studies. *Annals of Tourism Research, 37*(4), 1055–1075.

Randles, S. and Mander, S. (2009). Practice(s) and ratchet(s): A sociological examination of frequent flying. In S. Gössling and P. Upham (eds), *Climate change and aviation: Issues, challenges and solutions* (pp. 245–271). London: Earthscan.

Ryan, C. (1997). *The tourist experience: A new introduction.* London: Cassell.

Scott, D. (2011). Why sustainable tourism must address climate change. *Journal of Sustainable Tourism, 19*(1), 17–34.

Scott, D., Hall, C.M. and Gössling, S. (2012). *Tourism and climate change: Impacts, adaptation and mitigation.* London: Routledge.

Scott, D., Peeters, P. and Gössling, S. (2010). Can tourism deliver its "aspirational" greenhouse gas emission reduction targets? *Journal of Sustainable Tourism, 18*(3), 393–408.

Sedgley, D., Pritchard, A. and Morgan, N. (2012). "Tourism poverty" in affluent societies: Voices from inner-city London. *Tourism Management, 33*, 951–960.

Semenza, J.C., Hall, D.E., Wilson, D.J., Bontempo, B.D., Sailor, D.J. and George, L.A. (2008). Public perception of climate change: Voluntary mitigation and barriers to behavior change. *American Journal of Preventative Medicine, 35*(5), 479–487.

Sharpley, R. (2003). *Tourism, tourists and society.* Huntingdon: Elm.

Steg, L. and Vlek, C. (2009). Encouraging pro-environmental behaviour: An integrative review and research agenda. *Journal of Environmental Psychology, 29*, 309–317.

Sustainable Aviation. (2011). *Progress report 2011.* Retrieved 7 May 2013, from www. sustainableaviation.co.uk/progress-report/.

Tung, V.W.S. and Ritchie, J.R.B. (2011). Exploring the essence of memorable tourism experiences. *Annals of Tourism Research, 38*(4), 1367–1386.

Turner, V. (1977). Variations on a theme of liminality. In S.F. Moore and B.G. Myerhoff (eds), *Secular ritual* (pp. 36–52). Assen: Van Gorcum.

Turner, V. (1982). *From ritual to theatre: The human seriousness of play.* New York: PAJ.

Vaughan, G.M. and Hogg, M.A. (2002). *Introduction to social psychology* (3rd edn). Frenches Forest: Pearson Education.

Whitmarsh, L., Seyfang, G. and O'Neill, S. (2011). Public engagement with carbon and climate change: To what extent is the public "carbon capable"? *Global Environmental Change, 21*, 56–65.

Zeppel, H. (2012). Collaborative governance for low-carbon tourism: Climate change initiatives by Australian tourism agencies. *Current Issues in Tourism, 15*(7), 603–626.

6 ZMET

A psychological approach to understanding unsustainable tourism mobility

Catheryn Khoo-Lattimore and Bruce Prideaux

Introduction

As the scale of global tourism flows continues to grow and the impacts of climate change become more apparent, issues related to tourism mobilities and the ability of destinations to sustain long-haul markets will become increasingly important. Distance is a particularly important factor because, as the average distance of individual trips grows, there is an increase in the collective environmental impact of annual emissions. The magnitude of anticipated increases in global air travel is demonstrated by a recent report from Boeing that forecasts that the total demand for passenger jet aircraft is likely to grow from 20,000 aircraft in 2010 to 40,000 in 2030 (Boeing, 2013). One issue that is likely to affect the ability of destinations to maintain long-haul tourism flows in the future is the impact of legislation to reduce greenhouse gas emission. Actions of this nature are already occurring with carbon pricing schemes implemented in Australia and the EU (World Bank, 2013). Despite many protests, the UK government has imposed the Air Passenger Duty as a climate change mitigation strategy by increasing the price of air travel and in the process making the UK's air travel tax one of the highest in the world. The issue of sustainable mobilities will grow in importance as concerns over climate change begin to determine international policy agendas. The implications for tourism are obvious, particularly the long-haul aviation sector that is currently the tourism industry's major contributor to greenhouse gas emissions (UNWTO–UNEP–WMO, 2008). Any change in demand for long-haul travel due to climate change mitigation policies will have a knock-on effect on short-haul travel and in so doing change the demand patterns for domestic road, rail, air and sea transport. Changing patterns of demand of this nature will create new opportunities for destinations as well as close off existing opportunities.

A number of studies have noted the gap between awareness of climate change and personal actions undertaken to alter personal travel behaviours (Becken, 2004; Huebner, 2012; McKercher *et al.*, 2010). Several researchers have noted that engaging in air travel to distant destinations is seen as a right by many consumers (Becken, 2007). In some cases, lack of action stems from the belief that it is the role of other agencies, not the individual, to make changes that can reduce the impact of climate change (McKercher *et al.*, 2013). Changing attitudes is a complex task and as Pidgeon (2011) acknowledges, even in circumstances where accepting personal responsibility might save one's life, wearing seat belts for example, long-term communication strategies, backed by punitive action, were required to achieve the desired effects.

Understanding and then changing consumer attitudes is, therefore, a pressing issue. Achieving tourists' social and behavioural change could be a key to achieving responsible tourism, particularly where climate change is concerned. Research has demonstrated that tourists do not choose environmentally friendly transportation despite their declared positive attitudes towards sustainable tourism (Beudeanu, 2007). As Gössling *et al.* (2010) observed, technology and management alone cannot reduce emissions. To overcome the apparent reluctance to cut back on personal air travel (Higham *et al.*, 2013), there is a need to address existing discrepancies between tourist intention (what they say they would like to do) and behaviour (what they actually do). A recent paper by McNamara and Prideaux (2010) illustrates this gap, in that case between tourists' self-reported use of interpretation provided in a natural area and actual observed use. Although a large majority of respondents reported that the availability of environmental information at a forest site was important, only 24 per cent of visitors actually stopped to read interpretative signs, with the average time spent reading being 2.1 minutes. Recent evidence (Kollmuss and Agyeman, 2002; Sheeran, 2002) suggests that attitude–behaviour gaps present themselves as substantial psychological barriers to positive change. For an approach based on reducing these barriers to be effective, new psychological and behavioural approaches are required.

There is, therefore, an urgent need to understand responses at the individual level using techniques that reflect the complexities of a tourist's thoughts and behaviour. This chapter outlines a relatively new method to identify and evaluate varied deep-seated motivations in individual travel practices and those choices that may constitute barriers to increased sustainability. It outlines how the Zaltman metaphor elicitation method (ZMET) technique may be used to identify barriers to individuals adopting sustainable mobility practices. It draws on the psychology of travel and tourism, and on the growing use of photographs and photography as a tool for understanding travel psychology.

Researching barriers to change: reflecting on past and current practices

Thus far, research on behavioural change has been concerned with understanding either motivations for, or barriers to, sustainable mobility. On the former, much past research has advocated intervention from governments and corporations, with suggestions such as informal environmental education (Ballantyne and Packer, 2005), provision of green information on products (Oates *et al.*, 2008; Ottman *et al.*, 2006) and use of future threat indicators related to transport, such as traffic safety, crash cost and emissions of toxic substances (Gilbert and

Tanguay, 2000; Gudmundsson, 2001; Litman, 2003). Researchers who study motivation from a consumer perspective have identified other factors such as knowledge of environmental issues (Finger, 1994; Nilsson and Kuller, 2000), incentives (Young *et al.*, 2010), technological solutions, consumer values and identities (Skinner and Rosen, 2007; Steg and Gifford, 2005) and the belief that an individual can make a difference (Stern, 2000). In relation to barriers, the focus of this chapter, scholars have identified various factors such as misrepresentation of data by aviation industry (Gössling and Peeters, 2007), lack of information (Becken, 2007), inability to transfer environmental concerns across contexts (Higham *et al.*, 2013), lack of alternatives (Anable, 2005; Hunecke *et al.*, 2007), accessibility (Prillwitz *et al.*, 2007), perception of decreased comfort (Hensher and Reyes, 2000; Steg and Gifford, 2005) and habits (Eriksson *et al.*, 2008; Kenyon and Lyons, 2003).

While the methods used in the past have yielded insightful findings, the majority of studies have relied on traditional research methods. Lu and Nepal (2009), for example, content-analysed the first 15 volumes of the *Journal of Sustainable Tourism* (1993–2007) and argued that the data collection methods used for understanding tourist mobility issues have remained largely unchanged for the past 15 years. Although quantitative methods have been supplemented with more sophisticated analytical tools such as multi-dimensional modelling, geographical information systems and computer simulation, qualitative methods have largely remained with interviews and focus groups. Structured interviews have been criticised, with issues raised about the validity and reliability of their findings. It has been suggested that during a structured interview, individuals may not be capable of articulating the reasons for their choice since they have little awareness of the nature of the cognitive process that mediates complex behaviour (Lawson *et al.*, 1996). In addition, some people may only choose to express what they perceive as ideal because they are aware of being observed (Nakanishi, 1974). Methods that assist tourists to identify and express their own motivations and barriers, rather than those arising from a positivist framework, are therefore needed. In addition, use of textual analysis as a stand-alone technique has limited the ability of researchers to move beyond word-based descriptions of any given phenomenon.

The methods discussed above do not include the potential for investigation of unconscious needs and forces as possible barriers to sustainable tourist mobility. In response to this research gap a number of scholars (Kingsbury, 2005; MacCanell, 2002; Uriely *et al.*, 2011) have proposed psychoanalytic frameworks to identify deviant tourist behaviour but as Kingsbury observed, "engagements with psychoanalysis in tourism research are cursory at best" (p. 118). This could be due in part to biased assumptions about, and attitudes towards, the methodology of prevailing psychological approaches. Perhaps this is why Kingsbury is right in saying that while tourism scholars recognise the consequences of the unconscious process, they view the topic as too difficult to tackle and therefore ignore it completely. MacCanell (2002) who attempted to answer questions about complex tourist behaviour through the concept of ego and to a certain extent superego, has called for the application of psychoanalytic notions for robust analysis of their research areas. In the world of consultancy, Plog (1991) developed an understanding of the psychology of tourism and tourists as it relates to creating growth markets. More recently, Pearce *et al.* (2011) have produced a substantial review of relationships between tourism, psychology and well-being. Pearce and Packer (2013) describe further newly emergent links between psychology and tourism.

Given that the tourist experience is often of a visual nature it can be argued that in a tourism setting, techniques that emulate the way tourists see and engage with the world would be beneficial. An approach of this nature offers the possibility for new insights into the question of why tourists often display indifference to the impact on the environment in their everyday lives and when they make travel decisions.

Photography, a psychological tool

It is surprising that few of the methods used in the study of transport mobilities are borrowed from the discipline of psychology. Psychologists have often used techniques that are projective in nature to investigate relevant constructs in consumer decision-making. One such projective technique is the ZMET method. ZMET is derived from a synthesis of approaches drawn from the neurobiology, psychoanalysis and psychology literatures. It was developed for marketing research by Harvard Business School Professor Gerald Zaltman who first used this technique on a market research trip in the rural areas of Nepal (Khoo-Lattimore, 2008). Participants were given cameras and film and asked to take their own pictures of their everyday lives. After developing the films, he asked the participants to tell him what the pictures meant to them. The most striking observation was that most pictures did not show people's feet. During interviews with the participants Zaltman discovered that they did not aim the cameras wrongly, rather, being barefoot was a sign of poverty and people chose to hide their bare feet in the pictures. This discovery of a "hidden meaning" in pictures prompted Zaltman to develop a new approach to market research, where the primary objective was to gain an in-depth understanding of people's choices.

For tourism scholars, the use of pictures is a logical addition to the research toolbox when one considers how contemporary tourism at one level represents an amalgam of culturally, socially and psychologically derived mental images which are able to be portrayed by pictures and at another level are the result in part of using attractive pictures to create socially and psychologically derived images to sell destination experiences. As Larsen (2006, p. 241) observed, "it has become almost unthinkable to embark on a holiday without taking the camera along ... and return home with many snapshot memories". As such, a number of recent tourism researchers have used photography in their work. Kerstetter and Bricker (2009), exploring Fijian's sense of place, note that there are multiple approaches to photo elicitation – auto driving, reflexive photography and photo novella/photo voice – the latter two being most common in the social sciences. Cahyanto *et al.* (2013) use reflexive photography in their research into developing responsible rural tourism in Indonesia.

ZMET's use of pictures as the medium of data collection is based on the premise that most human communication is non-verbal (Zaltman, 1997; Zaltman and Higie, 1993). Previous research has shown that only a maximum of 30 per cent of the meaning in a social exchange is conveyed by words (Knapp, 1980; Mehrabian, 1971; Weiser, 1988). Zaltman on the other hand found that pictures contain metaphors that may be visual, verbal, mathematical or even musical (Zaltman, 1997). ZMET contends that these metaphors are instrumental in an attempt to understand the respondent's voice. A metaphor is the perception of one thing as if it were a different kind of thing (Lakoff, 1993). It compares seemingly unrelated subjects. For example, the dove is a metaphor for peace and freedom. The importance of metaphors in shaping people's thoughts and reality in their daily lives has been highlighted by many studies in the past (Gibbs *et al.*, 1996; Glucksberg, 1995, 2001; Glucksberg and McGlone, 1999; Kövecses, 2002; Short, 2001). The consensus of these studies has been that metaphors are able to assist researchers and psychotherapists to discover "important mental states that literal language might altogether miss or under represent" (Zaltman, 1997, p. 425). The following discussion outlines the potential for using ZMET as a new approach to understanding the reluctance of tourists to first, support and second, make holiday choices that support sustainable mobilities. As far as the authors are aware, no part of the manual has been previously published in an academic work.

The ZMET method

ZMET uses a ten-step one-to-one interview approach based on images collected or taken by participants. Prior to the interview, participants are instructed to collect or take between eight and twelve pictures that convey their thoughts and feelings about the topic under investigation (Coulter and Zaltman, 1994; Zaltman, 1997, 2003; Zaltman and Higie, 1993). If used to investigate aspects of transport mobilities the researcher could for example focus on the relationship between mode selection, cost of transport and expected benefits from travel. The results from the ZMET technique could be expected to show options that may exist for changing the mode choice (with the expectation of reducing climate change impacts) while retaining the expected benefits from travel.

The first stage of the ZMET interview is called storytelling, because participants are asked to tell the story behind each of the pictures that they have brought to the interview. Storytelling presents a good opportunity for uncovering relevant information about the topic. At this stage the researcher is mainly concerned with uncovering the reasons why the images were selected as well as the meanings behind each selected image. The ZMET interviewer training manual (2003, p. 17) advocates starting off the interview with the following statement:

> I am going to be asking some seemingly strange questions, asking for clarification of your comments and explanations of ideas that you think should

be obvious. Just imagine that I am from another country (or another planet) and I don't understand much about the topic or people. So I am going to ask a lot of questions. This is just part of our process, so please bear with me.

Following this statement, the interviewer refers to the first image with specific questions such as, "Could you describe this picture for me?", "What is this picture about?" or "How does this image relate to your thoughts and feelings about taking the train instead of the plane?" At this stage, the interviewer's goal is to identify the most interesting metaphors and probe for further elaboration and understanding of the core meaning within each metaphor. At the same time, the interviewer must become aware, spot interesting and relevant concepts and probe for associations with other concepts.

After all the pictures have been discussed, step 2 involves the interviewer asking if there were any images that participants sought but could not find. Alternatively, the interviewer could ask if the participants had any thoughts or feelings about the topics that were not represented in the images already discussed, and then ask the participants what image would express that idea. When participants describe a missed image, the interviewer follows up by asking the participants to expand upon their thoughts and feelings, treating the missing image as any other storytelling episode. Descriptions attached to missing images also enable the respondent to expand on his or her thoughts and feelings as if the images were available. This step ensures that the potential of a missing construct due to the respondent's inability to gain access to a particular image is ruled out.

In step 3, the participant is asked to sort the pictures into meaningful piles and provide a label or description to the pile. This step is useful because it establishes constructs or themes that are relevant to the consumer. If the probing undertaken in step 1 has already managed to explore the participant-selected images/constructs in detail, the third step tends to provide redundant information and the interviewer can move to the next step. However, if the interviewer has missed probing with some pictures, this third step is an efficient way to explore elicited concepts in depth. If new concepts are elicited, the interviewer will continue to probe and ladder all new concepts.

Step 4 uses a modified version of the Kelly repertory grid to identifying new concepts, meanings and distinctions. The Kelly repertory grid is described as a method for exploring personal construct systems (Fransella and Banister, 1977). Thus it is an attempt to see other peoples' worlds as they see them and to understand their situations and concerns. The interviewer selects three images to identify concepts, meanings and distinctions. Three images are necessary according to the Kelly repertory grid because this two-against-one question produces a bipolar scale for the identification of a construct. According to the ZMET interviewer manual (2003), images can be selected randomly but interviewers can also choose to intentionally select images that have similar meanings. The interviewer then asks, "Now, tell me how two of these pictures are similar yet different from the third ... with respect to your thoughts and feelings about not flying?"

In step 5 the participant is asked to select the picture that is most representative of the topic under investigation. Also included in this step is metaphor elaboration where the respondent is asked to imagine widening the frame of the picture in any direction or dimension to describe what would enter the picture, and in so doing reinforce the meaning for them. This step encourages the respondent to explore additional thoughts and feelings. In step 6, the participant is asked to describe pictures that might depict the opposite of the task assigned. This step of the ZMET interview is designed to use negative case analysis as a criterion for evaluating trustworthiness (Carson *et al.*, 2001) and is based on the view that while it is imperative to understand what is meaningful to the participants, it is also important to know what is not meaningful for them. Negative case analysis involves asking questions to find exceptions to a rule in a theory that therefore invalidate the rule. Used in a transport mobilities context this step could be expected to identify association between mode selection and expected opportunity costs of selecting near rather than distant destinations.

Step 7 is designed to bring the participant's unconscious thoughts to a level of awareness at which verbal articulation can occur (Zaltman, 1997). The participant is asked to use other non-visual senses (taste, sound, smell, colour, touch and feelings) to convey what does and does not represent the topic under investigation. This sensory image step offers an opportunity to elicit metaphors via the other senses besides the visual.

In step 8, the participant creates either a cognitive mental map or a vignette, depending on the project's focus. To create the vignette, the participant is asked to imagine a short movie that describes their thoughts and feelings about the topic. In contrast, for the mental map, the participant is asked to create a map using the constructs that have been elicited. During the construction of the cognitive map, the participant is asked to verify if the constructs recorded by the interviewer are accurate representations of what was said. The participant then writes down each accurate construct and is asked to draw lines from one construct to another showing how they are related. Here, the participant's viewpoint and confirmation is sought before a construct is identified in the map. Unlike the means–end chain hierarchical value maps, the cognitive map in ZMET and the links between constructs are constructed by the participants.

Using digital imaging techniques, step 9 involves the participant creating a summary image or montage expressing the topic under investigation. The purpose of this step is to help the participant stimulate and express thinking and not to develop aesthetically pleasing pictures (Zaltman and Higie, 1993). The participant's images are scanned and brought up on a computer screen for manipulation. With the help of the researcher, the participant decides on what picture to place where, what to crop in a picture, which picture to resize, to add colour, to change colour and so on until the participant is satisfied with the summary image they have created. When the summary image is created, the participant talks the interviewer through it. The relevance of a particular image editing decision is used to build a deeper understanding of their motives. Step 9 is essentially a step to reaffirm the constructs that have been uncovered and to ensure that a potential construct has not been missed.

The final stage of the ZMET interview entails data analysis and coding. In this stage the researcher develops key constructs by capturing common ideas, themes and concepts expressed by the participants. The constructs represent important aspects of the participants' cognitive maps such as core values, feelings, emotions, thoughts, ideas and themes. In this phase of the data analysis, interviews are transcribed verbatim, read and re-read. Audio files are listened to again and again until the researcher becomes familiar with the data after which coding may commence. As espoused by Zaltman, only key constructs mentioned by at least one-third of the participants are considered for analysis. Following this, transcripts are read and re-read while audio files, respondents' images, mental maps and the researcher's notes are scrutinised – this time, for construct pairs. A construct pair is found when participants associate one construct with another. As a rule of thumb, relationships indicated by one-quarter of the respondents can be considered important enough to be included in the consensus map (Christensen and Olson, 2002; Vorell *et al.*, 2003; Zaltman and Coulter, 1995). According to Zaltman (1997), a completed consensus map usually consists of approximately 25–30 constructs and represents 85 per cent of the constructs expressed by any one participant. Hence, data from four or five participants, randomly selected, are generally required to generate all of the constructs in a consensus map (Zaltman, 1997). The researcher analyses all participants' constructs collectively and builds a consensus map highlighting common themes and threads. The interaction between significant constructs as given by the participants is assessed. The result is a consensus map that diagrammatically portrays the relationships among the elicited constructs.

In a transport mobility experiment this stage may include identified relationships between time, specific modes, service standards and anticipated relationships with other tourists that share the mode, opportunity costs of mode–destination pairings and expectations of experiences. Collectively, data of this nature has the potential to reveal short-haul destinations that meet many if not all expectations that participants might have of long-haul destinations.

The evaluation

As the preceding discussion has highlighted, the ZMET technique can be used to understand deep-seated psychological factors that underlie behaviour. Examples of ZMET used in academic research include purchase of homes (Khoo-Lattimore, 2008), emotional responses of females towards intimate advertising (Siergiej and Eason, 2009), construction of experiences with the Internet (Annamma *et al.*, 2009), adoption of 3G mobile banking services (Lee *et al.*, 2003), understanding subcultures within an organizational context (Vorell *et al.*, 2003) and a study of brand images (Coulter and Zaltman, 1994). In all of these studies, ZMET produced useful results. For example, Khoo-Lattimore (2008), in her doctoral study on consumer home choice, found answers to why pre-purchase checklists used by many homebuyers and real-estate agents are an inaccurate representation of consumer home choice. The answers are illustrated in a ZMET consensus map which identified 24 motives influencing home choice. The ZMET map also highlighted interaction between motives and how 23 of the motives are underpinned by one motive of autobiographical memories which evokes a complex system of feelings, sensations and emotions that influence consumer home choice. The information generated by ZMET was able to assist homebuyers understand how their home choice is guided by internal images and deep-seated motives derived from many years of past experience. Based on this revealed understanding the participants were able to decide if these motives justified the price they paid for the property.

In one of the few studies in the tourism literature that used ZMET, Chen (2007) employed the technique to understand family vacationers. The research was published in a conference proceeding that is not readily available. Another tourism-related study by Christensen and Olson (2002) used ZMET to elicit consumers' motives for their involvement with mountain biking. Based on interviews with 15 committed mountain bikers, their study found four broad meaning themes associated with mountain biking: riding for challenge, thrill and a sense of accomplishment; sharing experiences and connecting with a group; seeking a transformation experience in their emotional and/or cognitive state; and escaping to nature. They reported that ZMET "is able to tap into consumers' knowledge that lies well below the surface of everyday conscious awareness" and elicited "vivid and detailed meanings in consumers' mental models, some of which are deep and perhaps unconscious" (Christensen and Olson, 2002, p. 497). Besides Chen (2007) and Christensen and Olson (2002), scholarly articles featuring ZMET are published in the Chinese language and are not widely available outside of China. Nevertheless, it is apparent that there is evidence of increased interest in using ZMET as a tool in tourism research. To date, ZMET has not been used to investigate issues related to sustainability or otherwise of mobility practices.

In the context of tourism mobilities ZMET has some potential to throw further light on the reluctance of consumers to select more environment-friendly transport options. Pictures play a deep symbolic role in helping to construct tourism and travel memories (Bærenholdt et al., 2004) and shape those memories into narratives or stories. ZMET, which uses participants' own pictures as the basic platform for research, can assist in understanding the attitude–behaviour gaps that exist in tourist mobility studies. Tourists, for example, can be asked to bring along pictures that represent their feelings about travel, their choice of travel mode or just flying per se. They can also be asked to bring pictures that signify reasons why they do not do what they say they are going to do. Although pictures have been used as a research method in many tourism studies (some examples include those of Albers and James, 1983; Lo et al., 2011; Markwell, 1997) as early as in the 1970s (Chalfen, 1979), the significance of ZMET for research into sustainable tourist mobility is that it integrates the use of pictures with psychological theories to build a consensus map. The advantage of a map of this nature is that it can provide insights into the purpose behind specific actions and decisions and may also be used to identify causal relationships in the decision-making process, and subsequently, the choice. Finally, the stories arising from the final collection of participants' pictures can be incorporated into communication strategies used by railways, buses and tour operators to affect behavioural change that will enhance the sustainability of tourist mobility.

The processes outlined above are missing from the most current research into behavioural change towards sustainable mobility. Many studies have concentrated on identifying factors but have not gone further to explain how these factors inter-relate to affect the tourists' ultimate choice. It has long been recognised that there is a fundamental need to understand destination choices, including the selection of the mode transport to be used. A number of researchers (Goodall, 1988; McGuiggan, 2000; Oppermann, 2000) have focused on specific aspects of choice including desires, motivation, image and loyalty. The implications of these factors for choice making are, however, unclear. As MacCanell (2002, p. 146) observes, "factors that motivate tourist desire are mysterious and illusive, even to the tourists themselves". MacCanell (2002)'s observation echoes the growing realisation that the act of choice involves both conscious and unconscious decisions. As Tran and Ralston (2006) note, most measures of motivation look only at conscious motives ignoring the possibility that unconscious motives also may affect destination choice. One approach to the assessment of unconscious motives is the thematic appreciation test that uses sets of photographs or images to identify unconscious motives. More recently, Uriely *et al.* (2011, p. 1051) demonstrated the utility of using psychoanalytic sociology "that focuses on the complex relationship between unconscious psychological forces and the social order" to identify the underlying reasons for the behaviour. It is apparent that the unconscious aspects of choice are important though not well recognised. ZMET offers researchers the opportunity to bridge the gap between the conscious and the unconscious when seeking to address issues such as mobilities choice and sustainability.

Finally it is important to note that there has been disapprobation of the ZMET technique that is attributed to its lack of novelty, likening ZMET to an amalgamation of established techniques. After all, photography, laddering and all of the steps laid out within ZMET are common knowledge. Related to this, other common earlier misunderstandings of ZMET have included the patenting of the technique for corporate use. For example, companies such as Coca Cola, Cadbury, Colgate, IBM, Samsung, Pfizer and even The World Bank have used ZMET to investigate both customers and employees. Some academics contend that this promotes unacceptable human control over material that is in the public domain. Furthermore, the detailed administrative procedures used in ZMET have not been readily available, leading many scholars to assume that ZMET cannot be easily adopted for their own research. However, what many do not realise is that ZMET was patented for use only in the USA and that researchers in other countries were free to use the technique. In a 1997 publication, Zaltman gave academics the right to use ZMET for scholarly purposes. Another possible reason for the low rate of adoption of the technique, despite it being available for academic use, may be the perceived need for researchers to have some training in psychology. However, this is not necessarily the case because Zaltman has developed a training manual that is most helpful and provides sufficient guidance for researchers with no specific training in consumer psychology to use the technique. In fact, the guidelines for academic researchers outlined in this chapter are based on recommendations extracted from the ZMET training manual (2003). Perhaps all it needs for scholars to begin considering ZMET is the expiration of the patent in the USA. Given that this happened in March 2013, it is hoped that there is now potential for its wider use.

Conclusion

As this chapter points out, it is likely that public sector measures to combat the threat of climate change will include measures that may include either financial disincentives to travel such as higher taxes or legal measures to limit travel that produces greenhouse gases or both. Given that while many consumers are aware of the links between greenhouse gas emission such as CO_2 and climate change but remain reluctant to change their travel patterns, there is an urgent need to identify strategies that may be used to change these behaviours. The ZMET approach is one option that might be used by scholars in this area of research, particularly as a bridge to connect the current research into unsustainable tourists' mobility behaviours with the limited body of knowledge about unconscious needs in the tourism literature. It may be that a simplified version of the ZMET approach could be used, perhaps entailing the use of photos and their interpretation to evaluate the travel options that consumers make based on the transport choices that are associated with selected destinations.

When compared to other conventional interviewing techniques, ZMET allows the researcher to cover more breadth and depth within the tourists' mobility behaviour. One strength of the technique is the incorporation of a variety of techniques from other disciplines such as psychology, art theory and marketing. Another advantage is that it facilitates the identification of both factors and the inter-relationships between these influences, rather than just the factors themselves in isolation. Arguably, ZMET is a powerful and, most importantly, a content-free procedure – that is, while the interviewer sets up the session to answer the research question, it is ultimately the respondent who provides most of the content. Finally, ZMET is based on the belief that each stage of the interview adds rigour to the one preceding it and as the researcher completes each stage, he/she is a step closer to understanding deep-seated psychological factors that underlie complex behaviour (Khoo-Lattimore *et al.*, 2009), including those relating to tourists' travel patterns and the sustainability of the environment.

Acknowledgements

This work/research was funded by Ministry of Higher Education (MOHE), Malaysia, under the Long Term Research Grant Scheme (LRGS) Programme (Reference no: JPT.S (BPKI) 2000/09/015JId.4(67)).

References

Albers, P.C. and James, W.R. (1983). Travel photography: A methodological approach. *Annals of Tourism Research, 15*(1), 134–158.

Anable, J. (2005). "Complacent car addicts" or "aspiring environmentalists"? Identifying travel behaviour segments using attitude theory. *Transport Policy, 12*, 65–78.

Annamma, J., Sherry, J. Jr., Venkatesh, A. and Deschenes, J. (2009). Perceiving images and telling tales: A visual and verbal analysis of the meaning of the Internet. *Journal of Consumer Psychology, 19*, 556–566.

Ballantyne, R. and Packer, J. (2005). Promoting environmentally sustainable attitudes and behaviour through free-choice learning experiences: What is the state of the game? *Environmental Education Research, 11*(3), 281–295.

Bærenholdt, J.O., Framke, W., Haldrup, M., Larsen, J. and Urry, J. (2004). *Performing tourist places*. Aldershot: Ashgate Publishing.

Becken, S. (2004). How tourists and tourism experts perceive climate change and carbon offset schemes. *Journal of Sustainable Tourism, 10*(2), 114–130.

Becken, S. (2007). Tourists' perception of international air travel's impact on the global climate and potential climate change policies. *Journal of Sustainable Tourism, 15*(4), 268–351.

Beudeanu, A. (2007). Sustainable tourist behavior – a discussion of opportunities for change. *International Journal of Consumer Studies, 31*, 499–508.

Boeing. (2013). *Long Term Market: Asia Pacific*. Retrieved from www.boeing.com/commercial/cmo/asia-pacific.html.

Cahyanto, I., Pennington-Gray, L. and Thapa, B. (2013). Tourist–resident interfaces: Using reflexive photography to develop responsible rural tourism in Indonesia, *Journal of Sustainable Tourism, 21*(5), 732–749.

Carson, D., Gilmore, A, Perry, C. and Gronhaug, K. (2001). *Qualitative marketing research*. London: Sage Publications.

Chalfen, R. (1979). Photography's role in tourism: Some unexplored relationships. *Annals of Tourism Research, 6*(4), 435–447.

Chen, P. (2007). Exploring unspoken words: Using ZMET to depict family vacationer mental models. In Joseph S. Chen, Thouraya Gherissi-Labben and Andrew Mungall (eds), *Proceedings of the First Hospitality and Leisure: Business Advances and Applied Research Conference* (pp. 39–49). Lausanne.

Christensen, G.L. and Olson, J.C. (2002). Mapping consumers' mental models with ZMET. *Psychology and Marketing, 19*(16), 477–501.

Coulter, R.H. and Zaltman, G. (1994). Using the Zaltman metaphor elicitation technique to understand brand images. *Association of Consumer Research, 21*, 281–295.

Eriksson, L., Garvill, J. and Nordlund, A.M. (2008). Interrupting habitual car use: The importance of car habit strength and moral motivation for personal car use reduction. *Transportation Research Part F, 11*, 10–23.

Finger, M. (1994). From knowledge to action? Exploring the relationships between environmental experiences, learning, and behavior. *Journal of Social Issues, 50*(3), 141–160.

Fransella, F. and Banister, D. (1977). *A manual for repertory grid*. London: Academic Press.

Gibbs, R.W., Jr., Colston, H.L. and Johnson, M.D. (1996). Proverbs and the metaphorical mind. *Metaphor and Symbol, 11*(13), 207–216.

Gilbert, R. and Tanguay, H. (2000). *Sustainable transportation performance indicators project. Brief review of some relevant worldwide activity and development of an initial long list of indicators*. Toronto: The Centre for Sustainable Transportation.

Glucksberg, S. (1995). Commentary on nonliteral language: Processing and use. *Metaphor and Symbol, 10*(11), 47–57.

Glucksberg, S. (2001). *Understanding figurative language.* New York: Oxford University Press.

Glucksberg, S. and McGlone, M.S. (1999). When love is not a journey: What metaphors mean. *Journal of Pragmatics, 31*(12), 1541–1558.

Goodall, B (1988). How tourists choose their holidays: An analytical framework. In B. Goodall and G. Ashworth (eds), *Marketing in the tourism industry. The promotion of destination regions* (pp. 41–60). London: Routledge.

Gössling, S. and Peeters, P. (2007). It does not harm the environment! An analysis of industry discourses on tourism, air travel and the environment. *Journal of Sustainable Tourism, 15*(4), 402–416.

Gössling, S., Hall, C.M., Peeters, P. and Scott, D. (2010). The future of tourism: Can tourism growth and climate policy be reconciled? A mitigation perspective. *Tourism Recreation Research, 35*(2), 119–130.

Gudmundsson, H. (2001). *Indicators and performance measures for transportation, environment and sustainability in North America (Research notes no. 148).* Roskilde: National Environmental Research Institute, Ministry of Environment and Energy. Retrieved from www.dmu.dk/1_viden/2_publikationer/3_arbrapporter/rapporter/AR148.pdf.

Hensher, D.A. and Reyes, A.J. (2000). Trip chaining as a barrier to the propensity to use public transport. *Transportation Research, 27,* 341–361.

Higham, J.E.S., Cohen, S.A. and Cavaliere, C. (2013). *Climate change and the "flyers' dilemma": A multi-national comparative analysis.* Paper presented at the international workshop on psychological and behavioural approaches to understanding and governing sustainable tourism mobility of Freiburg 2012 – NHTV Centre for Sustainable Tourism and Transport, Black Forest, Freiburg, Germany.

Huebner, A. (2012). Public perceptions of destination vulnerability to climate change and implications for long-haul travel decisions to small island states. *Journal of Sustainable Tourism, 20*(7), 939–951.

Hunecke, M., Haustein, S., Grischkat, S. and Bohler, S. (2007). Psychological, socio-demographic, and infrastructural determinants of ecological impact caused by mobility behaviour. *Journal of Environment Psychology, 27,* 277–292.

Kenyon, S. and Lyons, G. (2003). The value of integrated multimodal traveller information and its potential contribution to modal change. *Transportation Research Part F, 6,* 1–21.

Kerstetter, D. and Bricker, K. (2009). Exploring Fijians' sense of place after exposure to tourism development. *Journal of Sustainable Tourism, 17*(6), 691–708.

Khoo-Lattimore, C. (2008). *Home truths: Understanding the key motives that underlie consumer home choice* (Doctoral dissertation). University of Otago, Dunedin, New Zealand.

Khoo-Lattimore, C., Thyne, M. and Robertson, K. (2009). The ZMET method: Using projective technique to uncover hedonic factors underlying consumer home choice. *The Marketing Review, 9*(2), 139–154.

Kingsbury, P. (2005). Jamaican tourism and the politics of enjoyment. *Geoforum, 36*(1), 113–132.

Knapp, M.L. (1980). *Essentials of nonverbal communication.* New York: Holt, Rinehart and Winston.

Kollmuss, A. and Agyeman, J. (2002). Mind the gap: Why do people act environmentally and what are the barriers to pro-environmental behavior? *Environmental Education Research*, 8(3), 239–260.

Kövecses, Z. (2002). *Metaphor: A practical introduction*. New York: Oxford University Press.

Lakoff, G. (1993). The contemporary theory of metaphor. *Metaphor and Thought*, 2, 202–251.

Larsen, J. (2006). Geographies of tourist photography: Choreographies and performances. In J. Falkheimer and A. Jansson (eds), *Geographies of Communication: The Spatial Turn in Media Studies* (pp. 241–260). Göteborg: Nordicom.

Lawson, R., Tidwell, P., Rainbird, P., Loudon, D. and Bitta, A. (1996). *Consumer behaviour in Australia & New Zealand*. Sydney: McGraw-Hill.

Lee, M.S.Y., McGoldrick, P.J., Keeling, K.A. and Doherty, J. (2003). Using ZMET to explore barriers to the adoption of 3G mobile banking services. *International Journal of Retail & Distribution Management*, 31(16), 340–348.

Litman, T. (2003). *Sustainable transportation indicators*. Victoria Transport Policy Institute. Retrieved from www.vtpi.org.

Lo, I.R., McKercher, B., Lo, A., Cheung, C. and Law, R. (2011). Tourism and online photography. *Tourism Management*, 32(4), 725–731.

Lu, J. and Nepal, S. (2009) Sustainable tourism research: An analysis of papers. *Journal of Sustainable Tourism*, 17(1), 5–16.

MacCanell, D. (2002). The ego factor in tourism. *Journal of Consumer Research*, 29(1), 146–151.

Markwell, K. (1997). Dimensions of photography in a nature-based tour. *Annals of Tourism Research*, 24(1), 131–155.

McGuiggan, R. (2000). The Myers–Briggs type indicator and leisure attribute preference. In A. Woodside, G. Crouch, J. Mazanec, M. Oppermann and M. Sakai (eds), *Consumer psychology of tourism, hospitality and leisure* (pp. 245–267). Wallingford: CABI Publishing.

McKercher, B., Prideaux, B., Cheung, C. and Law, R. (2010). Achieving voluntary reductions in the carbon footprint of tourism and climate change. *Journal of Sustainable Tourism*, 18(3), 297–317.

McKercher, B., Prideaux, B. and Pang, S. (2013). Attitudes of tourism students to the environment and climate change. *Asia Pacific Journal of Tourism Research*, 18(1–2), 108–143.

McNamara, K.E. and Prideaux, B. (2010). Reading, learning and enacting: Interpretation at visitor sites in the wet tropics rainforest of Australia. *Environmental Education Research*, 16(2), 173–188.

Mehrabian, A. (1971). *Silent messages*. Belmont: Wadworth.

Nakanishi, M. (1974). Decision net models and human information processing. In G.D. Hughes and M.L. Ray (eds), *Buyer/consumer information processing* (pp. 75–88). Chapel Hill, NC: University of North Carolina Press.

Nilsson, M. and Kuller, R. (2000). Travel behaviour and environmental concern. *Transportation Research*, D(5), 211–234.

Oates, C.J., McDonald, S., Alevizou, P., Hwang, K., Young, W. and McMorland, L.A. (2008). Marketing sustainability: Use of information sources and degrees of voluntary simplicity. *Journal of Marketing Communication*, 14(5), 351–365.

Oppermann, M. (2000) Where psychology and geography interface in tourism research and theory. In A. Woodside, G. Crouch, J. Mazanec, M. Oppermann and M. Sakai (eds), *Consumer psychology of tourism, hospitality and leisure* (pp. 19–38). Wallingford: CABI Publishing.

Ottman, J.A., Stafford, E.R. and Hartman, C.L. (2006). Avoiding green marketing myopia: Ways to improve consumer appeal for environmentally preferable products. *Environment: Science and Policy for Sustainable Development, 48*(5), 22–36.

Pearce, P.L. and Packer, J. (2013). Minds on the move: New links from psychology to tourism. *Annals of Tourism Research, 40*, 386–411.

Pearce, P.L., Filep, S. and Ross, G. (2011). *Tourists, tourism and well being.* New York: Routledge.

Pidgeon, N. (2011). *Public understanding of and attitudes towards climate change (Report 5: international dimensions of climate change).* London: UK Government Foresight Office.

Plog, S.C. (1991) *Leisure travel: Making it a growth market … again!* New York: John Wiley.

Prillwitz, J., Harms, S. and Lanzendorf, M. (2007). Interactions between residential relocations, life course events, and daily commute distances. *Transportation Research Record, 2021*, 64–69.

Sheeran, P. (2002). Intention–behavior relations: A conceptual and empirical review. *European Review of Social Psychology, 12*(1), 1–36.

Short, D.C. (2001). Shining a torch on metaphor in HRD. *Advances in Developing Human Resources, 3*(13), 297.

Siergiej, E. and Eason, K. (2009). Intimate advertising: A study of female emotional responses using the ZMET explorations. *Explorations: The Journal of Undergraduate Research and Creative Activities for the State of North Carolina, 4*, 1–22.

Skinner, D. and Rosen, P. (2007). Hell is other cyclists: Rethinking transport and identity. In D. Horton, P. Rosen and P. Cox (eds), *Cycling and society* (pp. 83–96). Aldershot: Ashgate.

Steg, L. and Gifford, R. (2005). Sustainable transport and quality of life. *Journal of Transport Geography, 13*, 59–69.

Stern, P.C. (2000). Toward a coherent theory of environmentally significant behavior. *Journal of Social Issues, 56*, 407–424.

Tran, X. and Ralston, L. (2006). Tourist preferences influence of unconscious needs. *Annals of Tourism Research, 33*(2), 424–441.

UNWTO-UNEP-WMO. (2008). *Climate change and tourism: Responding to global challenges.* Madrid: UNWTO.

Uriely, N., Ram, Y. and Malach-Pines, A. (2011). Psychoanalytical sociology of deviant tourist behavior. *Annals of Tourism Research, 38*(3), 1051–1069.

Vorell, M.S., Theses, O.E. and Center, D. (2003). *Application of the ZMET methodology in an organizational context comparing black and white student subcultures in a university setting.* Ohio: Miami University.

Weiser, J. (1988). See what I mean? Photography as nonverbal communication in cross-cultural psychology. In E. Poyatos (ed.), *Cross-cultural perspectives in nonverbal communication* (pp. 245–290). Boston, MA: Hogrefe Publishers.

World Bank. (2013). Domestic carbon pricing initiatives offer hope for future market. Retrieved from www.worldbank.org/en/news.

Young, W., Hwang, K., McDonald, S. and Oates, C. (2010). Sustainable consumption: Green consumer behaviour when purchasing products. *Sustainable Development, 18*, 20–31.

Zaltman, G. (1997). Rethinking market research: Putting people back in. *Journal of Marketing Research, 34*(14), 424–437.

Zaltman, G. (2003). *How customers think: Essential insights into the mind of the market.* Boston, MA: Harvard Business School Press.

Zaltman, G. and Coulter, R.H. (1995). Seeing the voice of the customer: Metaphor-based advertising research. *Journal of Advertising Research, 35*(4), 34–51.

Zaltman, G. and Higie, R. (1993). *Seeing the voice of the customer: The Zaltman metaphor elicitation technique* (pp. 93–114). Cambridge, MA: Marketing Science Institute.

ZMET Interviewer Training Manual. (2003). Pittsburgh, PA: Olson Zaltman Associates.

7 The attitude–behaviour gap and the role of information in influencing sustainable mobility in mega-events

Acácia Cristina Mendes Malhado,
Lindemberg Medeiros de Araujo and
Rainer Rothfuss

Introduction

Mega-events provide unique opportunities and challenges for the provision of sustainable transport options (Davenport and Davenport, 2006; Dodouras and James, 2004). The sharp increase in demand for mobility from residents and tourists in the host cities provide significant opportunities to promote sustainable alternatives for local transport, changing attitudes and modifying behaviours (Lenskyj, 2000; Silvestre, 2009). Conversely, poor planning or investment before and during an event could cause negative social (Hall, 1992; Ritchie and Hall, 1999; Silvestre, 2009), economic (Matheson, 2002; Siegfried and Zimbalist, 2000) and environmental impacts (Ahmed and Pretorius, 2010; Davenport and Davenport, 2006; Koenig and Leopkey, 2009; Jago *et al.* 2010); damaging the image of the host city and country and reinforcing unsustainable travel behaviour patterns. Therefore, mobility planning is a critical aspect of mega-events, not only to justify immediate measures to minimise these impacts, but also to find solutions for the persistent mobility problems affecting major metropolitan areas globally (Gakenheimer, 1999; Moss and O'Neil, 2012; Vasconcellos, 2001).

However, the mobility dilemmas are not exclusively a consequence of physical limitations, but are powerfully related to behavioural and environmental issues (Silva *et al.*, 2008). The challenges of integrating these issues with respect to travel attitudes and behaviour have been gradually increasing in the transport and tourism literature (Kelly *et al.*, 2007), but more research addressing these challenges is needed in order to develop a sustainable approach for mega-event planning (Koenig and Leopkey, 2009).

Securing efficient mobility is a vital component for any event's success. Therefore, the research aims of this study are to examine the role of information in shaping the choices, attitudes and behaviour of tourists and residents, and to provide insights for transport and mega-event planners in developing appropriate measures for mega-event mobility management. There is a pressing need to fully understand how access to information can contribute (or even hinder) the massive demand for mobility associated with such events. However, it is important to note that research on these issues in the context of event management is still in its infancy (Robbins *et al.*, 2007). Indeed, one of the main benefits of a sports mega-event is the catalytic effect it can have on transport projects for host cities (Kassens, 2012).

Linking sustainability, transport and mega-events

According to the United Nations Environment Programme (UNEP, 2012) current global urban development trends are characterised by three fundamental changes: First, an era characterised by the second wave of urbanisation. Second, an era of resource scarcity and constraints. Finally, an era of increased uncertainty in terms of global changes. Therefore, one of the pre-eminent discussions around sustainable development is to figure out how these changes will impact mobility across the globe. More generally, transport plays a fundamental role in the global economy (OECD, 2009; Rodrigue *et al.*, 2006); and, controversially, provides an enormous challenge for sustainable development (Igwe, 2006; Yigitcanlar *et al.*, 2008). The consequences of the private transport use on the environment in terms of air quality and pollution, greenhouse gases, ozone depletion, use of non-renewable resources and noise are enormous (e.g. Dimitrov, 2004; Tertoolen *et al.*, 1998). According to UNEP (2009), the use of unsustainable transport modes have raised serious concerns about human health, productivity loss and climate change.

One of the attempts to address these mobility issues was presented in the work developed by Black *et al.* (2002, p. 186) who argue that to ensure sustainable mobility urban transport needs to

> provide access to goods and services in an efficient way for all inhabitants of the urban area, protect the environment, cultural heritage and ecosystems for the present generation, and ... not endanger the opportunities of future generations to reach at least the same welfare level as those living now, including the welfare they derive from their natural environment and cultural heritage.

In other words, a sustainable transport should balance the three dimensions of sustainable development; making transport socially accessible (meeting the basic transport needs of individuals and providing equal opportunities for humans), economically efficient (cost-benefit analysis, higher net value) and environmentally friendly (optimising the use of natural resources and mitigating climate change).

The environmental dimension of planning and implementing mega-events has become a growing area of research (Dredge and Whitford, 2010; Preuss, 2012; Van Wynsberghe *et al.*, 2012). The Sydney 2000 Olympic Games, recognised as the first "green games", and its impacts were extensively discussed in the academic community; for example, the studies of Cashman (2002), Cashman and

Hughes (1998, 1999) and Lenskyi (2002). Another range of publications can be found concerning the environmental issues of the Beijing 2008 Olympic Games (see Brajer and Mead, 2003; Mead, 2008; Xin *et al.*, 2012; Wu *et al.*, 2011). For example, Beyer (2006) investigated the effectiveness of Beijing's Olympic Action Plan and assessed the barriers and challenges to overcome the main environmental problems. The 2006 FIFA World Cup (FWC) in Germany was the first World Cup to have an environmental concept – the "Green Goal" programme – and was also analysed in terms of the sustainable benefits within the Green Goal framework (see BMU, 2007; Dolles and Söderman, 2010; Öko-Institut, 2006).

Mega-events inevitably produce both negative and positive effects for the host cities (Ahmed and Pretorius, 2010; Chappelet, 2012; Emery, 2002; Fredline, 2004; Hall, 2001). A mega-event can offer benefits for the host city/country, for example, in terms of improved transport infrastructure (Ritchie, 1984; Saunders, 2010) and tourism growth (Hinch and Higham, 2004). Contrarily, it creates considerable pressure on the urban infrastructure and services related to transport (Ahmed and Pretorius, 2010; Gössling, 2002; Jones, 2001; Schmied *et al.*, 2007). As Bovy (2006) points out, the increased transport demand causes, in the short-term, an intensification of already existing problems in most host cities, such as congestion and pollution.

Likewise, according to Konrad Adenauer-Stiftung (KAS) (2011, p. 25) the "infrastructure projects and the presence of a high number of tourists and spectators cause a significant increase in greenhouse gas emissions during the event". In an attempt to measure the impact of transport during a mega-event, Schmied *et al.* (2007) estimated that transport represented 90 per cent of the GHG of all large sporting events in Germany in 2005. Although this is mostly derived from international and domestic air transport to and from the host city, land transport (urban and inter-cities mobility) should not be neglected as the improvements made to urban transport design during mega-events may even impair the ability to foster sustainable development in the future if not well planned.

Undoubtedly, a considerable amount of infrastructure investment might be required to meet the increased demand for transport during the events, which afterwards may "result in underutilization of that capacity" (Robbins *et al.*, p. 304). As the keynoter speakers of the European Conference of Ministers of Transport (ECMT, 2003) remarked, a strategic decision is needed on the capacity of the transport infrastructure; the priority should lie in adapting the infrastructure to the event or, still, adapting the event to the available infrastructure.

Kassens (2009, p. 19) defines transport during mega-events as "a unique temporary combination of mass transport flows, requiring the involvement of all available metropolitan transport modes with different service levels and requiring temporary and long-term modifications of a transport system". It is therefore important to investigate how event-related transport investments and strategies enable environmental, social and economic sustainable development.

Although the literature generally cites transport as the cause of many negative impacts in a mega-event, there is a promising aptitude for benefits such as the increased usage of sustainable transport means. Mega-events create a special environment for promoting new ways of thinking, behaving and changing attitudes towards more sustainable ways of life (Deloitte, 2010; Pellegrino and Hancock, 2010; Preuss, 2012). As Burke and Woolcock (2009) point out, a positive user experience of event visitors concerning sustainable travel modes may bear the potential of stimulating changes in daily mobility behaviour beyond the mega-event among residents and tourists. Additionally, efficient mobility management can motivate increased use of public transport during mega-events, even improving air quality and road congestion during the event (e.g. Friedman *et al.*, 2001; Hensher and Brewer, 2002).

The attitude–behaviour gap

There is an enormous and varied literature on how attitudes and behaviours can be changed. One of the first types of behavioural change models to be developed was the rational choice model. The Theory of Reasoned Action (TRA), developed by Fishbein and Ajzen (1975), is one of the most commonly applied rational choice models. The TRA holds that the individual's intention to engage in a particular behaviour is basically constructed by the attitudes that influence that behaviour. Analogous to the TRA is the Theory of Planned Behaviour (TPB), which seeks to explain how the intentions to perform a given behaviour are formed (Ajzen, 1991).

Both theories have had some eminence in explaining and predicting behaviours, but have often been ineffective (Weinstein, 2007). This is because there is frequently a so-called "attitude–behaviour gap" between the attitudes of an individual and their actual behaviour (Kollmuss and Agyeman, 2002); meaning that, usually, attitudes are not the prior determinant for behaviour. This evidence was also noted by Eagly and Kulesa (1997) who added that although some individuals generally hold pro-preservation attitudes, they routinely engage in environmentally unfriendly actions, such as driving to work instead of using sustainable means of transport.

One potential barrier to achieve a behavioural change is defended to be habit (Triandis, 1977). Thus, repeated activities are very likely to become a habit rather than a reasoned action (Verplanken *et al.*, 1998). For example, Moller and Thøgersen (2008) found that the decision of using a car is made, usually, automatically; people do not even consider using another alternative. The Theory of Interpersonal Behaviour (TIB) is one of the most famous theories discussing the importance of habit into behavioural change. The TIB is an integrated behaviour choice model that combines habit and facilitating conditions as intervening between intention and behaviour (Triandis, 1977).

However, most of the theories only contemplate the internal determinants in explaining the transition towards a new behavioural pattern and, inevitably, do not consider some external conditions. These external factors may also have a significant influence in explaining the reasons why individuals' behaviours may not be consistent with their attitudes. This inconsistency between attitudes and behaviour is commonly explained in terms of lack of information on certain issues and has been addressed by a number of authors.

Blake (1999) and Burgess *et al.* (1998) defend the position that information generates knowledge which is then very likely to lead to behavioural change. In other words, people who have more information and knowledge about environmental issues are likely to act more ethically in reference to the environment (Owens, 2000). Dickson (2001) and Gale (2008) argue that education and knowledge are key elements to overcome the attitude–behaviour gap, further contributing towards behavioural change. Generally, the above authors defend the role of information in building a bridge between attitude and behaviour, especially in respect of environmental issues.

Arguably, the decision on the use of private or public transport may not be only related to the lack of information on environmental issues, but is very likely to be also connected with the lack of information on transport systems; which by its nature, mega-events have more agency to influence. Brög *et al.* (2002) defend that individuals are not going to change behaviour if they are not aware of the decent alternatives to car travel. Further, they add that lack of information about the transport services is the cause for the opposition to the use of public transport. Dziekan (2008) adds that the decision about travel mode is also influenced by traveller information and, ultimately, can be used as knowledge for future travel decisions. As a result, the promotion of sustainable mobility for mega-events should also be planned in a way that presents information to advance alternatives to car use.

Nevertheless, the impact of information on behavioural change is still being seen controversially. A number of authors argue that the provision of information is not sufficient to lead to a behavioural change which, in isolation, would close this gap (see Barr and Gilg, 1998; Myers and Macnaghten, 1998). In the travel behaviour field, Tertoolen *et al.* (1998) represent a similar position: when people were simply given information on the impact of their travel choice, and had no information on how to change it, they changed their attitudes (travelling by car is not that bad) rather than their behaviour.

Considering the special context of sports mega-events, most of the transport strategies usually rely on reducing the use of private cars through a range of economic and functional constraints (push measures) such as higher parking fees near to the stadium, and prohibitive measures to car circulation. However, these strategies very rarely provide people with information on how to use and benefit from sustainable transport modes (pull measures). Thus, an effective strategy for coping with increased traffic demand during mega-events would have the ability to reverse private car use by advancing walking, cycling and public transport (PT) use through the availability of reliable and efficient information.

Case-study: the 2010 FIFA World Cup in Johannesburg, South Africa

In 2010, the FIFA World Cup (FWC) was held in nine cities across South Africa. The city of Johannesburg was chosen as a case-study because it is a focus for debates about a range of issues concerned with transport and mobility. The city has a population of 3.8 million (City of Johannesburg, 2012) and its land area is very large – listed as 1.645 km^2 (Metropolis, 1998). Johannesburg is the capital of South Africa's smallest but wealthiest province, which generates about 38 per cent of the country's gross domestic product (GDP) (KPMG, 2012).

Johannesburg, especially in the post-apartheid era, has faced strong processes of urbanisation, modernisation, population growth and, ultimately, a substantial increase in unsustainable consumption. This has resulted in a steep increase in private car ownership, generally recognised as an important symbol of power and status globally (Bastos and Martins, 2012; Diekstra and Kroon, 1997; Mohamad and Kiggundu, 2007; Sheller and Urry, 2000).

South Africa's public transport system is generally perceived as unreliable and unsafe (UITP, 2010). In contrast to the inappropriate public transport, the country has an excellent road infrastructure (Ahmed and Pretorius, 2010). The combination of these two scenarios may be enforcing a pro-car culture, conse-quently collaborating to worsen the well-known transport problems. To help minimise this structural imbalance favouring private transport the Department of Transport (2006) of South Africa released the "Transport Action Plan" which clearly aimed at promoting and supporting a pro-PT strategy for the 2010 FWC.

The use of PT was highly encouraged during the 2010 FWC; for example, rail was the "mass mover" mode which assisted the stadiums to clear fans within the two-hour deadline mandated by FIFA. Another way to encourage the PT use during the event was through the implementation of a new transport system, BRT, which was considered an efficient travel mode for the mega-event (Innova-tion Transportation Solutions, 2010). Moreover, transport integration also played an important role in the management of transport during the event; including a number of innovative transport integration options such as "Park-and-Ride" and "Park-and-Walk". These new opportunities and experiences may, in the future, contribute to the promotion of behavioural change.

Another priority was to reduce regular daily travelling, for example, of com-muters and students. Therefore, travel demand management approaches to manage short-term travel were applied. For example, endorsement of school holidays, incentives for workers to take holidays during this period and the offer of flexible working hours on match days helped to avoid extra congestion gene-rated from regular daily travel.

However, while these options and strategies coped with the visitors' transport needs during the World Cup, the development of a sustainable public transport system legacy was not truly put into practice (Ahmed and Pretorius, 2010). Accord-ing to the same authors, the cumbersome investments in roads infrastructure (impro-vement and expansion) outweighed and undermined the investments in alternative travel modes, consequently reinforcing the sustained use of private transport. Death (2011) added that the greatest opportunities of the 2010 FWC could have been its communication (information) potential to catalyse greater environmental awareness and a powerful commitment to sustainable advancement. However, due to lack of coordination these opportunities were unfortunately missed.

Methodology

This case-study is based on the role of information in shaping attitudes and behaviours of residents and tourists to sustainable mobility during the 2010 FWC in Johannesburg, South Africa. Initially, a literature review addressing mega-events, sustainability, transport, mobility and behavioural models was carried out. This overview highlighted a gap between the attitudes and behaviours of individuals. The review also demonstrated that this gap is partly caused by a lack of information. Based on this finding, four hypotheses were constructed (defined and assessed in the following sections).

Data collection and sampling

The case study focuses on the 2010 FIFA World Cup in Johannesburg, South Africa, from 15 June to 5 July 2010. Most of the studies have focused on daily mobility. Thus, a case study involving a mega-event is a promising topic of research to understand and overcome the challenges/problems in such a context. A quantitative approach was adopted and 233 surveys were carried out. The questionnaire mainly contained closed questions, but also included a few open questions. The survey was based on self-completion questionnaires delivered in person. This approach has the advantage of increasing response rates and allowing respondents to clarify ambiguous questions (Bowling, 2005). A non-probability sampling technique, called convenience sampling, was used. According to the United Nations Educational, Scientific and Cultural Organization (UNESCO, 2005) a convenience sample is a survey technique where individuals are selected from the target population on the basis of their accessibility or convenience to the researcher. Therefore, the sampling frame consisted of the residents and tourists in Johannesburg who were visiting the fan-parks/public viewing or attending the football matches in the Soccer City Stadium. The reason for choosing such sites to sample the population was because tourists and fans were demanding transport from/to these places during the World Cup.

The questionnaire was designed to identify the attitude and behaviour of residents and tourists to transport and sustainability for mega-events. The variables were generated through an in-depth analysis of the behavioural theories (e.g. Theory of Planned Behaviour, Theory of Interpersonal Behaviour and Theory of Reasoned Action) and the role of information in explaining the so-called attitude–behaviour gap. The software utilised for the quantitative data analysis was the IBM SPSS (Statistical Package for the Social Sciences). Two main inferential statistics were performed. First, tests of group differences, mainly related to the group tourists and residents, group car users and non-car users, and the differences (within the group) to their mean scores on some response variables. Second, tests of association with studied variables. Four main hypotheses relating to attitude, behaviour and perception of respondents to a range of variables in transport were statistically tested.

Respondents' profile

Initial descriptive analysis was conducted to demonstrate the socio-demographic characteristics of the survey population (Figure 7.1). Most respondents had a high school certificate or university degree, and the majority belonged to the younger age group 18–35 (75 per cent). Almost half of respondents were classified in the upper social classes (A, B).[1] Considering the place of residence of the respondents, almost half of them were South Africans (44 per cent, being 40 per cent residents), 21 per cent Europeans, 15 per cent Latin Americans, 12 per cent Americans, 4 per cent African and the further 4 per cent from other continents.

Hypothesis 1: The attitude–behaviour gap is relevant in a sports mega-event context.

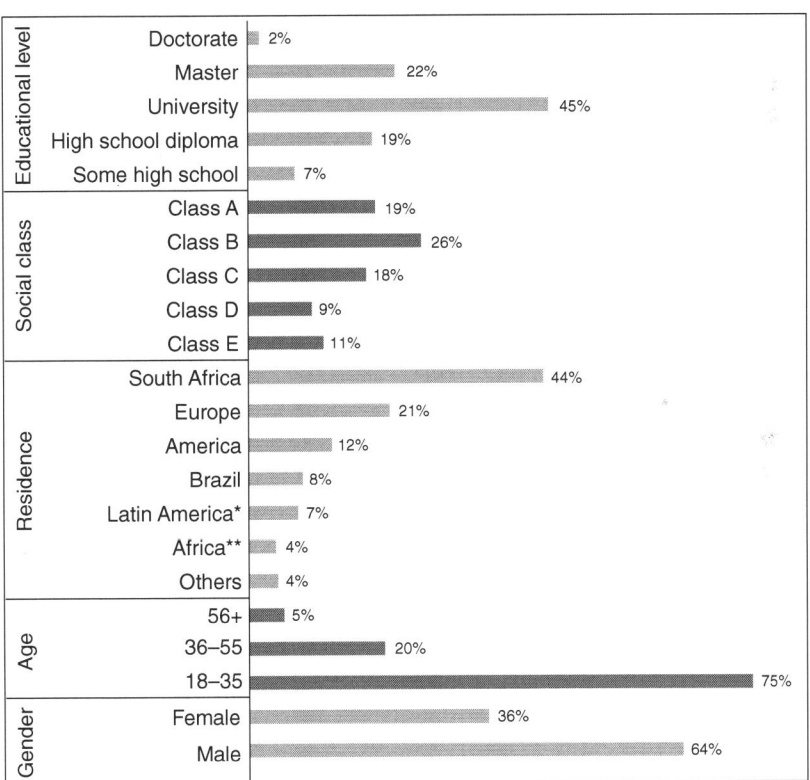

Figure 7.1 Socio-demographic characteristics of survey population (Johannesburg, South Africa).

Note
* Excluding Brazil ** Excluding South Africa.

Hypothesis 1 evaluated if the "attitude–behaviour gap" can be detected in a mega-event context. This hypothesis investigated if those respondents who had a positive attitude to transport improvements aimed at the 2010 FWC were also willing to change their behaviour towards a sustainable transport mode after the event. The two questions relating to the present hypothesis were only addressed to South Africans (including the residents of Johannesburg) who are daily car users. The selected sample population to this hypothesis was 68 respondents.

Figure 7.2 relates to a question about the personal perception of the improved transport system. The daily car users who also used their cars during the 2010 FWC valued the transport improvements higher (81 per cent) than those who were daily car users but who also tried an alternative travel mode during the 2010 FWC (72 per cent). The explanations given by the respondents in a qualitative response (closed and open question) outlined two patterns: (1) the car users had a bad experience using an alternative travel mode and (2) the transport infrastructure improvements had made the use of car favourable or more attractive.

Figure 7.3 relates to the respondents' willingness to change behaviour. The analysis here only considers daily car users who also used their car during the event, and who had a positive attitude to transport improvements but still declared no change in behaviour under any circumstance (a range of facilitating conditions was still provided as an alternative to the option: "I would not change my behaviour"). This group was found to be around 12 per cent and statistically significant. This finding suggests that even if people have a positive attitude to a sustainable travel mode, it does not mean that they are going to use it. Such a result would also support the notion that the "attitude–behavior" gap is also relevant in the singular occasion of a sports mega-event.

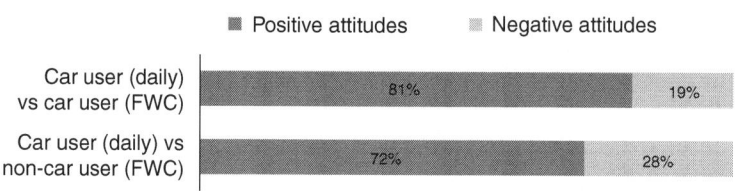

Figure 7.2 Car user and non-car user attitudes to transport improvement for 2010 FWC.

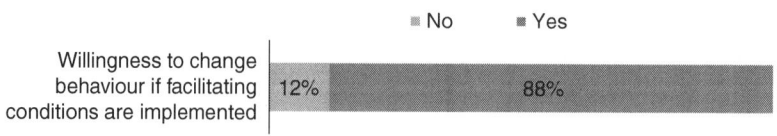

Figure 7.3 Car users with positive attitudes and their willingness to change behaviour.

Hypothesis 2: There is an association between information evaluation and transport mode satisfaction during a sport mega-event.

The traveller's level of satisfaction is an important determinant of modal choice (Gebeyehu and Takano, 2008). Information also plays a crucial role in supporting modal choice (Dziekan, 2008). Therefore, it can be assumed that the most satisfied transport users are those who have better information about their transport options and services. Both variables (satisfaction and information) are ordinal and suitable for correlation. The Pearson correlation coefficient ($r=0.303$, $N=232$, $p<0.001$) indicates that the more positive the information evaluation, the higher the user satisfaction. A coefficient considerably smaller than one can be interpreted as: while there is a clear and significant tendency that higher values for information go along with higher satisfaction, this does not hold true for every single individual. In this case, some respondents may have found the information provided on transport options helpful but have still not been satisfied with the choice or service and vice versa. The analysis is split to get separate results for car users and non-car users. For non-car users, the Spearman's Rho correlation coefficient is slightly higher than for car users ($r_s(79)=0.325$ and $r_s(149)=0.284$, respectively; $p<0.001$). This means that positive information evaluation brings a high value of satisfaction among the non-car users when compared to car users.

Hypothesis 3:"Lack of information" is the main factor keeping car users from a more sustainable modal choice.

The questions relating to the present hypothesis were only addressed to car users and assessed a number of factors to prevent sustainable transport use. As Figure 7.4 shows, the most important factors identified as preventing car users from using PT or NMT (Non-Motorised Transport) were "safety and security" ($M=4.43$, $SE=0.100$, $SD=1.070$) and "convenience" ($M=4.21$, $SE=0.098$, $SD=1.040$), with "travel cost" being the least important ($M=3.27$, $SE=0.140$, $SD=1.463$). Considering only the cluster tourists, a distinct pattern emerges: "lack of information" ($M=4.45$, $SE=0.139$, $SD=1.107$) with "safety and security" ($M=4.69$, $SE=0.099$, $SD=0.815$) were the most important reasons in preventing the use of NMT or PT. This feature is of fundamental importance in establishing the argument that the attitude–behaviour gap can be overcome by reliable and efficient information provision during a mega-event.

Hypothesis 4: Environmental importance is associated with modal choice at home and during mega-events.

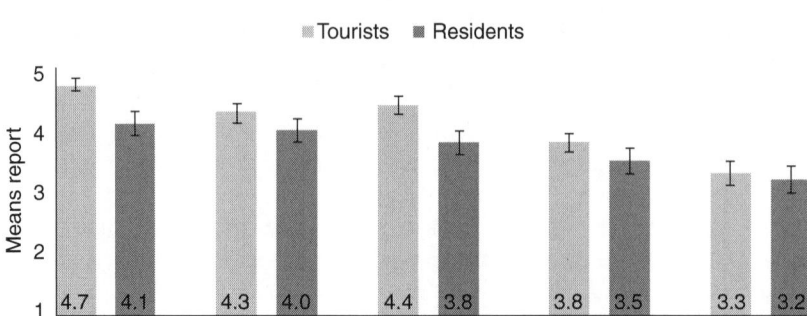

Figure 7.4 Factors preventing sustainable mobility (means report).

Note
Measurement: Likert scale ranging from 1 = not at all important to 5 = very important.

This hypothesis relates to a question measuring the importance that the environment has in travel mode decision-making during the 2010 FWC and travel mode choice in daily life. The regression coefficient which measures the association between environmental importance and travel mode choice during 2010 FWC is not statistically significant ($R^2 = 0.01$, $F(2, 222) = 0.41$, $p = 0.661$); for descriptive analyses see Figure 7.5.

The importance of the environment when choosing the daily travel mode was also assessed (Figure 7.6). The results show that this variable was also not significant, with a regression coefficient of $R^2 = 0.01$ ($F(2, 222) = 0.426$, $p = 0.654$). The model estimates that going by car corresponds to a lower environmental concern than choosing NMT and PT ($\beta_{car} = 2.9$ (constant), $\beta_{PT} = 0.06$, $\beta_{NMT} = 0.27$, $p = 0.654$); however, as already noted, the results are not significant. Considering

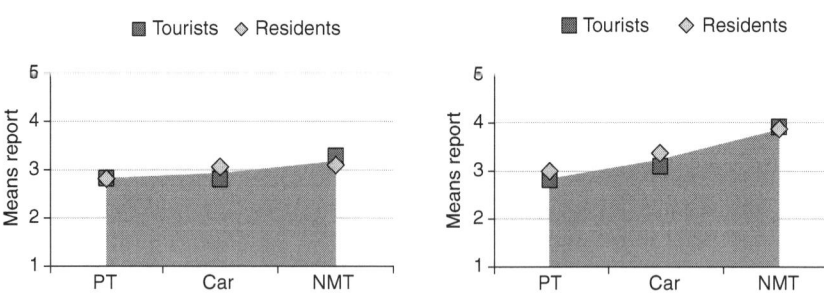

Figure 7.5 Environmental importance and 2010 FWC modal choice.

Figure 7.6 Environmental importance and daily modal choice.

the cluster (tourists vs residents) separately, the regression coefficient is still statistically not significant ($R^2 = 0.01$, $F(3, 222) = 1.13$, $p = 0.337$).

Concluding, it seems that environmental importance does not influence directly the modal choice in a mega-event or on a daily basis. According to the tests, there are no obvious association between respondents who consider the environment to be important and those who use a more sustainable mobility (PT and NMT). This seems to illustrate the common observation in psychological studies that attitude does not dictate behaviour directly – the often cited attitude–behaviour gap.

Discussion

The attitude–behaviour gap presented in the theoretical section was assessed statistically in four hypotheses. The results confirm that in a mega-event context there is also a divergence between the attitude people have towards transport and the behaviour they may adopt. For example, a number of tourists and residents who had a positive attitude towards transport improvement openly declared that they are not going to change their behaviour under any circumstance. As Butcher (2003) suggests, there is an increasing number of environmentally conscious tourists, even though they do not behave in an environmentally friendly way.

One of the main findings of the hypotheses was that information evaluation is correlated to transport satisfaction, meaning that the more information is available to a user, the higher the satisfaction level is. This correlation is stronger among non-car users. Many respondents highlighted qualitatively (open questions) that it was difficult to find efficient information about using PT and walking to the stadiums; and still, when the information was provided in the booklets, it was confusing and hard to follow. A study performed by King *et al.* (2009) concluded that access to information is therefore still perceived as a barrier to the use of public transport. The study of Deloitte (2010) emphasised higher quality of information as main areas requiring improvement for the 2010 FWC. Thus, the provision of efficient information is reflected in a higher level of satisfaction with the transport system, especially among public transport users; consequently, resulting in a higher willingness to use an alternative to a car.

The lack of information has also been identified as the second most important factor restricting the use of PT among tourists. Naturally, tourists need information on bus schedules, stops, stations and routes to be able to effectively and conveniently use this transport mode. In the same way, tourists need clear signposting and appropriate infrastructure to be able to move on foot around the city. Dziekan (2008) points out that reliable information provision is the basis for trust in the transport system. Thus, the use of sustainable transport can be sustained by providing information about available services and infrastructure (or its condition) and also ensuring satisfactory travel information to users (e.g. travel time, travel distance – especially for walking and cycling, travel mode integration, etc.). Individuals can only choose from options of which they have enough information to make an informed decision (Gärling and Golledge, 2000).

Lack of information may be also linked to the lack of environmental awareness. According to Brög *et al.* (2002) and Litman (2013), individuals are not aware about the environmental (e.g. mitigating climate change, resource conservation) benefits of using sustainable travel options.

Both tourists and residents scored (means report) astonishingly low concerning environmental importance in their personal travel mode choice, especially considering the event. Seemingly, most citizens are not particularly concerned about climate change or simply not aware of the association between their own travel behaviour and the broader environmental problems. This is consistent with studies which suggest that although knowledge and information about the negative environmental impact of car use raises some awareness, it is usually insufficient to change actual behaviour (Hagman, 2003). This suggests that environmentally friendly behaviour is often adopted for other reasons than environmental concern (Hergesell, 2011; Stern, 2000). Considering the special economic situation of South Africans, it is very likely that some of these reasons relate to travel cost and affordability. Accordingly, environmental concern does not mean that tourists and residents will automatically favour more sustainable travel behaviours. The study by King *et al.* (2009) produced similar results: participants were unaware of the extent of their personal contribution to climate change with respect to their modal choice. Therefore, the attributed environmental importance did not significantly influence modal choice. One of the explanations for this finding is that car use is still perceived as providing a socially desirable status, hence overriding any environmental concerns. Additionally, the positive perception of safety, security and convenience (a personal effect) by using a private transport may also undermine any responsibility associated with environmental impacts (usually assumed as a global effect).

The role of information and its impact on mobility during mega-events

There is increasing evidence that attitudes and behavioural issues need to be considered to ensure successful mobility management during mega-events and for long-term behavioural change, contributing further to the economic, social and environmental dimensions of the host city and country. This study generates primary data to inform the role of information in responding to the attitude–behaviour gap related to sustainable mobility during mega-events. Fundamentally, the study supports the influence of information in changing attitudes and behaviours for transport choice; especially in the face of the favourable improvements resulting from hosting the mega-event. However, no single factor can entirely provide the answer to the complexity of transport and sustainability in mega-events. Nevertheless, considerable understanding on the role of information in shaping travel mode choices was acquired through literature and empirical discussions.

There is a need for implementing a wide range of policies and strategies to reduce car use and encourage PT and NMT. As demonstrated in the study, by providing better information on PT and NMT, a temporary change in transport choice is both possible and important to adequately cope with increased transport demand during mega-events. Clearly, there is still a need for further empirical research to clarify the relationships among mega-events, sustainable mobility and the role of information in shaping travel behavioural change. A topic for further research could determine which categories of information (characteristics of services, IT, environmental campaigns) are most important for making sustainable transport modes more reliable and attractive.

Based on the case-study and literature review, information provision can be a potential deterrent to car use habit. In the same way, providing effective information about PT and NMT services may also contribute to change perceptions on a range of influencing factors (e.g. convenience, security and safety) and, finally, using information to generate positive attitudes to sustainable modals and other environmental issues. This in turn may determine new behavioural patterns. In summary, the overall effect of providing reliable, efficient and satisfactory information on these issues should be a net increase in the sustainability of mega-events, favouring also a long-term shift to sustainable travel modes in daily life.

Note

1 Social classes divisions based on net monthly income: class E (<US$370), class D (US$370–592), class C (US$592–2.550), class B (US$2.550–3.326) and class A (>US$3.326) (IBGE, 2011).

References

Ahmed, F. and Pretorius, L. (2010). Mega-events and environmental impacts: The 2010 FIFA World Cup in South Africa. *Alternation, 17*(2), 274–296.

Ajzen, I. (1991). The theory of planned behavior. *Organizational Behavioral and Human Decision Processes, 50*(2), 179–211.

Barr, S. and Gilg, A. (1998). Environmental communication and the cultural politics of environmental citizenship. *Environment and Planning A, 30*, 1445–1460.

Bastos, V. M and Martins, S. F. (2012). Automóvel versus bicicleta: Disparidade na sociedade de consumo. *Boletim Gaúcho de Geografia, 39*(1–2), 105–112.

Beyer, S. (2006). The Green Olympic Movement: Beijing 2008. *Chinese Journal of International Law, 5*(2), 423–440.

Black, J. A., Paez, A. and Suthanaya, P. (2002). Sustainable urban transportation: Performance indicators and some analytical approaches. *Journal of Urban Planning, 128*(4), 184–192.

Blake, J. (1999). Overcoming the value-action gap in environmental policy: Tensions between national policy and local experience. *The International Journal of Justice and Sustainability, 4*(3), 257–278.

BMU (Bundesministerium für Umwelt) (2007). *Green Champions für Sport und Umwelt.* Leitfaden für umweltfreundliche Sportgroßveranstaltungen. Berlin.

Bovy, P. (2006). Solving outstanding mega-event transport challenges: The Olympic experience. *Public Transport International, 6*, 32–34.

Bowling, A. (2005). Mode of questionnaire administration can have serious effects on data quality. *Journal of Public Health, 27*(3), 281–291.

Brajer, V. and Mead, R. (2003). Blue skies in Beijing? Looking at the Olympic effect. *The Journal of Environment & Development, 12*(2), 239–263.

Brög, W., Erl, E. and Mense, N. (2002, May). Individualised marketing: Changing travel behaviour for a better environment. Paper presented at the OECD Workshop: Environmentally Sustainable Transport, Berlin.

Burgess, J., Harrison, C. and Filius, P. (1998). Environmental communication and cultural politics of environmental citizenship. *Environment and Planning A, 30*(8), 1455–1460.

Burke, M. and Woolcock, G. (2009). Getting to the game: Travel to sports stadia in the era of transit-oriented development. *Sport in Society: Cultures, Commerce, Media, Politics, 12*(7), 890–909.

Butcher, J. (2003). *The moralisation of tourism: Sun, sand … and saving the world?* London: Routledge.

Cashman, R. (2002). Impact of the Games on Olympic host cities [Lecture notes]. Retrieved from http://olympicstudies.uab.es/lectures/web/pdf/cashman.pdf.

Cashman, R. and Hughes, A. (1998). *The Green Games: A Golden opportunity.* Sydney: UNSW Press.

Cashman, R. and Hughes, A. (1999). *Staging the Olympics: The event and its impact.* Sydney: UNSW Press.

Chappelet, J. (2012). Mega sporting event legacies: A multifaceted concept. *Papeles de Europa, 25*, 76–86.

City of Johannesburg (2012). The 2012 "State of the City" address. Johannesburg: Official website of the city of Johannesburg. Retrieved from www.joburg.org.za/index. php?option=com_content&id=7865:the.

Davenport, J. and Davenport, J. L. (2006). The impact of tourism and personal leisure transport on coastal environments: A review. *Estuaries Coastal and Shelf Science, 67*(1), 280–292.

Death, C. (2011). Greening the 2010 FIFA World Cup: Environmental sustainability and the mega-event in South Africa. *Journal of Environmental Policy and Planning, 13*(2), 99–117.

Deloitte (2010). *Transport survey: The road ahead.* South Africa.

Department of Transport (2006). *Action plan for ensuring operational success and establishing a legacy of improvement towards the 2010 World Cup.* Pretoria, Republic of South Africa. Retrieved from www.cityenergy.org.za/files/transport/resources/2010/2010_transportactionplan.pdf.

Dickson, M. (2001). Utility of no sweat labels for apparel consumers: Profiling label users and predicting their purchases. *The Journal of Consumer Affairs, 35*(1), 96–119.

Diekstra, R. and Kroon, M. (1997). Cars and behaviour: Psychological barriers to car restraint and sustainable urban transport. In R. Tolley (ed.), *The greening of urban transport* (pp. 147–157). London: John Wiley & Sons.

Dimitrov, P. (2004, October). *Overview of the environmental and health effects of urban transport in the Russian Federation and the other countries in Eastern Europe, the Caucasus and central Asia.* Report from the Conference on implementing sustainable urban travel policies in Russia and other CIS countries, Moscow. Retrieved from www. internationaltransportforum.org/IntOrg/ecmt/urban/Moscow04/Dimitrov.pdf.

Dodouras, S. and James, P. (2004, March). *Examining the sustainability impacts of mega-sports events: Fuzzy mapping as a new integrated appraisal system.* Paper presented at the 4th International Postgraduate Research Conference in the Built and Human Environment, Salford.

Dolles, H. and Söderman, S. (2010). Addressing ecology and sustainability in mega-sporting events: The 2006 football World Cup in Germany. *Journal of Management & Organization*, *16*(4), 587–600.

Dredge, D. and Whitford, M. (2010). Policy for sustainable and responsible festivals and events: Institutionalisation of a new paradigm – a response. *Journal of Policy Research in Tourism, Leisure & Events*, *2*(1), 1–13.

Dziekan, K. (2008). *Ease-of-use in public transportation: A user perspective on information and orientation aspects* (Unpublished doctoral dissertation). Royal Institute of Technology: Stockholm. Retrieved from http://kth.diva-portal.org/smash/get/diva2:13493/FULLTEXT01.

Eagly, A. H. and Kulesa, P. (1997). Attitudes, attitude structure, and resistance to change: Implications for persuasion on environmental issues. In M. H. Bazerman, D. M. Messick, A. E. Tenbrunsel and K. A. Wade-Benzoni (eds), *Environmental, ethics, & behavior: The psychology of environmental valuation & degradation* (pp. 122–153). San Francisco: The New Lexington Press.

Emery, P. R. (2002). Bidding to host a major sports event: The local organising committee perspective. *International Journal of Public Sector Management*, *15*(4), 316–335.

European Conference of Ministers of Transport (2003). *Transport and exceptional public events*. Report of the hundred and twenty second round table on transport economics. Paris, France. Retrieved from http://internationaltransportforum.org/pub/pdf/03RT122.pdf.

Fishbein, M. and Ajzen, I. (1975). *Belief, attitude, intention, and behavior: An introduction to theory and research*. Reading, MA: Addison-Wesley.

Fredline, L. (2004). Host community reactions to motorsport events. In B. W. Ritchie and D. Adair (eds) *Sport tourism: Interrelationships, impacts and issues* (pp. 155–171). Clevedon: Channel View Publications.

Friedman, M. S., Powell, K. E., Hutwagner, L., Graham, L. M and Teaque, W. G. (2001). Impact of changes in transportation and commuting behaviors during the 1996 Summer Olympic Games in Atlanta on air quality and childhood asthma. *Journal of American Medical Association*, *285*(7), 897–905.

Gakenheimer, R. (1999). Urban mobility in the developing world. *Transportation Research Part A*, *33*, 671–689.

Gale, H. (2008). How does drama work in environmental education? *Earth & Environment*, *3*, 159–178.

Gärling, T. and Golledge, R. G. (2000). Cognitive mapping and spatial decision-making. In R. Kitchin and S. Freundschuh (eds), *Cognitive mapping: Past, present and future* (pp. 44–65). New York: Routledge.

Gebeyehu, M. and Takano, S. (2008). Modeling the relationship between travellers' level of satisfaction and their mode choice behavior using ordinal models. *Journal of the Transportation Research Forum*, *47*(2), 103–118.

Gössling, S. G. (2002). Global environmental consequences of tourism. *Global Environmental Change*, *12*(4), 283–302.

Hagman, O. (*2003*). Mobilizing meanings of mobility: Car users' constructions of the goods and bads of car use. *Transportation. Research Part D*, *8*, 1–9.

Hall, C. M. (1992). *Hallmark tourist events: Impacts, management & planning*. London: Belhaven Press.

Hall, C. M. (2001). Imaging, tourism and sports event fever. In C. Gratton and I. Henry (eds), *Sport in the city: The role of sport in economic and social regeneration* (pp. 166–183). London: Routledge.

Hensher, D. A. and Brewer, A. (2002). Going for gold at the Sydney Olympics: How did the transport perform? *Transport Reviews, 22*(4), 381–399.

Hergesell, A. (2011). Climate friendly tourist behaviour. *Sustainable Tourism: Socio-Cultural, Environmental and Economics Impact, 1*, 95–105.

Hinch, T. D. and Higham, J. E. S. (2004). *Sport tourism development.* Clevedon, UK: Channel View Publications.

IBGE (Instituto Brasileiro de Geografia e Estatística) (2011). *Censo 2010: Estimativa da população residente nos municípios brasileiros.* Rio de Janeiro. Retrieved from www. ibge.gov.br/home/estatistica/populacao/estimativa2011/metodologia_08112011.pdf.

Igwe, A. (2006). The transport challenge in the sustainability of megacities. *Wessex Institute Transactions on the Built Environment, 89*, 23–32.

Innovation Transportation Solution (2010). *Transport event management: 2010 FIFA World Cup South Africa. Cape Town: ITS Ltd.* Retrieved from www.itse.co.za/content/ tinymce/plugins/openfile/uploads/files/Jhb2010WCAfterEventTranspManPresent.pdf.

Jago, L., Dwyer, L., Lipman, G., van Lill, D. and Vorster, S. (2010, February). *Tourism, sport and mega-events sustainability: Contributing to the roadmap for recovery.* Paper presented at the International Colloquium on Mega-event Sustainability, Johannesburg.

Jones, C. (2001). Mega-events and host-region impacts: Determining the true worth of the 1999 Rugby World Cup. *International Journal of Tourism Research, 3*, 241–251.

Kassens, E. (2009). *Transportation planning for mega events: A model of urban change* (Unpublished doctoral dissertation). Massachusetts Institute of Technology, Massachusetts.

Kassens, E. (2012). *Planning Olympic legacies: Transport dreams and urban realities.* Oxford: Routledge.

Kelly, J., Haider, W. and Williams, P. (2007). A behavioural assessment of tourism transportation options for reducing energy consumption and greenhouse gases. *Journal of Travel Research, 45*(3), 297–309.

King, S., Dyball, M., Webster T., Sharpe, A., Worley, A. and DeWitt, J. (2009). *Exploring public attitudes to climate change and travel choices: Deliberative research.* [Final report for Department for Transport]. Retrieved from www.dft.gov.uk/pgr/scienceresearch/social/climatechange/attitudestoclimatechange.pdf.

Koenig, S. and Leopkey, B. (2009, June). *Canadian sporting events: An analysis of legacy and sports development.* Paper presented at the Administrative Sciences Association of Canada Conference, Niagara Falls.

Kollmuss, A. and Agyeman, J. (2002). Mind the gap: Why do people act environmentally and what are the barriers to pro-environmental behavior? *Environmental Education Research, 8*(3), 239–260.

Konrad Adenauer-Stiftung (2011). *Sustainable mega-events in developing countries: Experiences and insights from host cities in South Africa, India and Brazil.* Germany. Retrieved from www.kas.de/wf/doc/kas_29583–1522–1–30.pdf?120124104515.

KPMG (2012). Johannesburg. KPMG International Cooperative: Johannesburg. Retrieved from https://static.kpmgglobalfrontiers.com/media/7/3/6/736.pdf.

Lenskyj, H. J. (2000). *Inside the Olympic industry: Power, politics and activism.* New York: SUNY Press.

Lenskyj, H. J. (2002). *The best Olympics ever? Social impacts of Sydney 2000.* Albany: SUNY Press.

Litman, T. (2013). *Well measured: Developing indicators for sustainable and livable transport planning.* Victoria Transport Policy Institute, Victoria. Retrieved from www. vtpi.org/wellmeas.pdf.

Matheson, V. A. (2002). Upon further review: An examination of sporting event economic impact studies. *The Sport Journal*, *5*(1), 1–3.

Mead, R. W. (2008). Environmental cleanup and health gains from Beijing's Green Olympics. *China Quarterly*, *194*, 275–293.

Metropolis (1998). *Metropolis: 097 Johannesburg*. World Association of the major Metropolises, South Africa. Retrieved from www.metropolis.org/sites/default/files/metropolitan_regions/442_097_johannesburg_eng.pdf.

Mohamad, J. and Kiggundu, A. T. (2007). The rise of the private car in Kuala Lumpur, Malaysia. *IATSS Research*, *31*(1), 69–77.

Moller, B. and Thøgersen, J. (2008). Car use habits: An obstacle to the use of public transportation? In C. Jensen-Butler, B. Sloth, M. M. Larsen, B. Madsen and O. A. Nielsen (eds), *Road pricing, the economy and the environment* (pp. 301–313). Berlin: Springer.

Moss, M. L. and O'Neill, H. (2012). Urban mobility in the 21st century: A report for the NYU BMW project on cities and sustainability. Retrieved from http://wagner.nyu.edu/rudincenter/publications/NYU-BMWi-Project_Urban_Mobility_Report_November_2012.pdf.

Myers, G. and Macnaghten, P. (1998). Rhetorics of environmental sustainability: Commonplaces and places. *Environment and Planning A*, *30*, 333–353.

OECD (Organisation for Economic Co-operation and Development) (2009). *Transport for a global economy: Challenges & opportunities in the downturn*. Paris: OECD Publishing.

Öko-Institut (2006). *Green goal legacy report*. Federal Ministry for the Environment, Nature Conservation and Nuclear Safety. Frankfurt: Organizing Committee (OC) 2006 FIFA World Cup. Retrieved from www.fifa.com/mm/document/afsocial/environment/01/57/12/66/2006fwcgreengoallegacyreport_en.pdf.

Owens, S. (2000). Engaging the public: Information and deliberation in environmental policy. *Environment and Planning A*, *32*, 1141–1148.

Pellegrino, G. and Hancock, H. (2010). *A lasting legacy: How major sporting events can drive positive change for host communities and economies*. United Kingdom: Deloitte Touch Tohmatsu.

Preuss, H. (2012). Green economy challenges for the FIFA World Cup and the Olympic Games. [Working Papers Series]. Retrieved from www.sport.uni-mainz.de/Preuss/Download%20public/Working%20Paper%20Series/Working_Paper_No_8_Version_2_Green_economy_challenges_for_the_FIFA_World_Cup_and_the_Olympic_Games.pdf.

Ritchie, B. and Hall, M. (1999, September). *Mega events and human rights*. Paper presented at Conference on Sports and Human Rights, Sydney.

Ritchie, J. (1984). Assessing the impact of hallmark events: Conceptual and research issues. *Journal of Travel Research*, *23*(1), 2–11.

Robbins, D., Dickinson, J. and Calver, S. (2007). Planning transport for special events: A conceptual framework and future agenda for research. *International Journal Tourism Research*, *9*(5), 303–314.

Rodrigue, J., Comtois, C. and Slack, B. (2006). *The geography of transport systems*. New York: Routledge.

Saunders, G. (2010, February). *South Africa's national tourism plan: Imperatives to manage tourism beyond 2010*. Paper presented at the International Colloquium on Mega-event Sustainability, Johannesburg.

Schmied, M., Hochfeld, C., Stahl, H., Roth, R., Armbruster, F., Türk, S. and Friedl, C. (2007). Green champions in sport and environment: Guide to environmentally-sound large sporting events. *German Olympic Sports Confederation* (DOSB), Frankfurt.

Sheller, M. and Urry, J. (2000). The car and the city. *International Journal of Urban and Regional Research*, *24*(4), 737–757.

Siegfried, J. and Zimbalist, A. (2000). The economics of sports facilities and their communities. *Journal of Economic Perspectives*, *14*(3), 95–114.

Silva, A. N., Costa, M. S. and Macedo, M. H. (2008). Multiple views of sustainable urban mobility: The case of Brazil. *Transport Policy*, *15*(6), 350–360.

Silvestre, G. (2009). *The social impacts of mega-events: Towards a framework* (Unpublished master's thesis). University of Westminster, London.

Stern, P. C. (2000). Toward a coherent theory of environmentally significant behaviour. *Journal of Social Issues*, *56*(3), 407–424.

Tertoolen, G., Van Kreveld, D. and Verstraten, B. (1998). Psychological resistance against attempts to reduce private car use. *Transportation Research Part A*, *32*(3), 171–181.

Triandis, H. C. (1977). *Interpersonal behaviour*. Monterey, CA: Brooks/Cole.

UITP (International Association of Public Transport) (2010). Public transport in Sub-Saharan Africa. Belgium: International Association of Public Transport. Retrieved from www.uitp.org/knowledge/pdf/PTinSSAfr-Majortrendsandcasestudies.pdf.

UNEP (United Nations Environment Programme) (2009). *Reducing emissions from private cars: Incentive measures for behavioural change.* Nairobi: UNEP. Retrieved from www.unep.ch/etb/publications/Green%20Economy/Reducing%20emissions/ UNEP%20Reducing%20emissions%20from%20private%20cars.pdf.

UNEP (United Nations Environment Programme) (2012). *Sustainable, resource efficient cities: Making it happen!* Nairobi: UNEP. Retrieved from www.unep.org/urban_ environment/PDFs/SustainableResourceEfficientCities.pdf.

UNESCO (United Nations Educational, Scientific and Cultural Organization) (2005). *Quantitative research methods in educational planning* [Series editor: Module 3]. Retrieved from www.iiep.unesco.org/capacity-development/training/training-materials/ quantitative-research.html.

Van Wynsberghe, R., Derom, I. and Maurer, E. (2012). Social leveraging of the 2010 Olympic Games: "Sustainability" in a City of Vancouver initiative. *Journal of Policy Research in Tourism, Leisure and Events*, *4*(2), 185–205.

Vasconcellos, E. A. (2001). *Urban transport, environment and equity: The case for developing countries.* London: Earthscan Publications.

Verplanken, B., Aarts, H., Van Knippenberg, A. and Moonen, A. (1998). Habit versus planned behaviour: A field experiment. *British Journal Social Psychology*, *37*(1), 111–128.

Weinstein, L. D. (2007). Misleading tests of health behaviour theories. *Annals of Behavioural Medicine*, *33*(1), 1–10.

Wu, D., Zhang, S., Xu, J. and Zhu, T. (2011). The CO_2 reduction effects and climate benefit of Beijing 2008 Summer Olympics Green Practice. *Energy Procedia*, *5*, 280–296.

Xin, J., Wang, Y., Wang, L., Tang, G., Sun, Y., Pan, Y. and Ji, D. (2012). Reductions of PM2.5 in Beijing-Tianjin-Hebei urban agglomerations during the 2008 Olympic Games. *Advances in Atmospheric Sciences*, *29*(6), 1330–1342.

Yigitcanlar, T., Fabian, L. and Coiacetto, E. (2008). Challenges to urban transport sustainability and smart transport in a tourist city. *The Open Transportation Journal*, *2*, 29–46.

Part II

Behavioural aspects of climate change and tourism mobilities

8 Carbon offsetting

Motives for participation and impacts on travel behaviour

Eke Eijgelaar and Danny de Kinderen

Introduction

Climate change has been described as the greatest environmental challenge of our time, not only for mankind (UN, 2011), but also for tourism (OECD-UNEP, 2011). Severe impacts of climate change, generally linked to exceeding 2°C global temperature rise, can only be prevented to some extent by drastically reducing the use of fossil fuels and thereby greenhouse gas (GHG) emissions within the next few decades. In this respect, an 80–95 per cent reduction of CO_2 emissions by 2050 compared to 1990/2000 levels is recognised as the minimum required effort (Allison *et al.*, 2009; Rogelj *et al.*, 2011). Even with a full implementation of these goals, an increase above 2°C is not unlikely (World Bank, 2012).

The contribution of global tourism to anthropogenic CO_2 emissions has been estimated at around 5 per cent for 2005, corresponding to 1,302 Mt CO_2, 75 per cent of which were from transport and 40 per cent from aviation alone. Tourism's CO_2 emissions are estimated to increase 135 per cent (to 3,000 Mt CO_2) by 2035, which includes the high efficiency gains forecasted for air transport (Peeters and Dubois, 2010; UNWTO-UNEP-WMO, 2008). The share of aviation will increase as air travel is expected to grow faster than overall tourism trips (ICAO, 2010; UNWTO, 2011). Emission scenarios for civil aviation vary from 1,034 to 3,105 Mt CO_2 for 2050 (Lee *et al.*, 2013). The further development of tourism CO_2 emissions is in stark contrast to the aforementioned global emission reduction needs. In fact, when assuming this business-as-usual growth path, tourism would exceed the global economy's reduced emission budget by mid-century on its own (Scott *et al.*, 2010). Given these developments it is not surprising that some regard the (mainstream) tourism industry as becoming less sustainable (Bramwell and Lane, 2012; Buckley, 2012; Gössling *et al.*, 2012).

In acknowledgement of the limited short-term energy reduction potential of technological improvements in aviation and the absence of short-term structural changes in travel behaviour, carbon offsetting has been accepted as an intermediate, albeit less effective solution for mitigating tourism emissions. This research aims to register the motives for buying offsets, but more particularly the effect of offsetting, as well as not offsetting, on the travel behaviour of Dutch tourists.

Mitigation approaches and barriers

Tourism transport emissions can be mitigated through a variety of approaches: technological, managerial, educational and behavioural. These approaches are ideally aimed at avoiding or reducing emissions, with offsetting as an alternative (Becken and Hay, 2007; Gössling, 2011). Technological improvements, including the switch to biofuels, are industry favourites, but their mitigation potential is not sufficient to reduce (passenger) transport emissions to the required levels, as has been analysed in various regional and modal contexts (Dray et al., 2012; Lee et al., 2013; Peeters, 2010; Schäfer et al., 2009; Skinner et al., 2010). The large-scale use of biofuels cannot be expected in the short-term due to economic and technological barriers (IEA, 2009; Timilsina and Shrestha, 2011; Sims et al., 2011), besides a range of other issues, like availability, indirect land-use change, social impacts and undesirable GHG balances (Ariza-Montobbio and Lele, 2010; Dray et al., 2012; Kant and Wu, 2011; Melillo et al., 2009; Searchinger et al., 2008). Without strong policy intervention and investment, high reductions from technology are unlikely (Dray et al., 2012; IEA, 2009). Improving management, for instance in aviation operations and air traffic control, requires changing and/ or harmonising regulations and procedures; frequently long-term processes. Here also, policy intervention seems essential, as self-regulation by the industry is not effective (Gössling et al., 2012). The mitigation potential of current policies that could impact tourism demand appears to be limited (Gössling et al., 2008; Mayor and Tol, 2007, 2010; Pentelow and Scott, 2011). For aviation emissions even a combination of new technology, biofuels and market-based measures (MBM) like the EU-ETS or a global MBM is not likely to lead to sufficient reductions to stay within 2°C targets (Lee et al., 2013). Behavioural change is therefore an essential additional approach for achieving required emissions targets in (tourism) transport (Dray et al., 2012; Dubois et al., 2011; Peeters and Dubois, 2010; Skinner et al., 2010). Structural behavioural change in the form of modal shift, i.e. to low-carbon transport modes, would be very effective in reducing emissions, though there is a lack of knowledge of the mitigation potential of behavioural measures (Dray et al., 2012).

Behavioural change

Creating a higher awareness of climate change and tourism/transport impacts has been seen as one way of raising environmental concern and inducing travellers to change their attitudes and subsequently their travel behaviour, following models such as that of the Theory of Planned Behaviour (Ajzen, 1991). Several studies report evidence of heightened awareness of aviation impacts (Brouwer et al., 2008; Gössling et al., 2009; Higham and Cohen, 2011; McKercher et al., 2010). The problem is that raising awareness, e.g. by providing information, is most effective when the behavioural change aimed at is not money- and time-consuming, and not too constraining (Steg and Vlek, 2009). This appears to be the bottleneck for a voluntary change towards pro-environmental travel behaviour, causing a discrepancy between awareness and behavioural change. This so-called "attitude-behaviour", "value-action", "awareness-action" or "knowledge-action" gap has been found in a large number of studies on climate and travel behaviour change (Anable *et al.*, 2006; Becken, 2007; Cohen and Higham, 2011; Günther, 2008; Hares *et al.*, 2010; Higham and Cohen, 2011; Kollmuss and Agyeman, 2002; McKercher *et al.*, 2010; Randles and Mander, 2009; Stoll-Kleemann *et al.*, 2001; Tiller and Schott, 2013; Whitmarsh, 2009). One explanation is that people simply "want holidays, and on holiday they act hedonistically" (Buckley, 2012, p. 535). Hence, any perceived behavioural constraints are undesirable and they are not willing to give up on flying (Becken, 2007; Hares *et al.*, 2010; Lorenzoni *et al.*, 2007), even if they are environmentally concerned (Barr *et al.*, 2010). Also, concern for more tangible issues of daily life is usually prioritised above climate change, which is felt as a rather abstract and remote issue (Lorenzoni *et al.*, 2007; Tiller and Schott, 2013).

Travel behaviour and carbon offsetting

Following the reasoning by Steg and Vlek (2009), it could be argued that carbon offsetting has fewer constraints, and hence offset uptake should be easier to increase than to change travel behaviour, as a means of mitigation. There is evidence from German travellers that this is indeed so. This group would rather pay more, for compensation or higher transport costs, than make radical changes in travel behaviour. Particularly air and long-haul travellers were far more willing to compensate travel emissions than to forgo on flying or fly less (Günther and Lohmann, 2008). It does appear as though awareness of the contribution of air travel on climate change and some degree of environmental concern have positive effects on the willingness to compensate (Brouwer *et al.*, 2008; van Birgelen *et al.*, 2011), though this still does not need to translate into widespread offset practices.

In view of the lack of short-term solutions, carbon offsetting has been accepted as an intermediate, albeit less effective solution for mitigating tourism emissions. Carbon offsets connect individuals and companies with climate change (Paterson and Stripple, 2010), as offset providers present climate change as an urgent problem that requires immediate action, be it directly in the form of buying offsets or by (gradually) changing behaviour (Lovell *et al.*, 2009). For the industry, a lack of political pressure may have contributed to the widespread use of offsetting as main mitigation effort (Schmücker, 2011). A key asset of offsetting is its direct availability, allowing individuals and companies to take some form of immediate action – a "quick fix" – instead of waiting for companies or politics to do so (Lovell *et al.*, 2009, p. 2365).

However, carbon offsetting has a long list of negative associations attached to it, such as transparency and additionality (the principle that after implementation of an offset project, emissions need to be lower than would have occurred in absence of the project) issues, differences in carbon calculator results and average offset costs, limited potential in terms of participation, insufficient permanence of forestry-related offsets and whether offsetting leads to emission reduction at all (see Eijgelaar, 2011; Gössling *et al.*, 2007; Hyams and Fawcett, 2013; Reijnders, 2009; Schmücker, 2011). A key point of critique about offsetting is whether it is unethical to pay others to reduce one's emissions and not taking responsibility by changing behaviour or adopting new technologies (Bumpus and Liverman, 2008). In the near future, there may not even be room for doing this. The effectiveness of offsetting, when the average person should only emit 0.5 tons of CO_2 annually in 2050 (Allison *et al.*, 2009) and this figure is already being exceeded by an average tourist trip by air, is hard to justify. In the longer run, offsetting may even contribute to a net growth of global emissions, leading Anderson (2012, p. 7) to comment that "offsetting is worse than doing nothing".

The motives for, and psychological and behavioural effects of offsetting have not been widely studied (House of Commons Environmental Audit Committee, 2007). Carbon offsetting has been linked to an increase in awareness (of climate change and the environmental impacts of behaviour), possibly even able to shape behaviour to a certain extent (Mair, 2011; Paterson and Stripple, 2010). There is some evidence that offsetting does not have any effect on changing carbon behaviour, but more research into this is required (House of Commons Environmental Audit Committee, 2007).

Research aim

This research aims to register the motives for buying offsets, but more particularly the effect of offsetting, as well as not offsetting, on the travel behaviour of Dutch holidaymakers. CO_2 emissions of Dutch holidaymakers have increased by 20 per cent between 2002 and 2011. Most of this increase is caused by a change in travel behaviour, with more intercontinental trips made, translating in an increase of distance travelled and higher use of the airplane as origin-destination transport mode (de Bruijn *et al.*, 2013). The effects of offsetting could be a reduction of (air) travel because of the cost involved with offsetting or awareness of the large share of (air) travel in individual carbon footprints. A different hypothesis sees the option to offset emissions and/or the relatively low cost of offsetting as an incentive to travel more, as first hinted at by Koens (2004 in Boon *et al.*, 2007). Results can be compared with offset behaviour related research in other countries (e.g. Whitmarsh, 2008). The research will also verify whether consumers who hold more sustainability-oriented ethics are also more inclined to behave more sustainably (e.g. Pereira *et al.*, 2012).

Methodology

For this purpose a survey has been set out among the clients and Facebook followers of a Dutch tour operator with an above-average uptake of offsets. Their portfolio mainly offers adventure-oriented travel packages in overseas destinations. Clients who had booked a trip in 2011 or the first three months of 2012 and compensated their flight received an invitation via email. A total of 1,376 emails were sent out. Furthermore, a link to the questionnaire was put on the company's Facebook page to obtain response from both offsetters and non-offsetters. Survey-Monkey was used to set up and distribute the questionnaire. The survey ran throughout April 2012. The response was 238 clients and 163 Facebook followers (401 in total), with 372 questionnaires fully completed (229 clients, 143 followers).

This sample is not representative for all Dutch holidaymakers. Thanks to the large number of offsetters linked to this tour operator, the sample does however enable a comparison between offsetting and non-offsetting travellers. Survey questions referred to travel and offset behaviour in 2011, the motivations for airplane usage and (non-) offsetting, daily environmental behaviour, attitudes towards climate change, acceptance of mitigation measures and mitigation responsibility. A holiday in 2011 was defined as a trip with at least one overnight and excluded business trips. Except for travel behaviour all questions were based on a five-point Likert scale, many in the form of acceptance of, or agreement with, a statement. The original survey was in Dutch; questions and statements have been translated for this chapter. Results were analysed with SPSS (PASW Statistics 18).

Results

Respondent characteristics

The mean age of the completed sample ($N=372$) was 45, with 19 per cent in the 16–30, 30 per cent in the 31–45, 36 per cent in the 46–60 and 14 per cent in the 60+ age group. A clear majority (62 per cent) of respondents was female. Three quarters (74 per cent) had a higher education, which is far above the national average of 33 per cent. The Facebook sample ($N=143$) was younger than the client ($N=229$) and overall sample, and has a higher female share (75 vs 55 per cent in the clients sample).

Travel and offset behaviour

All results concern holiday trips made in 2011. In that year, respondents ($N=401$) showed 99 per cent holiday participation, with an average of 3.72 trips per person. Over two-thirds of these trips were made within the Netherlands (1.3) and Europe (1.42), while 1.0 trip was made intercontinentally on average. For all of these travels, 232 respondents had bought one or more offsets, while 169 had not compensated. The average trip number was slightly higher among offsetters than non-offsetters (3.8 versus 3.5), but no significant difference was found. Both groups made equal amounts of domestic trips. On average, non-offsetters (mean 1.63; sd 1.73) do take more holidays within Europe than offsetters (mean 1.22; sd 1.08). For this type of holiday a significant difference was found ($t=2.696$; df$=261.24$; $p=\leq0.007$ in a two-tailed test). Vice versa, offsetters (mean 1.2; sd 0.92) take more intercontinental holidays than non-offsetters (mean 0.64; sd 0.8). This difference is also significant ($t=-7.25$; df = 493; $p=<0.001$). Offsetters compensated about one-third of their trips (1.36) on average. The majority (92 per cent) of all respondents had used the airplane for at least one of their trips in 2011. Nearly all offsetters had flown (99 per cent) and 82 per cent of non-offsetters had done so. A majority of air travellers (62 per cent) compensated at least one trip, while 38 per cent did not.

Offsetters and non-offsetters were compared with aspects of daily environmental behaviour. The answers of 15 statements on daily behaviour have been combined into a "green score", using equal weighting.[1] The result was a total score from 1 to 5, with 1 corresponding to very environmentally friendly daily behaviour (e.g. "always" recycle), and 5 the opposite ("never" recycle). On average, offsetters (mean 2.09; sd 0.47) score "greener" than non-offsetters (mean 2.4; sd 0.53) (see Figure 8.1). The difference is significant ($t=6.13$; df 395; $p<0.001$ in a two-tailed test). Offsetting can account for 8.7 per cent of the differences in total green score (d=0.62).

Awareness of climate change and transport impacts

The majority of all respondents acknowledge the seriousness of climate change and the contributing role of human activities. Only subtle differences could be found for offsetters and non-offsetters for some of the issues in Figure 8.2. Offsetters showed slightly higher climate change concern and higher agreement of the airplane's role in transport emissions. Non-offsetters showed a little less faith

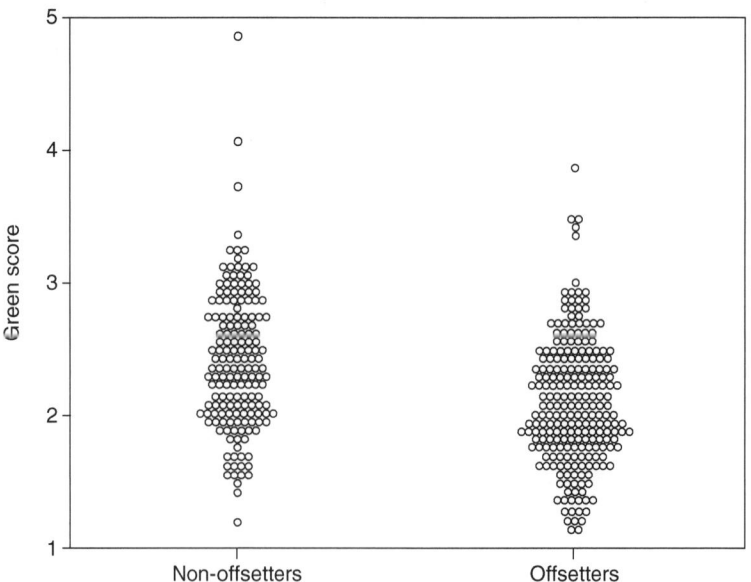

Figure 8.1 Environmental behaviour in daily life ("green score") per group.

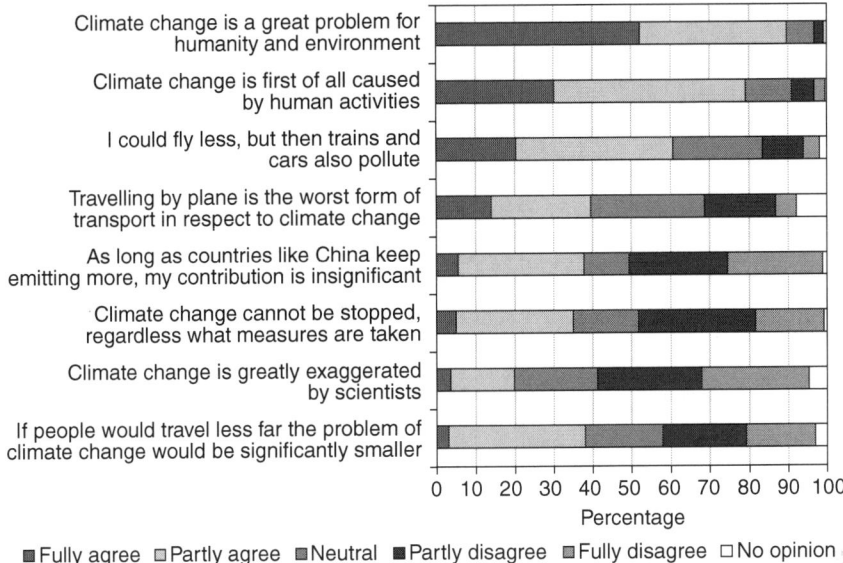

Figure 8.2 Knowledge and perception related statements about climate change.

in the possibilities of climate change mitigation, both through their own means and also regardless what measures are taken. The answers to several statements point towards limited knowledge of the impact of flying compared to other modes, and of the share of travel emissions in overall emissions.

Motives and responsibility

Respondents who used the airplane were asked for their motives for flying (see Figure 8.3). Accessibility, time and practicality are obvious top answers. Price does not seem to be an argument for this sample. Simple enjoyment of flying meets with agreement of 36 per cent of respondents.

Offsetting respondents were asked for their motives to purchase offsets (see Figure 8.4). Here, a mixture of concern, personal responsibility and emotions seem to dominate. Compensation as a means to keep flying meets with considerable approval too (58 per cent agreement), whereas the price of offsets appears less important.

Likewise, non-offsetters were asked for motives for not having compensated, but these answers are less pronounced. Figure 8.5 shows that there is still a lot of unawareness of the concept of offsetting and that for this group, the extra cost is a barrier. Issues of trust, ethics and responsibility also play a role. Climate change denial or concern were not strong reasons to not offset. However, strong agreement with any of the statements, as seen in Figure 8.4, is not seen here. Other motives could possibly apply.

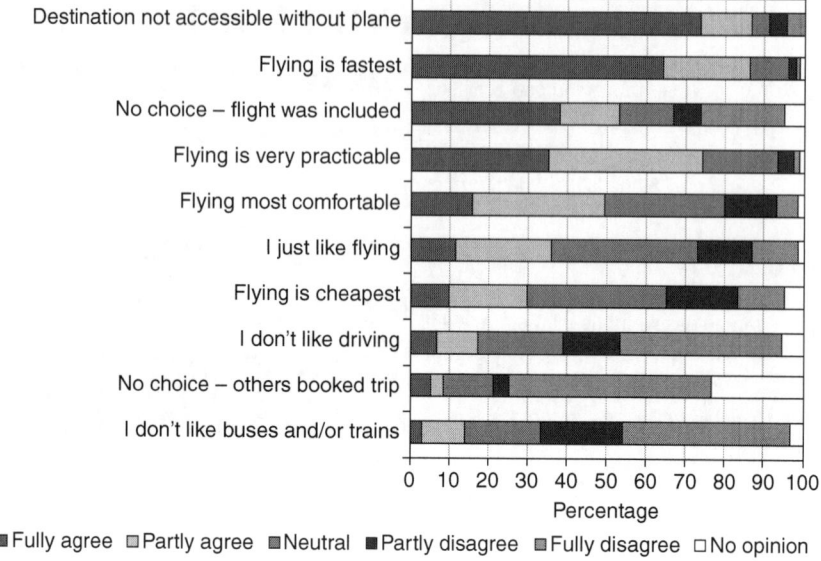

Figure 8.3 Motives for airplane usage.

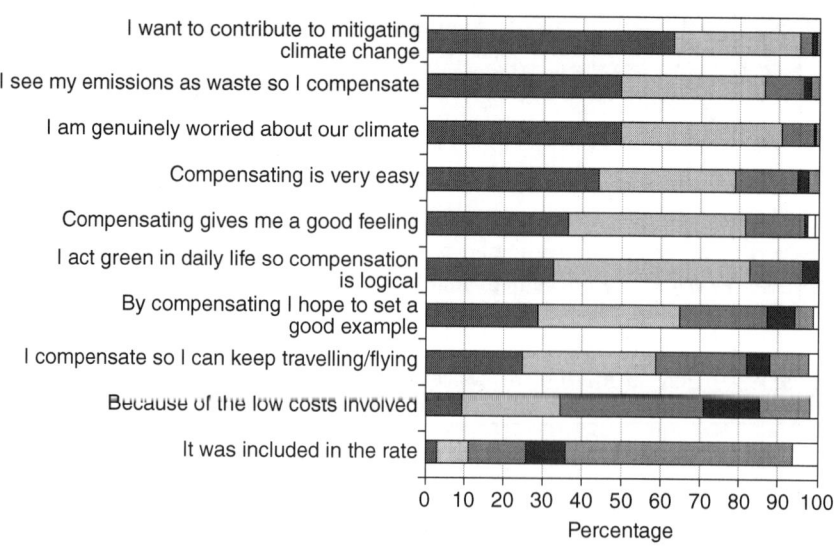

Figure 8.4 Motives for offsetting.

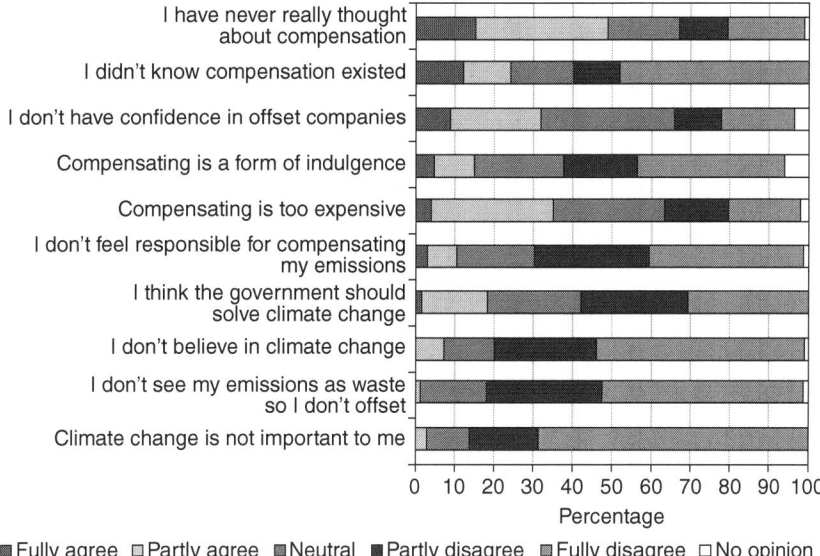

Figure 8.5 Motives for non-offsetting.

From the above, it is clear that responsibility for mitigating (own) impacts on climate change is an issue for respondents, albeit a varying one. Respondents were also asked to rate the responsibility of several actors for mitigating tourism impacts on climate change. Both offsetters and non-offsetters see airlines, the government, the tourism industry and international organisations all as similarly responsible to very responsible (92, 87, 82 and 75 per cent combined respectively). The difference is in consumer/traveller responsibility. Offsetters see this actor as more responsible than non-offsetters: 45 per cent responsible and 49 per cent very responsible versus 69 per cent responsible and 23 per cent very responsible. This correlation is significant (Chi^2=26.81; df=2; p<0.001). The cost of offsets was also defined as a motive for both groups. A separate statement on how important costs are for choosing to offset or not revealed that cost is important to very important for 75 per cent of non-offsetters against 62 per cent of offsetters. Again, the correlation is significant (Chi^2=12.36; df=4; p≤0.015).

Attitudes towards mitigation measures

Respondents were asked for their acceptance, and to rate the effectiveness, of several mitigation measures. When asked what airlines could do to mitigate their impact on climate change most effectively, little variance is shown for all respondents (see Figure 8.6). Only one significant difference was found between offsetters and non-offsetters: a weak association (Cramér's V−0.27) between

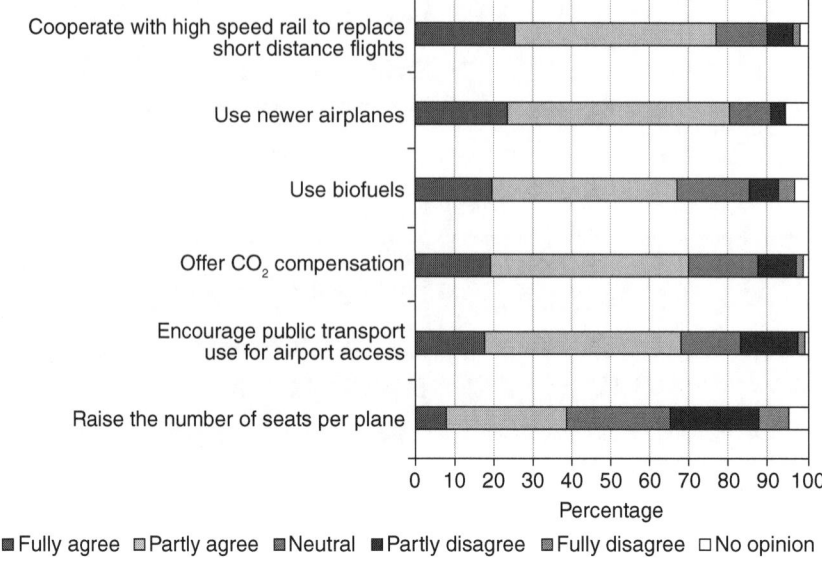

Figure 8.6 Perceived effectiveness of measures for airlines to mitigate their impact on climate change.

offsetting and "provide CO_2-compensation possibilities". Offsetters see this as a far more efficient manner of mitigation than non-offsetters (see Table 8.1; Chi^2=27.88; df=4; $p<0.001$).

Figure 8.7 shows all respondents' (dis)agreement with a range of statements on the effects of compensation on travel behaviour, and on attitudes towards changes in travel behaviour. Interestingly, differences between offsetters and non-offsetters were negligible, despite the subject of compensation. Clearly, personal travel freedom is extremely highly valued. Less or more nearby travel is hardly an option. Respondents confirm that offsetting does not change their travel behaviour nor the choice of their transport mode, though it has made them

Table 8.1 Perceived effectiveness of offering offsets for airlines to reduce their impact (%)

	Non-offsetters	*Offsetters*
	(N = 160)	*(N = 227)*
Very effective	12.5	24.7
Effective	43.1	54.6
Neutral	26.9	11.5
Hardly effective	14.4	6.6
Ineffective	1.3	1.3
No opinion	1.9	1.3

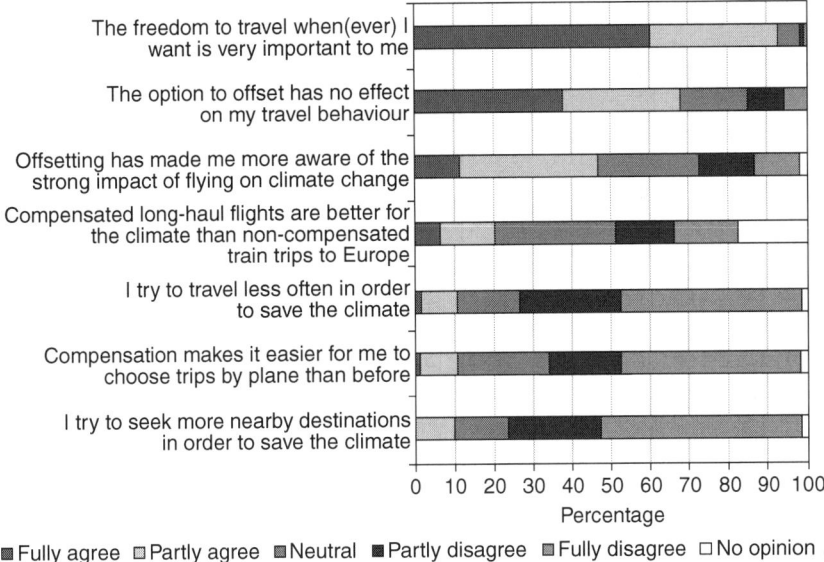

Figure 8.7 Compensation effectiveness and travel attitudes.

more aware of aviation impacts. The statement on comparing a compensated flight with a train trip apparently caused confusion, or shows that respondents are not entirely convinced of offset effectiveness after all.

As was to be expected in view of this desired freedom, acceptance of behaviour limiting measures to reduce the impact of flying (personal flight budget, no air travel under 1,000 km, one long holiday instead of multiple short ones) was very low for all respondents (see Figure 8.8). Cost-related measures met with higher acceptance, with significant differences between the two groups (see Tables 8.2 and 8.3). Offsetters were slightly more inclined to accept a worldwide tax (Cramér's V−0.20; Chi^2=15.50; df=4; $p \leq 0.004$), and much more acceptable to mandatory offsetting (Cramér's V−0.40; Chi^2=58.89; df=4; $p < 0.001$).

Table 8.2 "Flying more expensive through a worldwide flight tax" (%)

	Non-offsetters	Offsetters
	(N = 153)	(N = 220)
Fully acceptable	5.2	13.2
Acceptable	26.8	35.6
Neutral	16.3	17.8
Unacceptable	26.1	16.9
Fully unacceptable	25.5	16.4

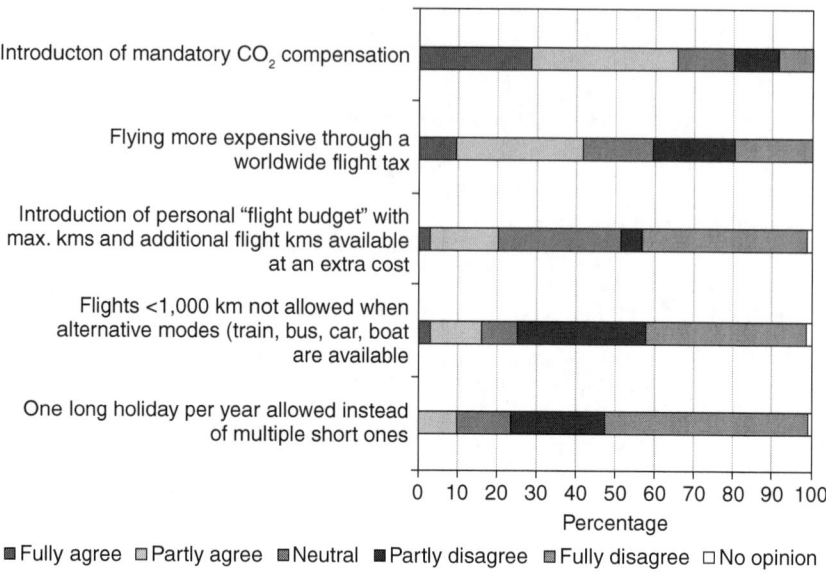

Figure 8.8 Acceptance of mitigation measures.

Socio-demographic differences between offsetters and non-offsetters

A weak association (Cramér's $V-0.36$) was found for different age categories and offsetting. The higher the age category, the more offsetters (in percentages). This correlation is significant; $Chi2=48.2$; $df=3$; $p<0.001$. To verify whether younger respondents were less "green" than older ones, a one-way ANOVA test was used. The "green score" previously mentioned differed significantly across the age groups, $F (3, 368)=16.5$; $p<0.001$. On average, the 16–30 group scores highest (mean 2.51; sd 0.45), followed by the 31–45 group (mean 2.31; sd 0.5), the 46–60 group (mean 2.1; sd 0.52) and the 60+ group (mean 1.96; sd 0.47). A post-hoc comparison in accordance with the Bonferroni method demonstrates that significant differences are shown to exist in all pair-based comparisons

Table 8.3 "Introduction of mandatory CO_2 compensation" (%)

	Non-offsetters	*Offsetters*
	(N = 153)	*(N = 220)*
Fully acceptable	14.4	38.8
Acceptable	31.4	40.2
Neutral	17.0	12.3
Unacceptable	22.9	4.1
Fully unacceptable	14.4	4.6

($p<0.05$) apart from the 46–60 years and 60+ groups. Furthermore, there is a significant difference between age groups and the amount of offsets bought (one-way ANOVA, $F (3, 368)=9.45$, $p<0.001$). There is a difference between the 16–30 years group (mean 0.33; sd 0.61) and all other groups, 31–45 (mean 0.89; sd 1.29), 46–60 (mean 0.9; sd 0.79) and 60+ (mean 1.91; sd 0.95), confirmed by a post-hoc comparison in accordance with the Bonferroni method ($p\leq0.001$).

When comparing educational level with offsetting, differences amongst these groups are smaller than those amongst the age categories. A chi-square test showed no significant difference between the educational level of respondents and whether they are offsetters or not. A causal relationship between the amount of offsets bought and level of education was also not found. Differences in sex in relation to offset behaviour were negligible.

Discussion

We found that offsetters tend to fly more often and further away than non-offsetters, but that they are also more focussed on being green in their daily life and see consumers (themselves) as more responsible for mitigating travel emissions than non-offsetters. Personal responsibility, however, does not translate into more sustainable travel behaviour. Travel behaviour restricting mitigation measures are not desirable, whereas financial ones are. We found some evidence that points at offsetting as a means to continue unsustainable travel behaviour. In this sample, the use of offsetting increases with age, but there is no relation with sex or education, though the sample showed well-above education levels overall. The convenience sample used is not representative for the Dutch population, as all respondents were customers or followers of a small, long-haul oriented tour operator. Hence, generalisations must be made with caution. However, results correspond to those of other work, signalling similarities in values, attitudes and travel behaviour in many Western countries.

The finding that offsetters fly more (often) than non-offsetters is supported by other research. In a UK study, more offsetters than non-offsetters used an airplane and also used it more often (Whitmarsh, 2008). Under German travellers a slightly higher share of airplane users and long-haul flyers were planning to offset than all travellers (Günther, 2008). Similar to our results, the UK study reports significantly higher pro-environmental values, a higher environmental self-identity, and more (regular) pro-environmental behaviour among offsetters than non-offsetters, except for flying (Whitmarsh, 2008). Ethical and environmental reasons were the strongest motivations to offset in our study, though obtaining a "good feeling" is also important. To some extent the latter can be linked to reducing guilt about flying, as mentioned by Kotchen (2009) and others. High pro-environmental values, expressed through a high moral obligation of offsetters to pay for their contribution to climate change, were confirmed by the international sample in Brouwer *et al.* (2008). As such, it appears that offsetters cannot translate their high environmental values or sustainability-oriented ethics into more sustainable travel behaviour, which confirms the results of other air travel behaviour related studies (e.g. Barr *et al.*, 2010). The desire to be able to travel where and whenever is a stronger motivation. In fact, a considerable amount of offsetters confirmed "compensating in order to be able to keep flying" as a valid motivation, though disagreement to a statement like "the option to offset has no effect on my travel behaviour" was less strong. Though not explicitly asked in this survey, we can draw a link to the UK study, where a same high percentage of offsetting and non-offsetting respondents (67 per cent) agreed that carbon offsetting does encourage continuation of unsustainable behaviour (Whitmarsh, 2008).

The aforementioned desire to travel freely is confirmed by refusal of behaviour-restricting mitigation measures and acceptance of financial ones, with the latter (taxes, mandatory offsets) meeting even higher acceptance with offsetters. German respondents were also most willing to accept higher transport costs and had no desire to forego on flying (Günther, 2008; Lohmann, 2007), an attitude shared by international travellers to New Zealand (Becken, 2007). Apart from age, socio-demographics were not related to offset behaviour. This may partly be due to the specific sample used. We have not investigated income levels. Whitmarsh (2008) linked UK offsetters to higher incomes, better education and higher knowledge and concern of climate change than non-offsetters. Others also found a higher willingness-to-pay (WTP) or intention to offset with higher incomes (Akter *et al.*, 2009; Brouwer *et al.*, 2008; MacKerron *et al.*, 2009).

Conclusion

In this chapter we have assessed the motives for offsetting holiday emissions, and the effects of offsetting and non-offsetting on attitudes and behaviour in a holiday context. Our sample was composed of clients and followers of a Dutch tour operator, with an above average education level. Offsets are bought for environmental reasons, but also for feeling good and for keeping on flying. We conclude that offsetters tend to fly more often and further away than non-offsetters, but that they are also more focussed on being green in their daily life and see consumers (themselves) as more responsible for mitigating travel emissions than non-offsetters. Personal responsibility and high environmental values, however, do not translate into more sustainable travel behaviour. Rather, it is passed on to other parties, for example in the form of mandatory, financial measures. Compensation appears to limit people's drive to contemplate other effective actions towards more sustainable travel. Measures that limit the freedom to travel are not accepted. Offsets do not lead to a reduction of travel, but do also not appear as an incentive to travel more. Rather, offsets support and sustain current travel behaviour.

This chapter provides more evidence that voluntary behaviour change in tourism is unlikely to yield large emission reductions (see also Becken, 2007), not even, or particularly amongst well-educated and environmentally aware travellers. The call for policy intervention is therefore increasing in tourism and transport emission issues (Dray *et al.*, 2012). For effective policy measures lessons can be learned from other areas where behavioural change is aimed at. In health sciences for example, the empirical evidence of effective obesity prevention approaches is limited but growing (Gortmaker *et al.*, 2011). Policy instruments like media campaigns on health-related behaviour are less successful when the behaviour is habitual or ongoing, like physical activity or food choices (Wakefield *et al.*, 2010), and there are some similarities here with the ineffectiveness of awareness raising by carbon offsetters in changing (habitual) travel behaviour. It is concluded that changing individual behaviour is hard to achieve without addressing the choice-related context, like making healthy foods available, identifiable and affordable to all people (Story *et al.*, 2008). In this context there is increasing indication that pricing instruments, e.g. raising the price of unhealthy food and beverages through taxes and lowering fruit and vegetable prices through subsidies, can be (more) effective policy instruments to induce people to change their food and drink behaviour. Even more so when the revenues are spent on prevention measures (Andreyeva *et al.*, 2010; Brownell and Frieden, 2009; Powell *et al.*, 2012). The relevance for tourism is easily made. Taxes, in the form of higher transport costs, are a more accepted approach for mitigating tourism emissions (Günther, 2008; Quack and Hallerbach, 2008) and the need for policy investment in (tourism) transport solutions is high (Dray *et al.*, 2012; IEA, 2009), given the ineffectiveness of industry self-regulation (Gössling *et al.*, 2012). Taxes would need to be spent on mitigation measures, for example compensating countries that lose from decreasing arrivals due to changing travel behaviour (see, e.g. Peeters and Eijgelaar, 2014). The choice-related context, i.e. making sustainable tourism products more available, identifiable and affordable, touches on subjects like labelling and marketing.

Acknowledgements

The authors would like to thank Sawadee Reizen for their willingness to participate in this independent study.

Note

1 Usage of energy-saving light bulbs, driving economy car, usage of public transport/ bicycle, glass recycling, paper recycling, purchase of fair trade products, consumption of vegetarian meals, purchase of certified fish, purchase of organic produce, purchase of A-labelled home appliances, effort in waste reduction, turning down heating at night, purchase of clean energy, low shower usage, home insulation efforts.

References

Ajzen, I. (1991). The theory of planned behavior. *Organizational Behavior and Human Decision Processes*, *50*(2), 179–211.

Akter, S., Brouwer, R., Brander, L. and van Beukering, P. (2009). Respondent uncertainty in a contingent market for carbon offsets. *Ecological Economics*, *68*(6), 1858–1863.

Allison, I., Bindoff, N. L., Bindschadler, R. A., Cox, P. M., de Noblet, N., England, M. H., Francis, J. E., Gruber, N., Haywood, A. M., Karoly, D. J., Kaser, G., Le Quéré, C., Lenton, T. M., Mann, M. E., McNeil, B. I., Pitman, A. J., Rahmstorf, S., Rignot, E., Schellnhuber, H. J., Schneider, S. H., Sherwood, S. C., Somerville, R. C. J., Steffen, K.,

Steig, E. J., Visbeck, M. & Weaver, A. J. (2009). *The Copenhagen diagnosis, 2009: Updating the world on the latest climate science.* Sydney, Australia: The University of New South Wales Climate Change Research Centre (CCRC).

Anable, J., Lane, B. and Kelay, T. (2006). *An evidence base review of public attitudes to climate change and transport.* London, UK: The Department for Transport.

Anderson, K. (2012). The inconvenient truth of carbon offsets. *Nature, 484*(7392), 7.

Andreyeva, T., Long, M. W. and Brownell, K. D. (2010). The impact of food prices on consumption: A systematic review of research on the price elasticity of demand for food. *Am J Public Health, 100*(2), 216–222.

Ariza-Montobbio, P. and Lele, S. (2010). Jatropha plantations for biodiesel in Tamil Nadu, India: Viability, livelihood trade-offs, and latent conflict. *Ecological Economics, 70*(2), 189–195.

Barr, S., Shaw, G., Coles, T. and Prillwitz, J. (2010). "A holiday is a holiday": Practicing sustainability, home and away. *Journal of Transport Geography, 18*(3), 474–481.

Becken, S. (2007). Tourists' perception of international air travel's impact on the global climate and potential climate change policies. *Journal of Sustainable Tourism, 15*(4), 351–368.

Becken, S. and Hay, J. E. (2007). *Tourism and climate change: Risks and opportunities.* Clevedon, UK: Channel View Publications.

Boon, B. H., Schroten, A. and Kapman, B. (2007). Compensation schemes for air transport. In P. Peeters (ed.) *Tourism and climate change mitigation: Methods, greenhouse gas reductions and policies.* Breda, Netherlands: NHTV Academic Studies.

Bramwell, B. and Lane, B. (2012). Getting from here to there: Systems change, behavioural change and sustainable tourism. *Journal of Sustainable Tourism, 21*(1), 1–4.

Brouwer, R., Brander, L. and Van Beukering, P. (2008). "A convenient truth": Air travel passengers' willingness to pay to offset their CO_2 emissions. *Climatic Change, 90*(3), 299–313.

Brownell, K. D. and Frieden, T. R. (2009). Ounces of prevention – the public policy case for taxes on sugared beverages. *New England Journal of Medicine, 360*(18), 1805–1808.

Buckley, R. (2012). Sustainable tourism: Research and reality. *Annals of Tourism Research, 39*(2), 528–546.

Bumpus, A. G. and Liverman, D. M. (2008). Accumulation by becarbonization and the governance of carbon offsets. *Economic Geography, 84*(2), 127–155.

Cohen, S. A. and Higham, J. E. S. (2011). Eyes wide shut? UK consumer perceptions on aviation climate impacts and travel decisions to New Zealand. *Current Issues in Tourism, 14*(4), 323–335.

de Bruijn, K., Dirven, R., Eijgelaar, E., Peeters, P. and Nelemans, R. (2013). *Travelling large in 2011: The carbon footprint of Dutch holidaymakers in 2011 and the development since 2002.* Breda, Netherlands: NHTV Breda University of Applied Sciences.

Dray, L. M., Schäfer, A. and Ben-Akiva, M. E. (2012). Technology limits for reducing EU transport sector CO_2 emissions. *Environmental Science & Technology, 46*(9), 4734–4741.

Dubois, G., Ceron, J.-P., Peeters, P. and Gössling, S. (2011). The future tourism mobility of the world population: Emission growth versus climate policy. *Transportation Research Part A: Policy and Practice, 45*(10), 1031–1042.

Eijgelaar, E. (2011). Voluntary carbon offsets a solution for reducing tourism emissions? Assessment of communication aspects and mitigation potential. *European Journal of Transport and Infrastructure Research, 11*(3), 281–296.

Gortmaker, S. L., Swinburn, B. A., Levy, D., Carter, R., Mabry, P. L., Finegood, D. T.,

Huang, T., Marsh, T. & Moodie, M. L. (2011). Changing the future of obesity: Science, policy, and action. *The Lancet, 378*(9793), 838–847.

Gössling, S. (2011). *Carbon management in tourism: Mitigating the impacts on climate change*. Abingdon, UK: Routledge.

Gössling, S., Broderick, J., Upham, P., Peeters, P., Strasdas, W., Ceron, J.-P. and Dubois, G. (2007). Voluntary carbon offsetting schemes for aviation: Efficiency, credibility and sustainable tourism. *Journal of Sustainable Tourism, 15*(3), 223–248.

Gössling, S., Haglund, L., Kallgren, H., Revahl, M. and Hultman, J. (2009). Swedish air travellers and voluntary carbon offsets: Towards the co-creation of environmental value? *Current Issues in Tourism, 12*(1), 1–19.

Gössling, S., Hall, C. M., Ekström, F., Engeset, A. B. and Aall, C. (2012). Transition management: A tool for implementing sustainable tourism scenarios? *Journal of Sustainable Tourism, 20*(6), 899–916.

Gössling, S., Peeters, P. and Scott, D. (2008). Consequences of climate policy for international tourist arrivals in developing countries. *Third World Quarterly, 29*(5), 873–901.

Günther, W. (2008). Die Sicht der Nachfrager: Akzeptanz von Handlungsoptionen der Urlauber in Bezug auf den Klimawandel. In DTV (ed.) *DTV Fachkonferenz: Klimawandel fordert Tourismuswandel.* Berlin, Germany.

Günther, W. and Lohmann, M. (2008). Klimaschutz und Reiseverhalten: Akzeptanz klimaschonender Verhaltensweisen im Urlaub. Kiel, Germany: F.U.R.

Hares, A., Dickinson, J. and Wilkes, K. (2010). Climate change and the air travel decisions of UK tourists. *Journal of Transport Geography, 18*(3), 466–473.

Higham, J. E. S. and Cohen, S. A. (2011). Canary in the coalmine: Norwegian attitudes towards climate change and extreme long-haul air travel to Aotearoa/New Zealand. *Tourism Management, 32*(1), 98–105.

House of Commons Environmental Audit Committee (2007). *The voluntary carbon offset market: Sixth report of session 2006–07.* London, UK: The House of Commons.

Hyams, K. and Fawcett, T. (2013). The ethics of carbon offsetting. *Wiley Interdisciplinary Reviews: Climate Change, 4*(2), 91–98.

ICAO (2010). *Environmental report 2010: Aviation and climate change*. Montréal: ICAO.

IEA (2009). Transport, energy and CO_2: Moving towards sustainability. Paris: OECD/IEA.

Kant, P. and Wu, S. (2011). The extraordinary collapse of jatropha as a global biofuel. *Environmental Science & Technology, 45*(17), 7114–7115.

Kollmuss, A. and Agyeman, J. (2002). Mind the gap: Why do people act environmentally and what are the barriers to pro-environmental behavior? *Environmental Education Research, 8*(3), 239–260.

Kotchen, M. J. (2009). Offsetting green guilt. *Stanford Social Innovation Review*, Spring 2009, 26–31.

Lee, D. S., Lim, L. L. and Owen, B. (2013). *Bridging the aviation CO_2 emissions gap: Why emissions trading is needed*. Manchester, UK: Manchester Metropolitan University.

Lohmann, M. (2007). *Akzeptanz klimaschonender Verhaltensweisen im Urlaub: Kommentar zu den Ergebnissen – F.U.R/Ipsos Befragung im April 2007*. Kiel, Germany: Forschungsgemeinschaft Urlaub und Reisen e.V.

Lorenzoni, I., Nicholson-Cole, S. and Whitmarsh, L. (2007). Barriers perceived to engaging with climate change among the UK public and their policy implications. *Global Environmental Change, 17*(3–4), 445–459.

Lovell, H., Bulkeley, H. and Liverman, D. (2009). Carbon offsetting: Sustaining consumption? *Environment and Planning A, 41*, 2357–2379.

MacKerron, G. J., Egerton, C., Gaskell, C., Parpia, A. and Mourato, S. (2009). Willingness to pay for carbon offset certification and co-benefits among (high-) flying young adults in the UK. *Energy Policy, 37*(4), 1372–1381.

Mair, J. (2011). Exploring air travellers' voluntary carbon-offsetting behaviour. *Journal of Sustainable Tourism, 19*(2), 215–230.

Mayor, K. and Tol, R. S. J. (2007). The impact of the UK aviation tax on carbon dioxide emissions and visitor numbers. *Transport Policy, 14*(6), 507–513.

Mayor, K. and Tol, R. S. J. (2010). The impact of European climate change regulations on international tourist markets. *Transportation Research Part D: Transport and Environment, 15*(1), 26–36.

McKercher, B., Prideaux, B., Cheung, C. and Law, R. (2010). Achieving voluntary reductions in the carbon footprint of tourism and climate change. *Journal of Sustainable Tourism, 18*(3), 297–317.

Melillo, J. M., Reilly, J. M., Kicklighter, D. W., Gurgel, A. C., Cronin, T. W., Paltsev, S., Felzer, B. S., Wang, X., Sokolov, A. P. & Schlosser, C. A. (2009). Indirect emissions from biofuels: How important? *Science, 326*(5958), 1397–1399.

OECD-UNEP (2011). *Climate change and tourism policy in OECD countries.* Paris: Organisation for Economic Co-operation and Development/United Nations Environment programme.

Paterson, M. and Stripple, J. (2010). My space: Governing individuals' carbon emissions. *Environment and Planning D: Society and Space, 28*, 341–362.

Peeters, P. (2010). Transport technology, tourism and climate change. In C. Schott (ed.) *Tourism and the implications of climate change: Issues and actions* (Vol. 3, pp. 67–90). Bingley (UK): Emerald.

Peeters, P. and Dubois, G. (2010). Tourism travel under climate change mitigation constraints. *Journal of Transport Geography, 18*(3), 447–457.

Peeters, P. and Eijgelaar, E. (2014). Tourism's climate mitigation dilemma: Flying between rich and poor countries. *Tourism Management, 40*, 15–26.

Pentelow, L. and Scott, D. J. (2011). Aviation's inclusion in international climate policy regimes: Implications for the Caribbean tourism industry. *Journal of Air Transport Management, 17*(3), 199–205.

Pereira, E. M. V., Mykletun, R. J. and Hippolyte, C. (2012). Sustainability, daily practices and vacation purchasing: Are they related? *Tourism Review, 67*(4), 40–54.

Powell, L. M., Chriqui, J. F., Khan, T., Wada, R. and Chaloupka, F. J. (2012). Assessing the potential effectiveness of food and beverage taxes and subsidies for improving public health: A systematic review of prices, demand and body weight outcomes. *Obesity Reviews, 14*(2), 110–128.

Quack, H.-D. and Hallerbach, B. (2008). *Sommerurlaub 2008: Repräsentativbefragung zum Sommerurlaub der Deutschen.* Trier, Germany: ETI.

Randles, S. and Mander, S. (2009). Aviation, consumption and the climate change debate: "Are you going to tell me off for flying?". *Technology Analysis & Strategic Management, 21*(1), 93–113.

Reijnders, L. (2009). Are forestation, bio-char and landfilled biomass adequate offsets for the climate effects of burning fossil fuels? *Energy Policy, 37*(8), 2839–2841.

Rogelj, J., Hare, W., Lowe, J., van Vuuren, D. P., Riahi, K., Matthews, B., Hanaoka, T., Jiang, K. & Meinshausen, M. (2011). Emission pathways consistent with a 2°C global temperature limit. *Nature Clim. Change, 1*(8), 413–418.

Schäfer, A., Heywood, J. B., Jacoby, H. D. and Waitz, I. A. (2009). *Transportation in a climate-constrained world.* Cambridge, MA: MIT Press.

Schmücker, D. J. (2011). Freiwillige Kompensation von Flugreiseneminssionen als nach-
fragerinduzierte Anpassungsstrategie: ein empirischer Anbietervergleich. *tw Zeitschrift
für Tourismuswissenschaft, 3*(2), 139–149.

Scott, D., Peeters, P. and Gössling, S. (2010). Can tourism deliver its "aspirational" greenhouse gas emission reduction targets? *Journal of Sustainable Tourism, 18*(3), 393–408.

Searchinger, T., Heimlich, R., Houghton, R. A., Dong, F., Elobeid, A., Fabiosa, J., Tokgoz,
S., Hayes, D. & Yu, T.-H. (2008). Use of US croplands for biofuels increases greenhouse
gases through emissions from land-use change. *Science, 319*(5867), 1238–1240.

Sims, R., Mercado, P., Krewitt, W., Bhuyan, G., Flynn, D., Holttinen, H., Jannuzzi, G.,
Khennas, S., Liu, Y., O'Malley, M., Nilsson, L. J., Ogden, J., Ogimoto, K., Outhred,
H., Ulleberg, Ø. & van Hulle, F. (2011). Integration of renewable energy into present
and future energy systems. In O. Edenhofer, Pichs-Madruga, R., Sokona, Y., Seyboth,
K., Matschoss, P., Kadner, S., Zwickel, T., Eickemeier, P., Hansen, G., Schlömer, S. &
C. von Stechow (eds) *IPCC special report on renewable energy sources and climate
change mitigation.* Cambridge, UK: Cambridge University Press.

Skinner, I., van Essen, H., Smokers, R. and Hill, N. (2010). *Towards the decarbonisation
of the EU's transport sector by 2050.* Didcot, UK; Brussels, Belgium: AEA/EC.

Steg, L. and Vlek, C. (2009). Encouraging pro-environmental behaviour: An integrative
review and research agenda. *Journal of Environmental Psychology, 29*(3), 309–317.

Stoll-Kleemann, S., O'Riordan, T. and Jaeger, C. C. (2001). The psychology of denial
concerning climate mitigation measures: Evidence from Swiss focus groups. *Global
Environmental Change, 11*, 107–117.

Story, M., Kaphingst, K. M., Robinson-O'Brien, R. and Glanz, K. (2008). Creating
healthy food and eating environments: Policy and environmental approaches. *Annu.
Rev. Public Health, 29*, 253–272.

Tiller, T. R. and Schott, C. (2013). The critical relationship between climate change
awareness and action: An origin-based perspective. *Asia Pacific Journal of Tourism
Research, 18*(1–2), 21–34.

Timilsina, G. R. and Shrestha, A. (2011). How much hope should we have for biofuels?
Energy, 36(4), 2055–2069.

United Nations (UN) (2011). Secretary-General calls climate change "quintessential
global challenge", citing also crime, pandemics, in Security Council meeting on new
challenges to peace [Online]. New York: United Nations. Retrieved from www.un.org/
News/Press/docs/2011/sgsm13964.doc.htm.

UNWTO (2011). *Tourism towards 2030: Global overview – advance edition.* Paper presented at UNWTO 19th General Assembly. Madrid: UNWTO.

UNWTO-UNEP-WMO (2008). *Climate change and tourism: Responding to global challenges.* Madrid: UNWTO-UNEP.

van Birgelen, M., Semeijn, J. and Behrens, P. (2011). Explaining pro-environment consumer behavior in air travel. *Journal of Air Transport Management, 17*(2), 125–128.

Wakefield, M. A., Loken, B. and Hornik, R. C. (2010). Use of mass media campaigns to
change health behaviour. *The Lancet, 376*(9748), 1261–1271.

Whitmarsh, L. (2008). Carbon offsetting: A way of avoiding emissions reductions?
Environment Research Web, 5 November. Retrieved from www.environmentalre-
searchweb.org/cws/article/opinion/36551.

Whitmarsh, L. (2009). Behavioural responses to climate change: Asymmetry of intentions
and impacts. *Journal of Environmental Psychology, 29*(1), 13–23.

World Bank (2012). *Turn down the heat: Why a 4°C warmer world must be avoided.* Washington, DC: International Bank for Reconstruction and Development/The World Bank.

9 Understanding temporal rhythms and travel behaviour at destinations

Potential ways to achieve more sustainable travel

*Janet E. Dickinson, Viachaslau Filimonau,
Tom Cherrett, Nigel Davies, Sarah Norgate,
Chris Speed and Chris Winstanley*

Introduction

Travel behaviour is inextricably linked to "time" in diverse ways with implications for sustainable mobility. At the most basic level, time is linked to travel through the speed equals-distance-divided-by-time equation. In this way natural laws govern the distance people may travel relative to the speed of movement and the time available. With less time available, distance decreases unless speed is increased. In general, increased travel speed is associated with higher energy intensity (Poumanyvong *et al.*, 2012). Given our current dependence on fossil fuel-based travel modes, this has led to higher greenhouse gas (GHG) emissions. Similarly, allocating more time to travel enables travel over a longer distance with greater GHG emissions even if speed is not increased. If more time to travel is available and there is access to faster modes, this has a twofold effect on increasing GHG emissions. However, the time allocated to travel in our daily lives has remained relatively constant and increased distance is a result of higher speed (Metz, 2008). Speed has increased in tourism through improved car and train infrastructure and greater use of aviation, and with it, both distance travelled and GHG emissions (Gössling *et al.*, 2009). The car and air travel dominate tourism transport modes (Scott *et al.*, 2010) and together account for 72 per cent of tourism GHG emissions (United Nations World Tourism Organisation – United Nations Environment Programme – World Meteorological Organisation, 2008). Though air travel plays a significant role in the GHG emissions of tourism (Becken, 2002), the focus of this study is on destination-based travel. The destination travel element is largely overlooked in existing sustainable tourism research (Hunter, 2002; La Lopa and Day, 2011) and the need for more research is well recognised (Warnken *et al.*, 2004). Therefore, the subsequent analysis considers land-based travel at a destination, especially car use.

In the tourist destination context, time is important to the transport demand management problem. Time-related visitation patterns generate peak transport demands through large numbers of people seeking to be at a specific place within a similar time frame. This has implications for all transport modes, though car travel presents particular temporal problems related to congestion and car park resource management (Mallet and McGuckin, 2000). Historically, solutions included improvements to road and car park infrastructure, offering alternative modes, such as buses or trains, and mechanisms to induce behaviour change in car users, either to encourage use of alternatives or to avoid peak times. While there have been some localised success stories (Page, 2005), the overall picture remains bleak, especially in rural destinations where car travel is a pervasive problem (Connell and Page, 2008).

Time also plays a role in individual mode choice decisions. The car is perceived to be convenient and offers individually tailored temporal flexibility (Dickinson and Robbins, 2008). In the tourism context, where trips are less predictable, subject to constant readjustment due to tourist preferences and often involve trip chaining, the car offers a unique "time shifting" device (Southerton *et al.*, 2001) that enables users to spontaneously adjust temporal plans to align with tourist opportunities. In rural destinations, in particular, there are often few alternatives available, and, where public transport is available, services are infrequent and often subject to delays due to traffic congestion. These cumulative temporal concerns favour car use and higher GHG emissions.

Time is also an important element that frames the tourist experience. Tourism represents a time to step outside the clock-time routines of day-to-day life (Elsrud, 1998; Richards, 1998; Stein, 2012) and research has explored the multiple temporalities of tourism and temporal rhythms that characterise destinations (Bærenholdt *et al.*, 2004; Germann Molz, 2010; Haldrup, 2004). Despite the importance of time in tourism and to the tourism transport problem, it has rarely been analysed from the perspective of the individual tourist experience and its role in tourist travel behaviour. Dickinson and Peeters (in press) suggest a need to better understand the role of time in the sustainable development of tourism. Tourist responses to time conditions influence travel behaviour and greater attention needs to be paid to this if we are to offer better insights to policy-makers.

To further highlight the significance of time, mobile media has emerged as a new sociotechnical substrate with ubiquitous capabilities that provide users with much enhanced space–time knowledge that is increasingly employed in travel contexts (Dickinson *et al.*, 2012).As a result, new travel tools, for example, those utilising real-time travel information, are emerging that have the potential to inform and guide new behavioural practice. Wajcman (2008, p. 67) suggests that such technology not only saves time but also provides users with a tool to mutually shape new material and cultural practices to actively "take more control of time". Therefore, this is a pivotal moment to examine temporal concepts in order to inform policy debate and the future governance of sustainable mobility.

This chapter draws on material collected within a wider research study, Sixth Sense Transport (sixthsensetransport.com), which focuses on decision-making in travel behaviour by using social networking principles to create visibility of potential transport options in time and space. The analysis presented here focuses on research conducted at a campsite in a UK rural tourism destination that explored a variety of temporal problems with respect to destination travel. The chapter's aim is to contribute an understanding of the role played by time in destination-based travel behaviour.

Time, tourism and travel behaviour

In clock-time cultures such as North America and Western Europe, time is regulated by the clock and it is difficult to conceptualise time in any other way. Relative to human existence, clock time is a recent phenomenon and one that is linked to travel. While time-keeping emerged from the Benedictine monasteries in the fourteenth century (Adam, 1995), it was only during the industrial revolution and the advent of transport structures that required coordination between different cities and countries that the time systems of different places became aligned. The result was the adoption of Greenwich Mean Time, a time standard across the globe (Speed, 2011). Contemporary society is governed by a variety of time schedules such as shop, office and attraction-opening hours. Transport infrastructures and travel patterns are to a large extent regulated by the clock-time system. For example, in tourism congestion builds up on motorways at the start and end of a holiday weekend, attraction car parks are busy on a Sunday and transport companies of all types charge peak rates at busy times of each day and at the start and end of a holiday period.

Clock time presents a linear view of time with a precisely defined measurement tool. This tool provides a mechanism to coordinate activities where people need to periodically meet. There are, however, challenges to the clock time brought about by socio-technological adaptations and other more experiential interpretations of time. An increasingly networked society has found some freedom from the clock-time bonds that once bound activities to distinct spatial and temporal settings (Couclelis, 2009). This is particularly evident in post-Fordist work environments where increasingly ubiquitous technology and global networks are changing work practices. Global communication technology can bring disparate parties to the immediate present (Frändberg, 2008; Klein, 2004). Castells (2000) describes this as the "Network Society" and suggests this is leading to a new paradigm of time that he refers to as "timeless time". In a similar vein, Klein (2004) describes the blurring of work time and social-life time and the disruption of time continuity as "fragmented time", while Urry (1994) refers to "instantaneous time". Dickinson and Peeters (in press) suggest that this more fluid experience of time has the potential to release tourists from day-to-day temporal and spatial constraints and yield opportunities for more sustainable tourist travel, though they are not optimistic that this path will emerge.

Our experience of time is not just linear and we are able to experience multiple times (Adam, 1995; May and Thrift, 2001). For instance, the leisure experience includes phases such as "anticipation" and "recollection" (Clawson and Knetsch, 1966) where we are looking forward to a future experience or remembering a past time (Adam, 1995). Csikszentmihalyi (2002) has also conceptualised "flow" where we become so immersed in an activity that we lose sense of time and it passes more quickly. Adam (1995, p. 12) also draws attention to the "recursiveness of daily existence" as we enact a pattern of rhythms relative to daylight hours, meal times and other bodily needs. Here clock time can be at odds with our bodies as contemporary society no longer modifies activity patterns relative to winter and summer. Merriman (2012) refers to the geographer's obsession with space–time and suggests a focus on the unfolding of events instead. He highlights that our experience of time is culturally specific and calls for more processual and relational accounts.

Within tourism, several authors have explored the multiple temporalities of tourism (Bærenholdt *et al.*, 2004; Germann Molz, 2010; Haldrup, 2004). Destinations exhibit their own temporal patterns that might be as distinctive as the landscape or architecture. Germann Molz (2010) suggests tourists actively seek out these different times. For instance, a destination may represent the past by reflecting a more traditional way of life (Dickinson and Peeters, in press) or enable a tourist to recapture some memory of past times based on repeat visitations (Bærenholdt *et al.*, 2004). Memories of past times are a very potent temporal force within tourism as they shape the future traveller identity (Hibbert *et al.*, 2013) and inform the information search process (Solomon *et al.*, 2010) and subsequent behaviour. The experience of the present time in tourism is therefore multiple and, in addition to past recollection, tourism involves anticipation of future activity.

A large degree of tourism is future-orientated as tourists not only imagine future holidays and future activities while staying at a destination, but also undertake a degree of planning, whether this is arranging time off work, booking transport or checking the opening hours of an attraction. With respect to this future orientation, contemporary westernised clock-time cultures view time as a scarce resource, a commodity to carve up into scheduled activities (Norgate, 2006). Scheduling tendencies are typical of cultures known for engaging in "monochromic" behaviours (Lindquist and Kaufman-Scarborough, 2007), where associated social norms prioritise the needs of the individual above the collective. In day-to-day life this can lead to car dependence since the car provides people with a "time shifting" device (Southerton *et al.*, 2001) that provides flexibility to deal with the subjective time pressure associated with meeting a series of scheduled tasks. Though tourists are arguably less time constrained, there is still much within tourism that binds tourists to clock-time regimes.

Tourism has been described as "time out" (Elsrud, 1998) that is generally interpreted as time away from work and the temporal constraints of our day-to-day existence. It is a chance to step out of our everyday routines (Richards, 1998) and find time for ourselves. A tourist's experience of time out is very varied as some choose to do very little while others pack in more activities than they do at home (Dickinson and Peeters, in press; Stein, 2012). Time out can also be interpreted as a stepping out of time as tourists may choose to ignore the clock time that governs their home life and enter an "extempore" existence (Kwan, 2007) where activities unfold on an "as and when required" basis. This is, however, contextual as in many respects clock time structures tourism just as much as other aspects of contemporary society. Most holidays have an inevitable time constraint linked to paid holiday from work and the temporal patterns of attraction opening hours, public transport, tour operators and meal times schedule the activity options of tourists day to day and hour by hour. In addition to these external temporal structures, tourists also have internal temporal concerns about the appropriate amount of time to allocate to activities and the best time to visit places (Germann Molz, 2010; Haldrup, 2004). This can be a significant logistical undertaking which Larson *et al.* (2007) suggest takes on work-like characteristics. Time allocation can generate anxiety (Germann Molz, 2010).

In addition to temporal problems such as congestion, time also presents travel opportunities in tourism. Tourists have a much higher degree of flexibility related to both when they travel and where (Gössling *et al.*, 2012). Assuming an awareness of localised traffic problems, tourists are better able to modify behaviour than working people. However, tourists often lack this local knowledge. Based on congestion it is self-evident that there is a degree of shared travel pattern among visitors, though this has yet to be captured and utilised in any meaningful way. Most attempts to coordinate travel and utilise spare vehicle capacity have been aimed at travel to work. However, there have been some attempts at travel collaboration through lift schemes mainly targeting young people. For example, Europe's www.carpooling.com, though aimed at short-distance commuting, is also actively used by non-residents and for longer trips. Glastonbury festival operates a lift-share scheme (Greener Festival, 2012) that actively utilises technology (for example, real-time travel updates and dedicated mobile app) to better visualise car-share opportunities (Ashden, 2012).

Space–time practices are increasingly modified by the mobile media and its location-based capabilities. For example, "micro-coordination" (Ling, 2004) has become well established as people are able to renegotiate meetings "on the go" through text and mobile phone calls. An accomplished micro-coordinator will make relatively few forward plans with friends, instead choosing to micro manage these relative to their evolving personal context. New forms of "synchronous-mediated communication" (Humphreys, 2010) are leading some to suggest a new transport paradigm is emerging (Couclelis, 2009) associated with new forms of anticipation and pace (Lemos, 2010).

This overview illustrates three significant strands of literature that point to time playing an important role in tourist travel behaviour and mobility: time is central to travel demand management and individual travel mode choices both in tourism and everyday life; time is conceptually important to tourism and is experienced in multiple ways; and new technology using mobile media is altering our conceptualisation of space–time travel behaviour. If we are to bring about a change in transport behaviours, understanding how tourists' perceived and the actual relationships with time impact on decision-making about transport behaviours is important.

Methodology

The study was interested in the temporal and spatial flow of people, objects, information and ideas and therefore required the application of mobile methods (Bærenholdt *et al.*, 2004; Buscher and Urry, 2009) to capture tourists not only on the move, but also over time. In other areas of transport research, data have been captured by methods such as accompanied trips (for example, "walking with", Sheller and Urry, 2006). However, in a tourist context, trips are rarely well contained, either spatially or temporally, and involve a high degree of spontaneity making accompanied trips time-consuming, difficult to organise and potentially intrusive for participants. Therefore alternative methods were needed that could capture data from participants on the move. Here data were captured using a one-day participant-generated photographic and diary record with a follow-up interview (the diary photograph, diary-interview method). This approach is similar to that applied by Line *et al.* (2011) to explore the use of Information Communication Technology in everyday travel and the diary-photograph, diary-interview method developed by Latham (2003). The research also sought to understand the spatial and temporal patterns of participants with data captured using a purpose-built smartphone app, Traverse, which employed remotely monitored trip research (Edwards and Griffin, 2013).

The study was based at a campsite in a rural destination on the UK south coast. Camping tourism provides a spatially bound community for study, but importantly also accounts for 17 per cent of the total overnight stays in the EU (EuroStat, 2012). The campsite was purposefully selected for the Sixth Sense Transport project. As well as being characteristic of UK campsites in rural areas, the site managers were willing to provide access to participants and support a variety of interventions over a two-year period. The campsite was medium size relative to campsites in rural destinations in the UK and located on a bus route, approximately 5 km from a seaside town and 1 km from coastal walks. A maximum variation sampling strategy was employed at the campsite to engage with a range of tourist types based on observed group and age characteristics (see Table 9.1).

Given that not all participants would have access to a smartphone for the use of the Traverse app, iPhones were loaned to participants. Participants were asked to carry the iPhone and launch the Traverse app whenever they left the campsite, even if this was just a short trip such as to walk the dog. Traverse was designed to record user location, using GPS and a time stamp, and, on each use, Traverse requested input of mode of transport. Data were therefore captured on the spatial location of users over time and by travel mode. To capture diary material participants were provided with a notebook and pencil to record notes and the iPhone provided for Traverse enabled capture of photographs. Participants were asked to reflect on the following in their diaries:

- Routines or repeated patterns.
- The order of events.

- The extent to which activities were planned.
- Rough allocation of time to activities, places, travelling.
- Any activities that were time-constrained.
- The time demands made by other people, places visited or things needed.
- Whether the participant felt in control of time.

Using this method, interviews should ideally be conducted as close as possible to the recorded events. In a tourism context it is difficult to arrange specific interview times due to tourists' tendency to ongoing renegotiation of activities. Experience showed that the best time to meet participants was the morning following their one-day diary activity. Interviews took place beside each

Table 9.1 Participant information

Participants (names are pseudonyms)	Gender	Age	Camping experience
Family groups			
Janet	F	30s	Repeat visitor
Sophie	F	40s	Repeat visitor
Sally	F	40s	Repeat visitor
Louise	F	40s	Repeat visitor
Teresa	F	40s	Repeat visitor
Jacob	M	30s	Repeat visitor
Graham	M	40s	First-time visitor
Carter	M	40s	First-time visitor
Oscar	M	40s	First-time visitor
Nicolas	M	40s	Repeat visitor
Carl	M	50s	First-time visitor
Couples			
Mary	F	20s	First-time visitor
Harriett	F	30s	First-time visitor
Julie	F	30s	Repeat visitor
Claire	F	40s	First-time visitor
Luke	M	20s	First-time visitor
Julian	M	20s	First-time visitor
Patrick	M	30s	First-time visitor
Marcus	M	30s	Repeat visitor
Mark	M	30s	First-time visitor
Saul	M	40s	Repeat visitor
Alex	M	40s	First-time visitor
Jack	M	50s	Repeat visitor
Singles			
Macy	F	50s	Repeat visitor
Darius	M	30s	Repeat visitor
Gareth	M	30s	Repeat visitor
Donald	M	40s	Repeat visitor
Large groups			
Jessica	F	Late teens	Repeat visitor
Gordon	M	20s	Repeat visitor

participant's tent. While each interview focused on one participant, the social nature of tourism often resulted in additional contributions from members of the participant's immediate party, most commonly from their partner. A semi-structured interview strategy was used beginning with a narrative approach, in conjunction with diaries and photos, to explore the temporal sequence and stories of the day. Twenty-nine interviews were conducted from July to September 2012. Each interview was recorded and interviews lasted around 40 minutes. Given the participant commitment, each participant received £10 in vouchers to spend at the campsite shop.

The analysis followed two strategies. The first strategy explored individual narrative structures to understand people's various movements and the temporal issues that impact on transport choices. The second strategy involved cross-case thematic analysis to identify generic concepts.

Findings

Participants found it hard to articulate the temporal nature of their travel. While people are well versed in talking about their tourist experiences, they are not used to reflecting on time, especially in a relatively abstract way. Following the diary instructions and narrative interview strategy, participants utilised a chronological reporting strategy, that involved reporting of the events as they unfolded, for example, "I did a walk by the seaside and I had some shopping to do, then I had lunch there and then I went to several shops" (Macy). Some respondents included much more specific references to clock time. For example: "1 o'clock, we had our lunch there … canoed out at about 3, no, half 2 maybe … we came back at half past 4" (Carter).

Both strategies reference past time in a linear way reflecting the dominant understanding of time in contemporary society, "this homogeneous and desacralised time has emerged victorious since it supplied the measure of the time of work" (Lefebvre, 2004, p. 73). Through a series of probes on the unfolding of time, time planning and time relative to other people, places and things, the interviews developed more depth to the temporal understanding. Conceptually the data orientate into three temporal themes: time fluidity; daily and place-related rhythms; and control of time.

Time fluidity

Without exception, participants experienced a degree of fluidity to their day. While some plans were made, there was little or no sense of having to do something or be somewhere at a specific time, and planned activities were open to amendment as other opportunities arose as Harriet describes:

> We were going to go over to Dancing Ledge and spend the day there but we just could not be bothered. It was sunny, so we just sat out here and chilled, so till the early afternoon and then we went out to Dancing Ledge …

because we can go along the flow a bit and change things, to adjust to how things are, then I feel we can dictate what we want to do and when we want to do it really.

Time fluidity was often organised around a notional "must do" list; however, participants felt no commitment to ticking off the items as "things may drop off the list" (Sally). Underlying this was a desire to "go with the flow", a term used by several participants, with some reflecting more deeply on this fluid time experience. For Mark it was about experience time or being in the present:

> Your body tells you everything. When you are hungry, you cannot think of anything else, when you need to go to the toilet, you cannot think of anything else, it is part of nature, is it not? You have got your own body clock and it tells you what to do, if you just listen to it, then it tells you what, you know, what you need. It is just about turning the timetable and trying to get rid of it, just letting things happen naturally ... it is all about control and we are trying, sort of, you know, having experience ... in a linear sense, of like going there and seeing that ta-da-da-da you know. Or you can actually just be and allow everything just to come to you, which it does, but in a different rhythm, it is not, you know, it is maybe a bit slower or whatever, but you know, things happen if I just stay still.

Mark acknowledges that being in the world is a bodily thing that does not always need ongoing cognitive monitoring of what happens next. His body "actively assumes" time and space (Cresswell, 2003, p. 276).

To articulate what they meant by fluid time, people referred to the structuring of everyday life and a desire to step away from the need to be somewhere by a certain time or a need to do so many activities.

> I'm a primary school teacher, so I have to be up relatively early, doing something at a specific time, it's just nice to just be able to take it at your own pace and relax ... it's nice to just be able to do things that I want to do and eat at the times that I want to rather than set slots.
>
> (Jim)

Clock time is at odds with the rhythms of our bodies (Adam, 1995) and it is evident in Jim's description that he is attempting to re-engage with a more natural rhythm relative to his body's needs. As Jim indicates, home life is very structured for most participants while tourism is very fluid. The accounts represent this tourism time as different to everyday life and a "time off" (Stein, 2012) that is not determined by others. This is a particular feature of rural tourism (Sharpley and Jepson, 2011) where time becomes less restricted and deceleration plays an important role (Matos, 2004). However, subsequent analysis demonstrates that the participants' use of time was not always as loose as they claim.

Some participants explicitly linked temporal fluidity with travel mode. Walking and car travel were dominant modes, both offering individual autonomy and a high degree of flexibility, albeit with distance constraints relative to walking. Flexibility is a significant factor in car travel behaviour since users can choose exactly when they want to leave and alter plans en route (Anable and Gatersleben, 2005). This flexibility is not afforded by public transport in rural areas:

> We get the bus from Swanage sometimes, I think the buses are a bit variable, I am not sure there are that many knocking around and it is quite easy if you have got a car, it is just easy to drive around.
>
> (Gordon)

> We live in London and we're so used to being able to go to any kind of restaurant and just not have to worry at all about being able to get public transport back, whereas obviously in rural areas public transport can be difficult anyway particularly after nine o'clock, ten o'clock or something.
>
> (Jim)

Here both Gordon and Jim indicate their temporal fluidity may be compromised by public transport. Gordon, in particular, notes the ease of car use.

Place rhythms

Despite the fluid nature of people's days and relatively ad hoc arrangements there was a daily rhythm and routine. With few exceptions, participants slept late and were in no rush to get going in the morning, much as you would expect in a tourism context.

> The way we always work is if the children are happy, then we have a good holiday, so yeah, normally we do not have a very early start … when they get up, they want to play with their friends, so you know, we are quite happy to sit here for a couple of hours, read, get a bit peace and quiet, you know, this makes a change.
>
> (Sally)

This led to a peak in departures from the campsite around 11.00, with few departures before 10.00. This was best captured by the Traverse data, a subset of which is presented in Figure 9.1. This shows people beginning to leave the campsite between 10.00 and 11.00. The patterns also begin to coalesce back around the campsite 17.00 to 18.00. Distinctive temporal and spatial patterns are also evident at places within the destination. For instance, Figure 9.1(b) illustrates a morning pattern when people head to a popular beach. Later in the day people return via the nearest town to the campsite, in order to buy food (Figure 9.1(e)). The data also reveal pub visits 21.00 to 22.00 (Figure 9.1(f)). Therefore, despite the apparent ad hoc arrangements, the tourists assumed distinctive place-related patterns. These reflect the multiple sustained time–space routines that Edensor describes as a kind of "place ballet" (cited by Cresswell, 2003, p. 279).

Germann Molz (2010) discusses how daily rhythms are both anticipated in tourists' imaginations of place and then enacted in tourist consumption of place. Destinations have temporal rhythms of place, a pace (speed and tempo) and flow that is characteristic (Bærenholdt *et al.*, 2004; Germann Molz, 2010; Haldrup, 2004) and creates a sense of place which Edensor (2010) likens to a heartbeat. Despite the churn of visitors, the individual patterns of people, which Häger-strand observed in his time geography analysis (Neutens *et al.*, 2011), coalesce with the place into daily rhythms and flows, that also embrace natural factors, such as the prevailing temperature and weather, and human factors like special events. Each new visitor adapts to the movements encoded in the destination (Edensor, 2010) and can become habitualised to temporal patterns they are unaware exist (Adam, 1995). These space–time routines, together with the availability of transport infrastructure, influence the transport choices available to tourists (Dickinson and Peeters, in press). Therefore, while tourists have time autonomy, the absorption of visitors into the flow of place generates destination congestion. Explanations for this are discussed in the next section. This flow of place, however, also yields a significant transport opportunity.

Both the temporal routines and the "must do" lists of participants exhibited similarities; however, there is currently little opportunity to harness this. Given campsites (and many other forms of tourism accommodation) host a spatially bound community, albeit temporary in nature, there is scope for travel collaboration. Preliminary analysis suggests scope for collaboration among visitors to campsites (Filimonau *et al.*, 2013). An example of this is illustrated in Figure 9.2, a storyboard developed with, and for, project participants to explain concepts. This illustrates how campsite visitors might tap into the knowledge that other members of the campsite community are visiting shops on the way home to avoid a 10-km round trip for a forgotten item. This moves away from the atomised individual to a network of actors who are able to realise space–time opportunities, especially given mobile communications technologies.

Control of time

Despite the time fluidity described by participants it was evident that there were a variety of temporal control elements. There was relatively little planning with respect to tourism activities on a day-to-day basis and, aside from the notional list of things to do, participants engaged in fairly minimal planning either the night before or on the actual morning.

> It was dictated by the weather, so we were choosing whether to do this walk yesterday or do it today, depending on whether we wanted to do it in the sun or possibly overcast conditions, so it was literally a decision over breakfast to go.
>
> (Marcus)

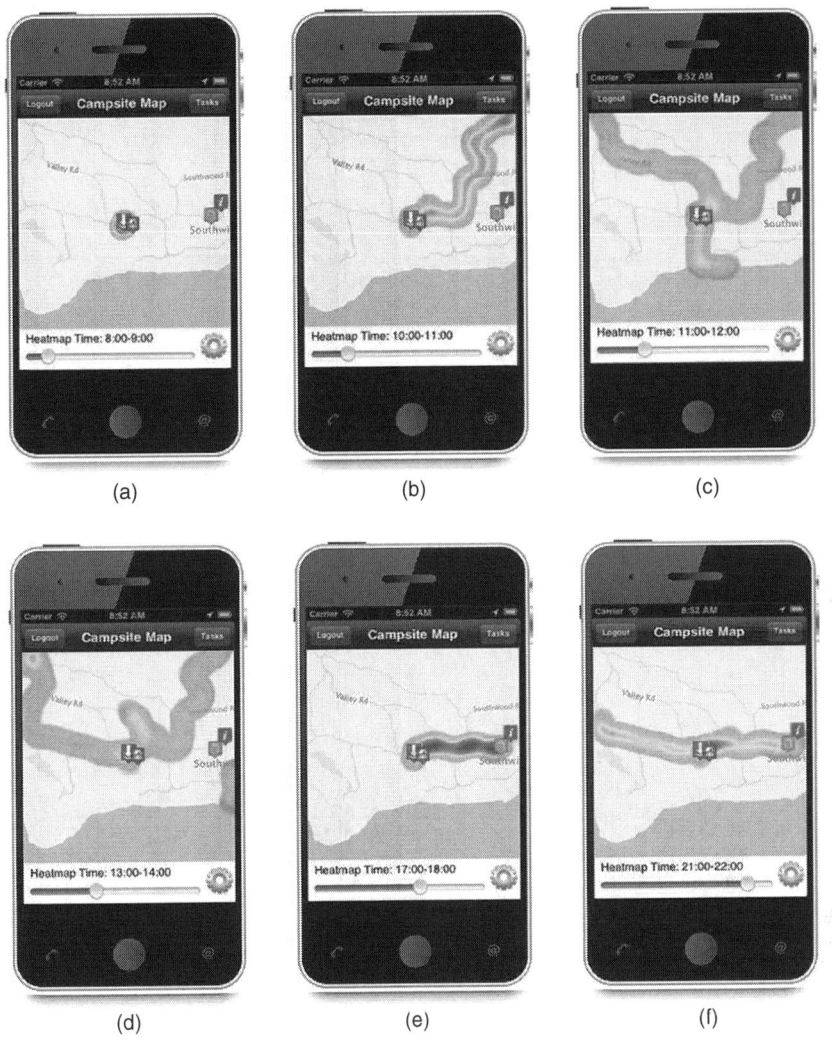

Figure 9.1 Traverse data visualised through a smartphone app. © Welf Aaron – Fotolia.com.

Some participants insisted there was no planning at all. For example, as Mark describes, "because things happen by themselves, you do not need to [plan], you know, you just experience the moment instead of trying to control the moment". Mark was particularly keen to immerse himself in experience time.

Participants reflected on the time demands made by other people, things and places. Participants took a flexible approach to place and acknowledged that sometimes they might move on more quickly if the opportunities afforded were exhausted, but more often participants seemed to find places demanded more of

Figure 9.2 Storyboard illustrating a collaborative shopping concept.

their time. Sophie describes an unexpected opportunity: "We hooked into this story-telling and I was thinking it would be half an hour but it ended up being an hour and twenty minutes which was lovely … we had to continue and finish it." Place experience and knowledge gained by word of mouth from local people or other visitors also played a role. In this way the "must do" list evolves throughout the stay, contingent on experience, actual events and, as discoveries are made, relative to the time needs of places. Accumulated place experience and knowledge also enable participants to avoid temporal problems, such as attraction crowding or road congestion. Given the patterns of repeat visitation, some participants had good local knowledge. In addition, even first-time visitors exhibited an intuitive familiarity given their experience in similar UK destinations. Therefore most participants were very aware that certain attractions would be crowded under particular conditions.

> If we are going to town, we would go in the morning because we know that in the afternoon it is really busy and you cannot park … if we are going to go somewhere where we know it is going to be busy, we avoid business by going in the morning, by going first thing.
>
> (Saul)

> If it was a really hot day, we would possibly go there [a popular beach] later on because it is a lively beach but, yeah, I did not really like it last year because of that, the packed in sardines [laughs].
>
> (Harriet)

In the context of round-the-world travel, Germann Molz (2010) discusses how tourists wish to allocate the right amount of time to place. Participants in this study were relaxed in their approach that reflects a much greater understanding of the place visited, as most participants were domestic tourists. Given the list of alternative activities, participants simply made substitutions as appropriate. However, as Saul and Harriet's quotes illustrate, participants did have internal temporal concerns about the appropriate time to visit places reflecting the findings of Germann Molz (2010) and Haldrup (2004).

More overt than the time demands of places were the time demands made by other people within the participants' immediate group. For example: "I think we have to allow a bit more time because it takes longer for eight people to mobilise and get moving" (Nicholas). Relationships with other people were one of the several factors that were attended to when planning. The presence of children in a group often shifted the focus of planning attention to the child's needs. For example, Louise suggests "if she's [Louise's daughter] enjoying it, I tend to be". Therefore plans often revolved around children's meal times, energy needs (how long they could walk without a break), or attention space (places that would absorb children's attention). For instance, Jacob describes accepting time would be spent on the beach despite poor weather conditions:

It was horrific, it was really, really blowing a gale on the beach and we said, regardless of what the weather was like the next day, we were going to go on the beach anyway because that is what the kids wanted to do.

The items needed during the stay also influence time planning. For instance, Sally describes: "If we go anywhere, we have got to take enough food", and Janet describes driving to shops to buy children's shoes:

Charlie then declared that the two sets of shoes that we'd packed were causing him problems and he sort of unleashed this enormous blister, so we called in on the way back through Swanage, which we had to go through anyway, to pick up some "crocs" for him.

Planning was also weather-related given the outdoor nature of rural tourism and Jim illustrates why planning is relatively immediate and requires alternative options:

We'd probably choose like one thing that we really wanted to do and base the day around that, but then we have in mind that there is other stuff that we may or may not want to do and particularly dependent on the weather … we tend to have a couple of like emergency wet-weather plans and then stuff that if it's a really nice day, we do that instead.

(Jim)

Travel planning is crucial to reduce uncertainty and enhance tourist satisfaction (Zalatan, 1996) and has been studied from a number of disciplinary angles (Jeng and Fesenmaier, 2002). Research has concentrated on consumer purchasing decisions, intentions and motives behind the destination, transport mode and holiday activities selection (Wong and Yeh, 2009). Once tourists have arrived, the travel plans are short term and more fluid (Stewart and Vogt, 1999) and multiple factors influence planning at the destination (Bansal and Eiselt, 2004).

Aspects such as weather changes, place-related discoveries and things suddenly needed are "incidents in time", unexpected moments that caused people to re-evaluate their plans. For example, Mary describes needing to find a bike shop to get parts for her partner's bike. This could be conceived as a negative event, however, it was not construed as such by Mary who took the opportunity to escape the heat (it was an unusually hot day) and read a book. It was just something that happened and they melded their day around this. Similarly, most parents acknowledged temporal impacts made by children. Sophie describes how "it would have been good to know if there was a lift going from the pub to here because by that time she [daughter] was exhausted".

Being able to maintain fluidity was very important to participants with several drawing attention to scheduled activities causing a loss of control of time:

I think there may be a time stress tonight. We were going to try and go to Weymouth for the opening Olympic ceremony so, we're thinking traffic's going to be [bad] ... that could be a stress because of the fact that it is going to be the fireworks or something at 7 o'clock and we'd quite like to be there to see the opening ceremony.

(Janet)

You lose control if you start saying I have got to do this and this perhaps.

(Mary)

This scheduled time replicates the temporal experience in day-to-day life and is known to generate a degree of time stress (Roxburgh, 2004). Participants also discussed other holiday contexts that they would avoid, such as a package holiday, which they associated with more organised and structured time. "I did not enjoy that so much because I almost felt I was on a schedule" (Carl).

Participants also described a variety of "control points in time" which sometimes presented space–time constraints. For example, the campsite shop closes at 19.00 or the last bus leaves at a specific time. Participants were particularly mindful of the need to factor in food, especially with children in the party and since it is a perishable resource:

If you go to the beach, say, or go somewhere first and then go late afternoon to get your shopping, there is nothing in the shop. You have to go in the morning to get reasonable things but then if you buy it in the morning, with it being hot, you have not got a lot of places to store things.

(Sally)

At home refrigerators and freezers provide "time storage" devices (Southerton *et al.*, 2001) that have radically altered our shopping patterns (Watkins, 2003). However, camping brings people back to a more basic relationship with produce and the purchase of food makes a routine temporal demand. These temporal control points align tourists with the "everyday time" that is still evident around them as they interact with tourism employees and local residents (Stein, 2012).

The combination of "incidents in time" and "control points in time" led participants to make ongoing adjustments in the present time, responding to a varied set of individually relevant contextual factors. In this sense the participant appeared to adopt a "responsive time" strategy.

Discussion

The tourist day unfolds in relation to the overall rhythms of place; tourists practice and reinforce these rhythms and the rhythms, in turn, reinforce behaviour. "Because people act in certain preconscious ways, any given order tends to get reestablished and reproduced owing to the 'naturalisation of its own arbitrariness'" (Bourdieu cited in Cresswell, 2003, p. 277).

Everyday life remains shot through and traversed by great cosmic and vital rhythms: day and night, the months and the seasons, and still more precisely biological rhythms. In the everyday, this results in the perpetual interaction of these rhythms with repetitive processes linked to homogeneous time.

(Lefebvre, 2004, p. 73)

The individual tourist succumbs to a variety of temporal forces that reinforce travel behaviour decisions that are not always sustainable. The natural temporal rhythms of people on holiday add to peak traffic flows. While these are less marked than the peaks associated with commuting and work, they exert pressure in a rural environment with limited infrastructure.

As this analysis illustrates, time fluidity was important to people, but as the interviews unfolded it was clear that time incidents and control points also influenced travel behaviour decisions. Opening times of shops and attractions, and scheduled events determine access to key resources and influence visitor flows. Where time appears short, visitors resorted to car use even when their intention was to avoid this. Cass *et al.* (2004) argue that some forms of transport are only possible for those with "time to spare". While tourists inherently have more time, incidents and temporal control points can lead tourists to trade off a quality experience for speed. This is best explained through the experience of multiple competing forms of time. The idea that time is multiple is not new (see Adam, 1995; Middleton, 2009), however, it is difficult for people to conceptualise this. The participants demonstrated the existence of multiple times by reference to: my time; my child's time; the food's time; the attraction's time; the place time; the bus time; clock time; the weather's time. Where multiple times briefly coexist they compete for attention with each individual determining which receives most attention. "Time is experienced differently by individuals depending on the importance of the experience" (Jäckel and Wollscheid, 2007, p. 86) and problems arise due to time–space coordination (Jarvis, 2005; Southerton, 2006). For parents "my child's time" dominates. For women in particular, their time sovereignty can be compromised by time demands of others (Davies, 2001). The personal experience of these multiple temporalities will have a strong influence on individual travel needs and behaviour.

The desire to embrace temporal fluidity embeds that concept with some value and meaning. The participants' experiences of time reflect slow travel where taking time out is a positive feature in which pauses, inactivity and the ability to respond to place encounters is valued (Dickinson *et al.*, 2011). Palmer (cited in Mullins, 2009) argues that higher travel speeds negatively affect tourism's sustainability since learning about unique places and establishing personal relationships with destinations lie at the heart of sustainable tourism. However, the car provides flexibility and speed and while participants were keen to walk, cycle or use public transport, the benefits of the car can outweigh other modes.

The findings highlight a number of temporal situations within a destination area that interact with place locations in a variety of ways to influence travel behaviour. Within rural destinations attractions are spatially dispersed and often

Figure 9.3 Storyboard illustrating a congestion avoidance concept.

require visitor travel to places off the main arterial routes and public transport networks. This is largely determined by natural features and entrepreneurial business opportunities historically developed with little regard to transport access. In the UK, the weather is relatively unpredictable and plays a significant role in temporal flows of participants given the outdoor context of rural tourism. In countries with more stable weather, it will play less of a role, though there may be predictable weather events, such as late afternoon thunderstorms in mountain regions, that impact on traffic flows. Short-term planning often reflected uncertainty about the weather leading to last minute decisions and changes to plans.

While many of these temporal situations would be difficult to address there is scope to better manage the visitors' understanding of the local temporal rhythms. Mobile technology is developing context-based systems that are able to deliver unique personalised information to users referenced to their current location in both space and time. This idea was explored with project participants and led to the emergence of a storyboard (Figure 9.3) that outlines how a context-based system might deliver relevant tourist travel advice to enable users to avoid temporal congestion through modified travel behaviour. The scenario in Figure 9.3 initially provides the user with advice to avoid congestion and, ultimately, by suggesting an alternative attraction closer to the current location, avoids a car trip altogether. The ideas described in Figures 9.2 and 9.3 are currently being developed in a collaborative travel app that has been field-tested in 2013. Technological interventions such as this introduce a temporal juxtaposition since many people choose camping in order to engage with a simpler, "past time", that is devoid of the technological devices that proliferate in their day-to-day lives. Despite this, many participants utilised mobile media to organise short-term planning and in this way two time frames overlap adding to the multiple experience of time.

Conclusion

This chapter has argued that temporal issues should not be overlooked in an analysis of travel behaviour within tourist destination areas. It shows that there are a variety of issues that present challenges for more sustainable travel. Time-fluid tourists desire flexible travel opportunities and this is best achieved through personal modes of transport. While walking was popular in the rural setting studied, car travel provides a high degree of convenience to visitors and is perceived to afford flexibility (Anable and Gatersleben, 2005). Despite the desire for fluidity, tourists were absorbed into place rhythms and flows and thus reproduce congestion and exceed local car park capacity. A variety of temporal incidents and control points aligned to linear clock time present a challenge to tourists. While the desire to go with the flow mediated the direct impact of unexpected incidents, there was often the need for transport flexibility and speed to achieve objectives. The car provides a useful time-shifting device that enhances opportunities in this respect.

The analysis draws attention to clock-time tensions embedded in contemporary society as people see tourism as a space to escape from their usual time-bound routines. Clock time is a relatively recent phenomenon in human society and other commentators (see Adam, 1995) have noted that this linear regime, with a series of scheduled events, is at conflict with a more humanistic approach to time. Transport systems are rooted in clock-time regimes through the need to coordinate arrivals, departures and capacity across space–time. Car travel provides a degree of individual freedom that is highly valued, though still bound to a variety of external time structures such as facility opening times. Given the temporal freedom sought by tourists and the opportunity afforded by car travel, it is not surprising that car travel is so routinely practised in rural tourism. Aside from the spatial constraints within rural tourism, time is a significant barrier to behaviour change from car to other modes of transport that cannot be overlooked.

Tourists' temporal freedom is also compromised by a multitude of competing forms of time. While the concept of multiple times and the possibility that individuals might experience time in more than one way is widely discussed in the geography and sociology literature (see Merriman, 2012; Wajcman, 2008), it has rarely been examined with respect to transport (see Middleton, 2009 for an exception). The transport literature is increasingly aware that our understanding of time is changing but has yet to grasp the travel implications of multiple temporal reference frames for individuals. The analysis presented here demonstrates how participants experience multiple times, each with attendant demands and some with specific transport needs. In addition, visitor travel is embodied and derived from rhythms that are absorbed from the destination, determined by biological routines and altered by other people, place encounters and the things needed. These rhythms reinforce travel behaviour choices.

Given that we are at a pivotal moment with respect to the emergence of a new sociotechnical substrate, in the form of mobile technology that has significant space–time capabilities, there is a key emerging research opportunity. If we are to harness technology to assist in more sustainable forms of travel then we need to better understand the role played by time. From this exploration of time in tourism destination travel, three key messages emerge that should influence the design of tourist destination-based travel systems. First, given the desire for temporal fluidity, transport systems need to evolve beyond clock-time regimes. Tourists seek respite from scheduling demands and tourism and travel information should seek to better manage this. Second, temporal forces favour personal modes of transport (car, walk, cycle), especially in rural areas where public transport is unlikely to provide adequate flexibility. Greater attention therefore needs to be paid to these modes. In this respect, the car is highly personalised and perceived to optimise travel fluidity and speed. This is currently unsustainable and, given the poor success rate of travel behaviour initiatives designed to reduce car dependence, there is a need for imaginative initiatives. Sustainable travel strategies are needed that adopt a positive and proactive stance to car travel that realise opportunities to utilise spare capacity in both public and private vehicles across the transport network. Research is needed to explore how highly individualised space–time scheduling and vehicle use might embed more collective strategies. Third, in order to realise network potential, mechanisms are needed to visualise travel opportunities and constraints in both real time and the immediate future based on historic data feeds representing destination rhythms and routines. New technology has the capacity to reveal highly personalised travel information to tourists to enable them to access local-ised space–time opportunities that demand less travel and are responsive to ongoing adjustments.

Mobile social networking tools are emerging to influence behaviour change in transport leading to new forms of transport networks that make visible other people, their transport plans, objects in transit and modes of transport (Davies *et al.*, 2012). Such systems will have implications for the temporal planning of travel and subsequent travel behaviour. To this end we need to better understand how time perception and use, which varies across cultural and demographic groups (Adams and van Eerde, 2010; Spears and Amos, 2012), impacts on sus-tainable mobility patterns. The data also raise questions about a more structured time encountered in other tourist settings that requires further research.

Acknowledgements

This work was partially funded by RCUK as part of the Sixth Sense Transport (6ST) Project.

References

Adam, B. (1995). *Timewatch: The social analysis of time*. Cambridge: Polity Press.

Adams, S.J.M. and van Eerde, W. (2010). Time use in Spain: Is polychronicity a cultural phenomenon? *Journal of Managerial Psychology*, 25(7), 764–776.

Anable, J. and Gatersleben, B. (2005). All work and no play? The role of instrumental and affective factors in work and leisure journeys by different travel modes. *Trans-portation Research Part A: Policy and Practice*, 39(2–3), 163–181.

Ashden. (2012). *Liftshare.com, UK. Ashden Case Study. Summary*. Retrieved from: www.ashden.org/files/Liftshare%20full%20winner.pdf.

Bærenholdt, J.O., Haldrup, M., Larsen, J. and Urry, J. (2004). *Performing tourist places*. Aldershot: Ashgate.

Dansal, H. and Eiselt, H.A. (2004). Exploratory research of tourist motivations and plan-ning. *Tourism Management*, 25, 387–396.

Becken, S. (2002). Analysing international tourist flows to estimate energy use associated with air travel. *Journal of Sustainable Tourism*, 10(2), 114–131.

Buscher, M. and Urry, J. (2009). Mobile methods and the empirical. *European Journal of Social Theory*, 12(1), 99–116.

Cass, N., Shove, E. and Urry, J. (2004). Transport infrastructures: A social-spatial-temporal model. In D. Southerton, H. Chappells and B. Van Vliet (eds), *Sustainable consumption: The implications of changing infrastructures of provision* (pp. 113–129). Cheltenham: Edward Elgar.

Castells, M. (2000). *The rise of the network society* (2nd edn). Oxford: Blackwell.

Clawson, M. and Knetsch, J.L. (1966). *Economics of outdoor education*. Baltimore: The Johns Hopkin University Press.

Connell, J. and Page, S.J. (2008). Exploring the spatial patterns of car-based tourist travel in Loch Lomond and Trossachs National Park, Scotland. *Tourism Management*, *29*(3), 561–580.

Couclelis, H. (2009). Rethinking time geography in the information age. *Environment and Planning A*, *41*, 1556–1575.

Cresswell, T. (2003). Landscape and the obliteration of practice. In K. Anderson, M. Domosh, S. Pile, N. Thrift (eds), *Handbook of cultural geography* (pp. 269–281). London: Sage.

Csikszentmihalyi, M. (2002). *Flow: The classic work on how to achieve happiness*. London: Random House.

Davies, K. (2001). Responsibility and daily life: Reflections over timespace. In J. May and N. Thrift (eds), *Timespace: Geographies of temporality* (pp. 133–148). London: Routledge.

Davies, N., Lau, M., Speed, C., Cherrett, T., Dickinson, J. and Norgate, S.H. (2012). Sixth sense transport: Challenges in supporting flexible time travel. *The 13th international workshop on mobile computing systems and applications*, 28–29 February, San Diego, California. www.hotmobile.org/2012/.

Dickinson, J.E. and Peeters, P. (in press). Time, tourism consumption and sustainable development. *International Journal of Tourism Research*. doi:10.1002/jtr.1893.

Dickinson, J.E. and Robbins, D. (2008). Representations of tourism transport problems in a rural destination. *Tourism Management*, *29*, 1110–1121.

Dickinson, J.E., Ghali, K., Cherrett, T., Speed, C., Davies, N. and Norgate, S. (2012). Tourism and the smartphone app: Capabilities, emerging practice and scope in the travel domain. *Current Issues in Tourism*. doi:10.1080/13683500.2012.718323.

Dickinson, J.E., Lumsdon, L. and Robbins, D. (2011). Slow travel: Issues for tourism and climate change. *Journal of Sustainable Tourism*, *19*(3), 281–300.

Edensor, T. (2010). Introduction: Thinking about rhythm and space. In E. Edensor (ed.), *Geographies of rhythm: Nature, place, mobilities and bodies* (pp. 1–8). Farnham: Ashgate.

Edwards, D. and Griffin, T. (2013). Understanding tourists' spatial behaviour: GPS tracking as an aid to sustainable destination management. *Journal of Sustainable Tourism*, *21*(4), 580–595.

Elsrud, T. (1998). Time creation in traveling: The taking and making of time among women backpackers. *Time and Society*, *7*(2), 309–334.

EuroStat (2012). *Tourism statistics at regional level*. http://epp.eurostat.ec.europa.eu/statistics_explained/index.php/Tourism_statistics_at_regional_level#Camping.

Filimonau, V., Dickinson, J.E., Cherrett, T., Davies, N., Norgate, S. and Speed, C. (2013, January). Rethinking travel networks: Mobile media and collaborative travel in the tourism domain. *Universities Transport Studies Group Conference Proceedings*, Oxford.

Frändberg, L. (2008). Paths in transnational time-space: Representing mobility biographies of young Swedes. *Geografiska Annaler*, *90*(1), 17–28.

Germann Molz, J. (2010). Performing global geographies: Time, space, place and pace in narratives of round-the-world travel. *Tourism Geographies*, *12*(3), 329–348.

Gössling, S., Ceron, J.P., Dubois, G. and Hall, C.M. (2009). Hypermobile travelers. In S. Gössling and P. Upham (eds), *Climate change and aviation. Issues, challenges and solutions* (pp. 131–150). London: Earthscan.

Gössling, S., Scott, D., Hall, C.M., Ceron, J.P. and Dubois, G. (2012). Consumer behaviour and demand response of tourists to climate change. *Annals of Tourism Research*, *39*(1), 36–58.

Greener Festival. (2012). *Traffic congestion and travel.* www.agreenerfestival.com/traffic-congestion-travel/.

Haldrup, M. (2004). Laid-back mobilities: Second-home holidays in time and space. *Tourism Geographies*, *6*(4), 434–454.

Hibbert, J.F., Dickinson, J.E. and Curtin, S. (2013). Understanding the influence of interpersonal relationships on identity and tourism travel. *Anatolia*. doi:10.1080/13032917.2012.762313.

Humphreys, L. (2010). Mobile social networks and urban public space. *New Media and Society*, *12*(5), 763–778.

Hunter, C. (2002). Sustainable tourism and the touristic ecological footprint. *Environment, Development and Sustainability*, *4*, 7–20.

Jäckel, M. and Wollscheid, S. (2007). Time is money and money needs time? A secondary analysis of time-budget data in Germany. *Journal of Leisure Research*, *39*(1), 86–108.

Jarvis, H. (2005). Moving to London Time: Household co-ordination and the infrastructure of everyday life. *Time & Society*, *14*(1), 133–154.

Jeng, J. and Fesenmaier, D.R. (2002). Conceptualizing the travel decision-making hierarchy: A review of recent developments. *Tourism Analysis*, *7*(1), 15–32.

Klein, O. (2004). Social perception of time, distance and high-speed transportation. *Time & Society*, *13*(2/3), 245–263.

Kwan, M.-P. (2007). Mobile communications, social networks, and urban travel: Hypertext as a new metaphor for conceptualizing spatial interaction. *The Professional Geographer*, *59*(4), 434–446.

La Lopa, J.M. and Day, J. (2011). Pilot study to assess the readiness of the tourism industry in Wales to change to sustainable tourism business practices. *Journal of Hospitality and Tourism Management*, *18*(1), 130–139.

Larson, J., Urry, J. and Axhausen, K.W. (2007). Networks and tourism: Mobile social life. *Annals of Tourism Research*, *34*(1), 244–262.

Latham, A. (2003). Research, performance, and doing human geography: Some reflections on the diary-photograph, diary-interview method. *Environment and Planning A*, *35*(11), 1993–2017.

Lefebvre, H. (2004). *Rhythm analysis: Space, time, and everyday life.* London: Continuum.

Lemos, A. (2010). Post-mass media functions, locative media, and informational territories: New ways of thinking about territory, place, and mobility in contemporary society. *Space and Culture*, *13*(4), 403–420.

Lindquist, J.D. and Kaufman-Scarborough, C.J. (2007). The polychromic-monochronic tendency model PMTS scale development and validation. *Time & Society*, *16*, 253–285.

Line, T., Jain, J. and Lyons, G. (2011). The role of ICTS in everyday mobile lives. *Journal of Transport Geography*, *19*(6), 1490–1499.

Ling, R. (2004). *The mobile connection: The cell phone's impact on society.* San Francisco: Morgan Kaufmann.

Mallet, W.J. and McGuckin, N. (2000). Driving to distractions: Recreational trips in private vehicles. *Transportation Research Record*, *1719*, 267–272.

Matos, R. (2004). Can slow tourism bring new life to alpine regions? In K. Weiermair and C. Mathies (eds), *The tourism and leisure industry. Shaping the future* (pp. 93–103). Binghamton, NY: Haworth Hospitality Press.

May, J. and Thrift, N. (2001). *Introduction.* In J. May and N. Thrift (eds), *Timespace: Geographies of temporality* (pp. 1–46). London: Routledge.

Metz, D. (2008). *The limits to travel. How far will we go?* London: Earthscan.

Merriman, P. (2012). Human geography without time–space. *Transactions of the Institute of British Geographers, 37*(1), 13–27.

Middleton, J. (2009). "Stepping in time": Walking, time, and space in the city. *Environment and Planning A, 41*, 1943–1961.

Mullins, P.M. (2009). Living stories of the landscape: Perception of place through canoeing in Canada's north. *Tourism Geographies, 11*(2), 233–255.

Neutens, T., Schwanen, T. and Witloz, F. (2011). The prism of everyday life: Towards a new research agenda for time geography. *Transport Reviews, 31*(1), 25–47.

Norgate, S.H. (2006). *Beyond 9 to 5: Your life in time.* London: Weidenfeld & Nicolson.

Page, S.J. (2005). *Transport and tourism: Global perspectives* (2nd edn). Harlow: Pearson.

Poumanyvong, P., Kaneko, S. and Dhakal, S. (2012). Impacts of urbanization on national transport and road energy use: Evidence from low, middle and high income countries. *Energy Policy, 46*, 268–277.

Richards, G. (1998). Time for a holiday? Social rights and international tourism consumption. *Time and Society, 7*(1), 145–160.

Roxburgh, S. (2004). There just aren't enough hours of the day: The mental health consequences of time pressure. *Journal of Health and Social Behaviour, 45*, 115–131.

Scott, D., Peeters, P. and Gössling, S. (2010). Can tourism deliver its aspirational greenhouse gas emission reduction targets? *Journal of Sustainable Tourism, 18*, 393–408.

Sharpley, R. and Jepson, D. (2011). Rural tourism. A spiritual experience? *Annals of Tourism Research, 38*(1), 52–71.

Sheller, M. and Urry, J. (2006). The new mobilities paradigm. *Environment and Planning A, 38*, 207–226.

Solomon, M.R., Bamossy, G., Askegaard, S. and Hogg, M.K. (2010). *Consumer behaviour: A European perspective.* Harlow: Financial Times/Prentice Hall.

Southerton, D. (2006). Analysing the temporal organization of daily life: Social constraints, practices and their allocation. *Sociology, 40*(3), 435–454.

Southerton, D., Shove, E. and Warde, A. (2001). *"Harried and hurried": Time shortage and the co-ordination of everyday life.* Manchester: Centre for Research on Innovation and Computing, University of Manchester.

Spears, N. and Amos, C. (2012). Revisiting western time orientations. *Journal of Consumer Behaviour, 11*, 189–197.

Speed, C. (2011) Kissing and making up: Time, space and locative media. *Digital Creativity, 22*(4), 235–246.

Stein, K. (2012). Time off: The social experience of time on vacation. *Qualitative Sociology, 35*, 335–353.

Stewart, S.I. and Vogt, C.A., (1999). A case-based approach to understanding vacation planning. *Leisure Sciences, 21*(2), 79–95.

United Nations World Tourism Organisation – United Nations Environment Programme – World Meteorological Organisation. (2008). *Climate change and tourism: Responding to global challenges.* Madrid: UNWTO.

Urry, J. (1994). Time, leisure and social identity. *Time and Society, 3*(2), 131–149.

Wajcman, J. (2008). Life in the fast lane? Towards a sociology of technology and time. *The British Journal of Sociology, 59*(1), 59–77.

Warnken, J., Bradley, M. and Guilding, C. (2004). Exploring methods and practicalities of conducting sector-wide energy consumption accounting in the tourist accommodation industry. *Ecological Economics, 48*(1), 125–141.

Watkins, H. (2003). Fridge stories: Three geographies of the domestic refrigerator. In M. Hard, A. Losch and D. Verdicchio (eds), *Transforming spaces. The topological turn in technology studies*, Darmstadt, Germany.

Wong, J.Y. and Yeh, C. (2009). Tourist hesitation in destination decision making. *Annals of Tourism Research, 36*, 6–23.

Zalatan, A. (1996). The determinants of planning time in vacation travel. *Tourism Management, 17*, 123–131.

10 Individual lifestyle as a determinant for sustainable tourism mobility

A transport planning perspective

Werner Gronau

Introduction: the relevance of re-thinking tourism-related transport needs

In the past decades we have seen a constant increase in transport demand, at the same time the share of tourism and leisure-related transportation has grown considerably over the last decades both in absolute and relative terms (Schiefelbusch *et al.*, 2007). Taking Germany as an example around 48 per cent of passenger transport performance is caused by leisure activities (Bundesregierung, 2001). Beside the simple increase of demand one can witness a shift towards more unsustainable transport modes, as for example the increasing air travel or as Chapmann (2007, p. 357) puts it: "All transport sectors are experiencing expansion and unfortunately there is a general trend that the modes which are experiencing the most growth are also the most polluting." However, this growth is a result of the simple fact that holidays and short breaks have become a part of today's societies. Whereas in the past travelling used to be a privilege, nowadays tourism has become a mass phenomenon of the western world. Exactly this increase has caused several debates and fostered a broad academic discussion on how to establish a more sustainable tourism industry. However, most studies focus on a local destination level and do not consider the related transport demand. Even studies examining the environmental impact of tourism frequently only consider the effects of tourism at the holiday destination. Their main objective is the development of a sustainable, green or gentle tourism and the protection of ecologically sensitive regions (e.g. Griffin, 2003); usually, the effects of travelling to these places are neglected. Therefore, the starting point for debating more sustainable tourism mobility is relatively complex. There are numerous studies on the environmental impact of transport and on the environmental impact of tourism, but as Schiefelbusch *et al.* suggest (2007, p. 94), "For some time transport research has considered leisure travel as a 'residual' part of mobility". Therefore only little attention and resources have been spent to understand this aspect of people's mobility. As a consequence there is a clear lack of understanding of what the determinants for transport needs in leisure and tourism actually are.

In early tourism literature, transport was understood as an accessibility provider for a destination, in other words a tool to support tourism development.

Therefore there is a clear deficit in understanding the tourism-related transport demand, as well as the specifics of travel mode choice in leisure and tourism. Both transport volume and mode choice are crucial when it comes to influencing transport demand in leisure and tourism, and aiming for a more sustainable tourism mobility. Reorganising the way people travel in their leisure and tourism time towards a more sustainable way is therefore a key challenge. The definition of transportation problems, the influencing factors and the possible solutions for this task have to be reconsidered. The new challenge is to set focus on the psychology of the transport user who has to be perceived as an active agent in the transport system. Thus, transport policy measures will be more successful when taking into account user capabilities and habits as done by transport psychologists. Psychological studies on travel mode choice focus on the influence of psychological determinants of mobility behaviour that are active in the situation when the decision is made, such as personal norms or travel mode choice habits (Haustein *et al.*, 2009). In order to understand and better manage this complex task, one has to understand the way transport planning has historically explained travel mode choice and the rather mechanical understandings of transport demand in general.

The "individual" in transport planning

In the 1980s transport planning mainly focused on transport infrastructure planning. During this time, transport demand was supposed to be managed by a steadily increasing amount of infrastructure, and by doing so, increasing transport capacities. Rye (1998) describes this phase in the case of the UK as a "predict and provide" mentality of officials aiming at building enough roads to satisfy demand. Little effort was spent on analysing transport behaviour. Therefore, transport planning focused on highly aggregated quantitative statistical models. Traffic was understood as a flow of vehicles, so transport planning was expected to follow the rules of fluid mechanics to solve existing problems concerning the "particle flow" of existing transport infrastructure. Transport demand subsequently was not seen as a result of individual activities, but as a static system. Despite of the fact that already in the 1970s Fishbein and Ajzen (1975) developed the attitude theory within the context of social psychology there was only limited response in the field of transport. Nevertheless it was the starting point for travel behaviour analysis and research (e.g. Golob *et al.*, 1979; Koppelman and Lyon, 1981).

The 1990s were characterised by an increasing number of models aiming at a better understanding and into deep analysis of transport behaviour, such as transport mode choice. One example of such an approach is the development of stated preference methods (e.g. Hensher, 1994; Louviere *et al.*, 2000). In many cases the common theoretical foundation was random utility theory (Ben-Akiva and Lerman, 1985), an extension of the expected utility theory that was the foundation of micro-economic choice theory. Ben-Akiva *et al.* (1999), Gärling *et al.* (1998) and Svenson (1998) argue that more recent developments of attitude theory, as well as of behavioural judgement and decision-making research, should also be included in transport planning discussions. Li and Hensher (2011) and van de Kaa (2010) provide reviews of such recent developments. In addition to these approaches and finalising the work on behaviourally realistic computational process models of activity-based travel choice (e.g. Axhausen and Gärling, 1992; Gärling, 2004; Ettema and Timmermans, 1997; Jones *et al.*, 1983) was a parallel development starting in the mid-1990s.

Scholars outside the transport engineering field have endeavoured to develop a broader more sociological, cultural and psychological understanding of mobility. Amongst these approaches the most relevant may be the extent to which the wider social context of mobility can be examined through the lens of research on lifestyles (Anable, 2005). Through such research, the notion of "mobility styles" has emerged as a pivotal approach to examining the motivations and barriers for the adoption of alternative travel modes. There are several examples of the mobility styles approach, usually focusing on a social-psychological understanding of travel behaviour (Anable, 2005). Such studies stress the attitudinal and/or behavioural differences between individuals, in order to distinguish clusters or groups of travellers (Anable, 2005; Dallen, 2007; Götz *et al.*, 2003; Gronau, 2005). The study of Anable (2005) with National Trust visitors in the United Kingdom examined four car-owning and two non-car owning groups with attendant names such as "complacent car addicts" and "car-less crusaders". The names refer to the social-psychological characteristics of each of the identified clusters. The obvious value of such approaches is the clear description of the characteristics of identifiable clusters of individuals, the motivations for travel behaviour and the opportunity to influence the travel mode choice of each cluster. Indeed, as indicated previously, the mobility styles perspective is one amongst numerous segmentation approaches which have been developed by environmental social scientists in recent years, to explore the ways in which scholars and practitioners might identify and develop social marketing strategies to promote behavioural changes (Darnton and Sharp, 2006).

Towards a "subject-oriented" transport-demand analysis

When taking into account the two previously outlined existing approaches which follow either quantitative computerised process models or highly individualised qualitative mobility style models, the question arises as to how to merge these two concepts. Without a doubt bringing together those two approaches would create numerous opportunities. The weakness of the qualitative models, in terms of quantifying demand or forecasting possible changes, is the strength of quantitative models. At the same time qualitative models provide valuable information on the influencing factors of transport behaviour, which are crucial for managing and planning transport demand. These aspects are of enormous relevance today, when it comes to the development of more sustainable transport behaviours in the leisure and tourism context. Present political and social challenges towards more sustainable tourism mobility pay an increasing attention to "soft" measures to encourage a modal shift from the private car to other more sustainable forms of transport. This is often referred to as travel demand management or mobility management (Lumsdon *et al.*, 2006). Realising all contradictions of today's society, as for example the contradictory relationship between everyday behaviour and existing environmental concerns, one has to conclude that understanding present-day travellers' attitude towards sustainable travel is a critical issue. The modern traveller is described as a hybrid, multi-optional and multi-mobile human being whose behaviour oscillates between rational aspects such as cost-awareness and emotional aspects like pleasure (Pikkemaat, 2004, p. 104 as cited in Prillwitz and Barr, 2012). Therefore any policy aiming at more sustainable travel behaviours has to accept the role of mobility style models and plan its activities accordingly.

This indicates there is a clear need to develop a "subject-oriented" transport planning framework; finding a compromise between quantitative process models and qualitative mobility style approaches. The way towards such an approach has to understand the "subject" as a "collective of aggregated individuals", which means the quantitative utilisation of qualitative mobility style groups. Such an integration of mobility style groups into a quantitative forecasting model first of all asks for a quantification of the share of mobility styles within a given population. The principle compatibility of these two approaches, even in the context of tourism, was already proven by the German research project INVENT – Sustainable Management in the Tourism Sector from 2002 to 2005 (see Götz and Seltmann, 2005), which aimed at a broader understanding of German holidaymakers in order to develop innovative travel that promotes sustainable tourism and fits easily into the mass- and high-volume markets. In order to be able to quantify the potential for these emerging products it was crucial to also answer the question concerning the share of each individual target group within the overall market. In order to find out the preferences of holidaymakers with regards to innovative tourism products, the project collected empirical data on the main motivations of tourists. In an initial phase in-depth interviews were conducted, followed by a representative study including 2,000 interviewees. Based upon this empirical data a target group typology of holidaymakers, interested in innovative and sustainable offerings, was formed. This typology formed the base for designing marketable products and marketing materials. The latter were then tested by the team, using methods taken from qualitative market and social research, before being included in the respective product portfolios of the business partners concerned. Öko-Institut calculated the carbon footprint of different holidaymaker types, which highlighted the potential savings to be made in mass-market travel offerings (see Götz and Seltmann, 2005). Beside the fact that this project bridged the gap between quantitative and qualitative models in the field of transport behaviour it also applied this approach within the field of leisure and tourism and, furthermore, it even utilised the established target groups for forecasting possible carbon reduction in order to quantify possible savings amongst the varying groups through the implementation of more sustainable tourism products. Unfortunately, what the project did not do was to integrate several transport modes within the specification of the target groups; its aims were to create new tourism products and therefore did not focus on travel mode choice and attitudes of the given groups concerning certain interventions.

Conceptualising a "subject-oriented" transport model in the context of leisure and tourism

Investigating travel mode choice is a critical tool for identifying ways to create more sustainable tourism mobilities. Different interventions have to be identified based on the attitude of travellers, but at the same time their impact has to be modelled in order to evaluate their specific effectiveness. Based upon a number of qualitative studies concerning different mobility patterns (see for example Bamberg, 2004; Freitag and Kagermeier 2002; Gärling *et al.* 1998; Heath and Gifford, 2002; Hunecke *et al.* 2001) today's knowledge of the different qualitative factors influencing the mobility behaviour in leisure time has improved steadily. Therefore, the already existing wide knowledge on the quantitative factors has to be supplemented by the qualitative findings in a subject-oriented model. To be able to decide on how to combine these two approaches, the following three pre-conditions and aims are defined: (1) the model has to combine mobility styles groups and classical quantitative forecasting methods; (2) it has to be able to forecast transport demand in the form of differentiated transport modes for a specific site or destination; (3) it has to provide the opportunity to change framework conditions to forecast their impacts.

First, following these specifications there is a need to identify site-related aspects aiming at a description of the mode-specific accessibility of a site or destination. Such aspects include the location situated within local, regional and national transportation networks, the type of given transportation modes available, mode-specific infrastructure (for example availability/amount/type of parking lots) and the frequency of given modes in case of public transport or airline supply. These aspects follow the classical transport planning approach in terms of "objective" quantitative measures influencing travel behaviour; they can easily be operationalised, quantified and measured as demonstrated in many previous cases. In addition to these aspects there has to be a second component to ensure a more disaggregated approach, meaning that the objective aspects are reflected and interpreted by individuals in diverse ways, leading to an individual specific behaviour. That is why these "objective" aspects have to be filtered through a "subjective" component allowing a more disaggregated understanding of the specific situation at certain sites through the eyes of certain target groups. Keeping these preconditions in mind this subjective component has to be introduced by mobility-style groups allowing a certain disaggregation while still being able to provide a quantification of the group-specific demand at certain sites or destinations. Consequently, it is important to identify the lifestyle-based mobility groups that combine the general lifestyle attributes with transport-orientated ones. Due to the specific situation of leisure and tourism-related demand – being rather different from everyday mobility patterns – the targeted groups have to focus on leisure and tourism dimensions, as well as leisure and tourism-specific transport needs. The result of such a course of action will be lifestyle-based mobility groups combining leisure and tourism interests with specific transport behaviour profiles during leisure time and therefore work as a collective of aggregated individuals for a quantitative forecasting model.

Following up on the second aim, lifestyle-based mobility groups offer the opportunity to forecast the frequency of each group at a specific site or destination based upon their specific leisure and tourism profile. Furthermore, it allows us to forecast the specific transport demand and style at a particular site/destination, and the specific volumes based upon the frequency of each group and their transport behaviour there. Each leisure and tourism site or destination can be affiliated with a specific share of each group, depending on its specific product portfolio. Therefore, the character of the specific site or destination has to be evaluated in order to evaluate its attractiveness for the different lifestyle-based mobility groups. At the same time, the transport profile of each group can be paired with site-specific aspects creating a probability for each group to appear within the given framework conditions at each site or destination. This site or destination-specific probability can then be multiplied by the overall share of the group, leading to a quantified demand forecast for a site or destination taking into account specific site characteristics such as accessibility, and at the same time including the subjective interpretation of those site-specific aspects on an aggregated level in form of the mobility groups.

This approach also ensures the third condition by utilising site-specific aspects in order to forecast the frequency of groups at a certain site; of course these site specific aspects might be changed to adapt to a changing situation at the site, such as an increased accessibility.

To sum up this section Figure 10.1 provides an overview on the way to join site-specific objective aspects with group-specific individual aspects. The figure illustrates the inter-relation of the main three dimensions, the "Site" on the left bottom, the "Lifestyle-based mobility groups" on the right bottom and the resulting modal split on the top. The site-specific aspects on the left side, such as accessibility measured in form of a transport index or the attractiveness of a site for specific target groups in form of the content index form the first component of the transport-demand model. The second component, on the right-hand side represented by lifestyle-based mobility groups, contributes to the model a group-specific leisure and tourism orientation and certain transport behaviour. As described before, based on the products provided by a certain destination and the interests of the different group forecasts, one can determine a group-specific frequency at a specific site or destination. Utilising the specific transport framework conditions at a site or destination, the transport behaviour and the specific frequency of the different groups, one can calculate a theoretic site or destination-specific transport demand in terms of a specific modal split.

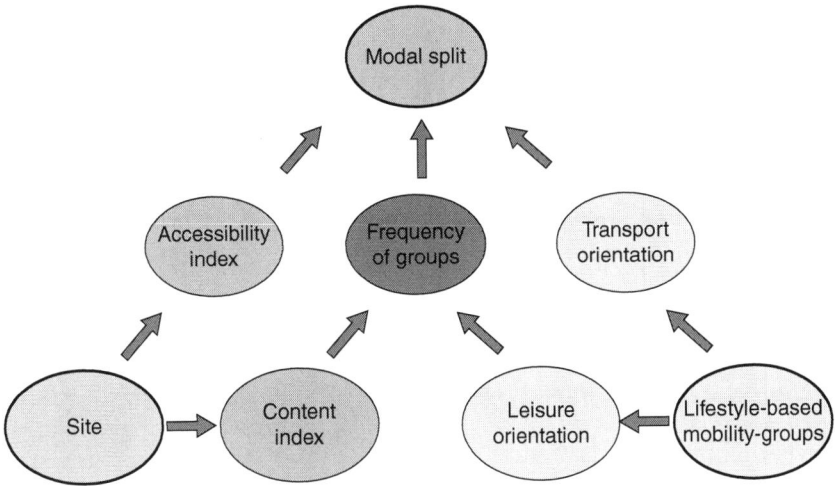

Figure 10.1 Transport-demand simulation based upon mobility groups.

Towards an empirical pilot implementation of a subject-oriented transport model

The following section will provide a brief insight into an empirical pilot implementation of the proposed model in the previous section. This first empirical implementation was part of a research project by the author granted by the German Research Foundation. Basically the project aimed at the development of a transport-demand-forecasting model for large-scale leisure facilities or urban entertainment centres. In more detail, the main objective was to proof the applicability of social-psychological mobility groups in the context of a quantitative transport-demand model. A more detailed description of the project is provided by Gronau (2005). Recognising the limited framework of this chapter, the following aspects of the implementation will be briefly explored:

- creation of leisure mobility groups;
- evaluation of the selectivity of the groups at varying sites;
- forecast of the transport behaviour in this case in the form of a theoretic modal split at given sites; and
- empirical evaluation of the theoretic modal split by an onsite survey.

The first mission to construct leisure mobility groups was already rather challenging due to the question of which dimensions describe the transport behaviour in leisure and tourism in the best way. Building on several empirical studies undertaken in Germany (see Bamberg, 2004; Freitag and Kagermeier, 2002; Gärling *et al.*, 1998; Hunecke *et al.*, 2001), the fun and function aspects were identified as the most important ones in judging transport alternatives for leisure time. Taking these findings into consideration, seven lifestyle-oriented, so-called "mobility groups" were defined according to similarities in their rationale for choosing a transport alternative in leisure time (see Lanzendorf, 2001). The construction of these groups is based on a household survey of 2,000 persons. The mobility groups were created as a result of cluster analysis using a wide variety of different indicators for respondent leisure interests, but also for their preferences towards a transport alternative in leisure time.

Figure 10.2 shows the different mobility groups within a grid system outlining their perception of the relevance of the fun and function factors when choosing a transport alternative. The two dimensions forming the axis of the grid were the result of specifically designed indices summing up several indicators. In the case of the dimension function aspects such as flexibility, reliability, security, speed, etc. were summed up. With reference to the Likert scale utilised during the data collection the dimensions range from –2 to +2, where the x-axis shows the "Function" factor and the y-axis the "Fun" factor. The figure illustrates that even in leisure time, the function factor has the greatest influence on choosing a

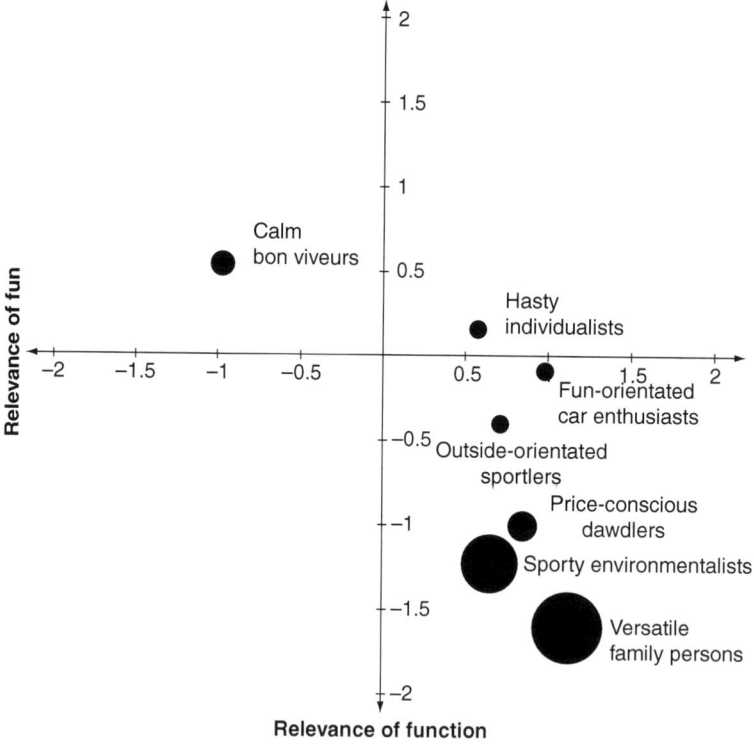

Figure 10.2 Relevance of "fun" and "function" differentiated by mobility groups

transport mode but, for certain groups, the fun factor also plays an important role (see Gronau, 2005). Although there are seven different mobility groups one can easily identify three main types:

- the first type emphasises only the function factor when it comes to choosing a transport alternative in leisure time;
- the second type somehow balances the two factors out; and
- the third type consists of just one group, clearly prioritising the function factor.

Speaking about the share of the different groups, the size of the various circles representing the groups within Figure 10.2 indicates the size of each group, highlighting that groups belonging to functional travellers represent a clear majority.

In order to forecast a specific transport behaviour, as for example in the case of travel mode choice in the leisure and tourism context, the interviewees were also asked to rate different transportation modes by the same aspects considering fun and function. In order to facilitate the comprehension of the reader the results

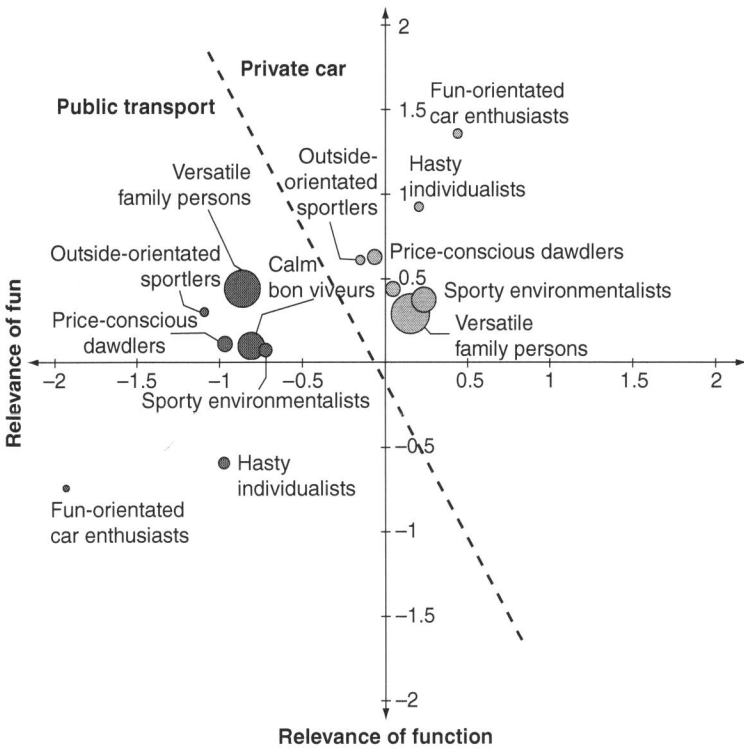

Figure 10.3 Connotations by "mobility groups" on transport alternatives.

were paired for the two main modes, the private car and public transport, illustrated in Figure 10.3. The figure indicates the group-specific association of the transport alternative car (light grey colour) and public transport (black colour). Public transport is perceived by all groups as less functional than the private car, but with reference to the fun factor, both transport modes are perceived quite similar by the majority of the groups. The overall relatively poor result for public transport in the leisure and tourism context is mainly caused by the predominant perception of the lacking function factor. Just two groups, made up of a very small number of persons, assume public transport is also less competitive with regards to the fun factor.

Accepting the perception of the various groups towards different transport alternatives one may assume a certain affinity towards specific transport modes. In the following the term affinity is understood as the combination of the importance of several transport-related factors on the one hand and the group-specific perception of those aspects for given transport alternatives on the other hand. Figure 10.3 therefore also indicates the group-specific affinity. Based on the fact that the affinity towards the private car is not overwhelming for the main part of the groups and the poor perception is mainly due to the lack of the function factor, one can conclude that, given an improvement in public transport provision, the attitude towards both alternatives could be almost the same. Two of the groups, however, show a strong resistance towards public transport use and it is likely that even a highly improved system would not be a real alternative to the private car for these groups.

By constructing the seven leisure mobility groups and briefly outlining their specific transport profile, the second question concerning the everyday relevance of those groups has to be answered. Therefore the evaluation of the selectivity of the groups at varying sites has to be analysed. In order to evaluate this selectivity in the context of real-life situations, an empirical study was performed on eight different leisure facilities. The sites varied by their accessibility and by their character in terms of a pure fun or an education-orientation. In the case of Figure 10.4 the two sites varying in their accessibility are presented for three leisure mobility groups. The Munich zoo and the thermal spa in Erding show different levels of accessibility, while the zoo has a direct subway connection but at the same time a restrictive parking lot management; the thermal spa offers a huge amount of free parking and a shuttle service to the nearby light-rail station every 20 minutes. It does not come as a surprise that throughout all groups the better accessibility of the zoo (first bar) is reflected by an overall higher share of public transport when compared to the thermal spa (second bar). At the same time the figure indicates the relevance of group affinity towards certain transportation modes. Even the poor supply of public transport at the thermal spa generates mode a share of 20 per cent among the "sporty environmentalists" due to their positive attitude towards the green mode. Of course related to a rising public transport quality level, this share rises even more visibly at the zoo. On the other hand the group "Fun-Orientated Car Enthusiasts" shows a significantly lower share of public transport even if there is a high quality service and if there are

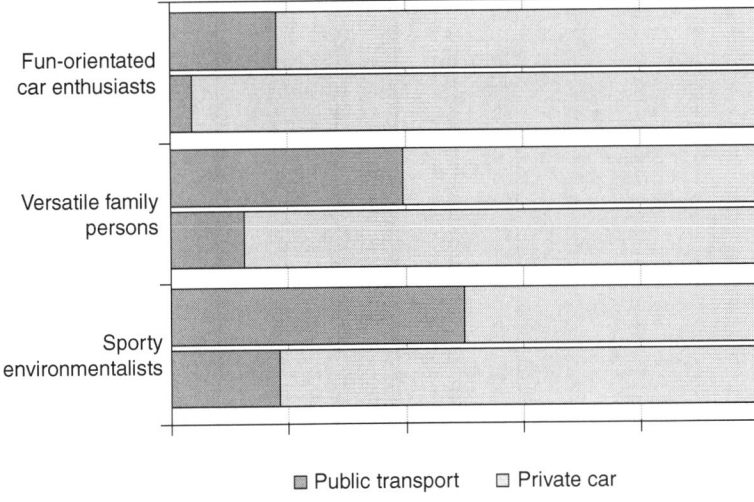

Figure 10.4 Modal split by mobility groups at site.

clear restrictions for parking, which of course fits very well to their specific transport affinity.

Figure 10.4 therefore clearly indicates the expected selectivity with reference to the specific affinity of each leisure mobility group. The use of public transport at both sites is almost twice as high for "Sporty Environmentalists" as it is for "Fun-Orientated Car Enthusiasts".

The additional travel time and effort needed due to the poor public transport supply is only accepted by people showing a clear affinity towards public transport. This selective effect in the use of public transport with reference to the different groups and their specific affinity can be identified in different intensities, but is still significant for all groups at all other locations. These findings lay the basis for the third step to utilise the empirically evaluated mobility groups in the context of a transport-demand model. Therefore, based upon the share of each mobility group at a given site a theoretic modal split was calculated bringing together the volume of a specific group with the site index, providing information on the existing framework conditions such as accessibility and the affinity towards transport alternatives, private car and public transport of each group. This theoretic modal split was finally compared with the one being measured during the empirical research at the given sites. In Figure 10.5 below one can see the deviation of the forecasted and the measured modal split at the given sites. Of course it has to be considered that based on the utilisation of only two transport alternatives the probability of a high difference is limited, but the deviation of maximum 4 per cent is still considerable.

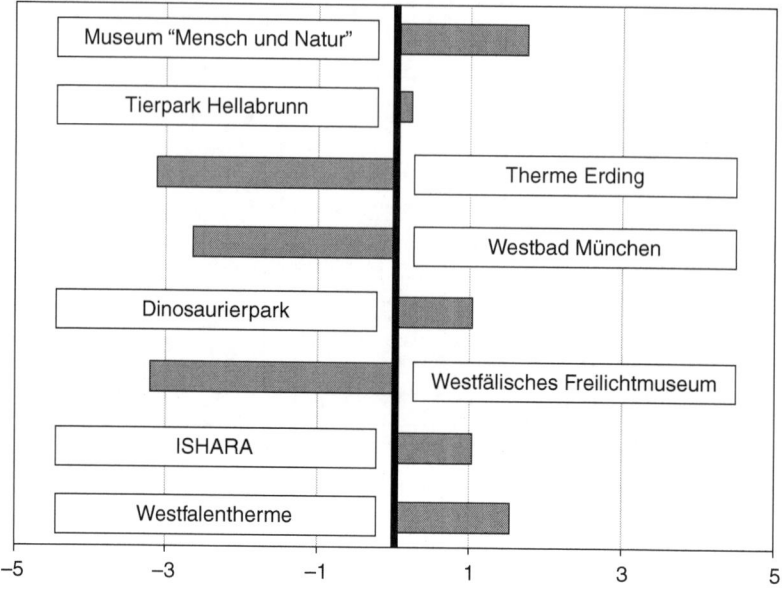

Deviation of modal split in percent

Figure 10.5 Deviation of modal split at all study sites.

Conclusion

The chapter outlined the need for a new understanding of transport demand and individual transport behaviour in the leisure and tourism context by reflecting on traditional approaches in transport planning. Elaborating on the highly aggregated transport models for explaining transport behaviour, the steadily increasing role of individual behaviour models in today's society was also outlined. By doing so, the growing gap between the qualitative individualised behaviour frameworks, especially in the leisure and tourism context, and the quantitative travel-mode-choice models was stressed. By presenting research results interpreting transport demand through the lens of lifestyle-based approaches the author outlines the opportunity for a "subject-orientated" transport-demand model, joining the highly aggregated quantitative models with the individualised qualitative ones. By outlining the main focus of the early mobility-style approaches to focus on a better understanding of transport behaviour, in order to develop target group specific transport products, the author stresses the need for a more analytical utilisation of mobility styles. Elaborating on the work of Lanzendorf (2001), Götz and Seltmann (2005) and Gronau (2005), the article argued that the utilisation of lifestyle-based mobility groups is an appropriate way to create quantitative transport-demand models including individual attitudes and therefore meet the expectations of today's heterogeneous transport behaviour in the leisure and tourism context. The lifestyle-based mobility groups combine leisure and tourism interests with a specific transport behaviour profile, and therefore work as a collective aggregate of individuals for a quantitative forecasting model.

The second part of the chapter concentrated on possible frameworks to operationalise such lifestyle-based mobility groups, as well as ways to utilise them in a quantitative demand model for the leisure and tourism context. Beside a brief overview on the construction of such groups, including the indicators used in the theoretic framework, conditions for the utilisation of such groups in a quantitative forecasting model for a specific site or destination were described. The selectivity of the lifestyle-based mobility groups on a given site is introduced as the possible link to connect the transport affinity of the various groups determining their specific modal split with the modal split on a given site. To evaluate the presented conceptual approach the results of an empirical pilot study were presented. The empirical evidence clearly indicates the selectivity of the lifestyle-based mobility groups at eight different sites. Furthermore the theoretically forecasted modal split rooting from the specific affinity and the frequency of the groups proved to represent the reality to a large degree. Therefore this chapter concludes that future transport-demand modelling in the leisure and tourism context should adhere to subject-orientated transport-demand models including lifestyle-based mobility groups.

References

Anable, J. (2005). "Complacent car addicts" or "aspiring environmentalists"? Identifying travel behaviour segments using attitude theory. *Transport Policy*, *12*(1), 65–78.

Axhausen, K.W. and Gärling, T. (1992). Activity-based approaches to travel analysis: conceptual frameworks, models, and research problems. *Transport Reviews*, *12*(4), 323–341.

Bamberg, S. (2004). Sozialpsychologische Handlungstheorien in der Mobilitätsforschung. In H. Dalkmann, M. Lanzendorf and J. Scheiner, J. (eds), *Verkehrsgenese* (pp. 51–70). Mannheim: Metagis.

Ben-Akiva, M. and Lerman, S.R. (1985). *Discrete choice analysis*. Cambridge: MIT Press.

Ben-Akiva, M., McFadden, D., Gärling, T., Gopinath, D., Bolduc, D., Borsch-Supan, D., Delquie, P., Larichev, O., Morikawa, T., Polydoropoulou, A., Rao, V. (1999) 'Extended framework for modeling choice behavior'. *Marketing Letters* 10, pp. 187–203.

Bundesregierung (2001). *Entwurf für die nationale Nachhaltigkeitsstrategie der BRD*. Government publication. Berlin, Germany.

Chapmann, L. (2007). Transport and climate change: A review. *Journal of Transport Geography*, *15*(5), 354–367.

Dallen, J. (2007). Sustainable transport, market segmentation and tourism: The Looe Valley branch line railway, Cornwall, UK. *Journal of Sustainable Tourism*, *15*(2), 180–199.

Darnton, A. and Sharp, V. (2006). *Segmenting for sustainability reports 1 and 2*. Didcot: Social Marketing Practice.

Ettema, D. and Timmermans, H.P.J. (1997). *Activity-based approaches to travel analysis*. Oxford: Pergamon.

Fishbein, M. and Ajzen, I. (1975). *Belief, attitude, intention, and behavior: An introduction to theory and research*. Reading, MA: Addison-Wesley.

Freitag, E. and Kagermeier, A. (2002). Multiplex-Kinos als neues Angebotselement im Freizeitmarkt. In A. Steinecke (ed.), *Tourismusforschung in Nordrhein-Westfalen: Ergebnisse – Projekte – Perspektiven* (pp. 43–55). Paderborn: Paderborner Geographische Studien zu Tourismusforschung.

Gärling, T. (2004). The feasible infeasibility of activity scheduling. In M. Schreckenberg and R. Selten (eds) *Human behaviour and traffic networks* (pp. 231–250). Springer: Berlin Heidelberg.

Gärling, T., Gillholm, R. and Gärling, A. (1998). Re-introducing attitude theory in travel behaviour research: The validity of an interactive interview procedure to predict car use. *Transportation, 25*, 147–167.

Golob, T.F., Horowitz, A.D. and Wachs, M. (1979). Attitude–behavior relationships in travel demand modeling. In D.A. Hensher and P.R. Stopher (eds) *Behavioral travel demand modelling* (pp. 739–757). London: Croom Helm.

Götz, K. and Seltmann G. (2005). Urlaubs- und Reisestile – ein Zielgruppenmodell für nachhaltige Tourismusangebote. In *ISOE-Studientexte*, Nr. 12. Frankfurt am Main.

Götz, K., Loose, W., Schmied, M. and Schubert, S. (2003). *Mobility styles in leisure time*. Paper presented at the 10th International Conference on Travel Behaviour Research, Lucerne (10–15 August).

Griffin, T. (2003). An optimistic perspective on tourism's sustainability. In R. Harris, T. Griffin and P. Williams (eds) *Sustainable tourism: A global perspective* (pp. 24–34). Oxford: Elsevier–Butterworth–Heinemann.

Gronau, W. (2005). *Freizeitmobilität und Freizeitstile. Ein praxisorientierter Ansatz zur Modellierung des Verkehrsmittelwahlverhaltens an Freizeitgroßeinrichtungen*. Mannheim: Metagis.

Haustein, S., Klockner, C.A. and Blobaum, A. (2009). Car use of young adults: The role of travel socialization. *Transportation Research Part F, 12*(2), 168–178.

Heath, Y. and Gifford, R. (2002). Extending the theory of planned behaviour: Predicting the use of public transport. *Journal of Applied Social Psychology, 32*(10), 2154–2189.

Hensher, D.A. (1994). Stated preference analysis of travel choices: The state of the practice. *Transportation, 21*(2), 107–133.

Hunecke, M., Blobaum, A., Matthies, E. and Höger, R. (2001). Responsibility and environment: Ecological norm orientation and external factors in the domain of travel mode choice behavior. *Environment and Behavior, 33*(6), 845–867.

Jones, P., Dix, M.C., Clarke, M.I. and Heggie, I.G. (1983). *Understanding travel behavior*. Aldershot, UK: Gower.

Koppelman, F. and Lyon, P.K. (1981). Attitudinal analysis of work/school travel. *Transportation Science, 15*(3), 233–254.

Lanzendorf, M. (2001). *Freizeitmobilität: Unterwegs in Sachen sozial ökologischer Mobilitätsforschung*. Materialien zur Fremdenverkehrsgeographie, Trier.

Li, Z. and Hensher, D.A. (2011). Prospect theoretic contributions in understanding traveller behaviour: A review and some comments. *Transport Reviews, 31*(1), 97–115.

Louviere, J.J., Hensher, D.A. and Swait, J. (2000). *Stated choice methods: Analysis and applications*. Cambridge: Cambridge University Press.

Lumsdon, L., Downward, P. and Rhoden, S. (2006). Transport for tourism: Can public transport encourage a modal shift in the day visitor market? *Journal of Sustainable Tourism, 14*(2), 139–156.

Prillwitz, J. and Barr, S. (2012). Green travellers? Exploring the spatial context of sustainable mobility styles. *Applied Geography*, *32*(2), 798–809.

Rye, T. (1998). *Can we make a business case for employer transport plans? Reducing traffic in cities: Avoiding the transport time bomb.* Paper presented at the Third Car Free Cities Conference, Edinburgh.

Schiefelbusch, M., Jain, A., Schäfer, T. and Müller, D. (2007). Transport and tourism: Roadmap to integrated planning developing and assessing integrated travel chains. *Journal of Transport Geography*, *15*(2), 94–103.

Svenson, O. (1998). The perspective from behavioural decision theory on modeling travel choice. In T. Gärling, T. Laitila and K. Westin (eds) *Theoretical foundations of travel choice modelling* (pp. 141–172). Oxford: Pergamon.

Van de Kaa, E.J. (2010). Applicability of an extended prospect theory to travel behavior research: A meta-analysis. *Transport Reviews*, *30*(6), 771–804.

11 Developing a long-term global tourism transport model using a behavioural approach

Implications for sustainable tourism policy making

Paul Peeters

Introduction

Tourism emits 5 per cent of anthropogenic carbon dioxide emissions of which about 75 per cent is caused by tourism transport (Scott *et al.*, 2010). If historic developments continue, it will be very difficult for tourism to significantly reduce its emissions to a sustainable level (Scott *et al.*, 2010). Technology-based efficiency improvements have so far been outpaced by volume and demand growth (Ch'eze *et al.*, 2011; Owen *et al.*, 2010; Sgouridis *et al.*, 2010). Therefore, changes in demand and travel behaviour will be inevitable to achieve sustainable tourism development with respect to climate change. Further, most current tourism studies cover only international trips, just 16 per cent of all global tourism trips (Peeters and Dubois, 2010). Finally, a long-term horizon is needed, up to at least the year 2100 in most climate scenarios (Girod *et al.*, 2009; Girod *et al.*, 2012; IPCC, 2000; Rogelj *et al.*, 2011) and even up to 2300 (Moss *et al.*, 2010). The main reasons for such a long-term focus in climate change scenarios are the "long-term (decades to centuries) trends in energy- and land-use patterns" and because of "the slow response of the climate system (centuries) to changing concentrations of greenhouse gases" (Moss *et al.*, 2010, p. 748). Most existing tourism demand models (Lim, 1997) and many tourism scenario studies cover only time horizons of 15–20 years (e.g. Forum for the Future, 2009; Schwaninger, 1984; UNWTO, 2011; WTO, 1998). Only a few studies take wider horizons, all dedicated to tourism and climate change (Ceron and Dubois, 2007, Mayor and Tol, 2010; Müller and Weber, 2007). Suitable system-based models for global tourism do not exist. Econometric models find increasing validity problems when describing longer-term futures; the coefficients defining such models are statistically derived, but not necessarily founded in the real world mechanisms of behaviour.

This chapter's goal is to create a tourism travel behaviour model founded in system dynamics, product diffusion and psychological mechanisms. System dynamics can model systems that lack data, proven theoretical foundations and need longer simulation periods (Sterman, 2000). The chapter explores a way to develop a novel tourism behaviour model that describes travel behaviour in terms of trips and distances travelled per transport mode at the global scale. Tourism's CO_2 emissions are, for a given level of technology, determined by trip numbers, distances travelled and transport mode (Peeters and Dubois, 2010). Therefore the model must provide estimates of trip numbers and distances per transport mode. Important model inputs are travel cost, travel time, income distribution, GDP/capita and population. Secondary inputs are transport infrastructure and technology that will affect both travel cost and travel time. The behavioural model has been created and tested with a dynamic version of the Global Tourism Transport Model (GTTM[dyn]). Two versions of GTTM[dyn] preceded the dynamic version: a basic version, GTTM[bas] programmed with Excel, with linear extrapolations and an advanced version, GTTM[adv], programmed in Powersim Studio (version 7), mainly based on linear projections but with automatic scenario generation capabilities used for back-casting towards certain emission goals. The GTTM[bas] and GTTM[adv] models are described by Dubois *et al.* (2011) and Peeters and Dubois (2010).

The ultimate goal of the GTTM[dyn] model is to provide insights into the impacts of tourism on greenhouse gas emissions and the effectiveness of policies to mitigate those emissions. The model will cover the period up to the year 2100. A consequence of that long time span is that we will need to calibrate the model over a similar period, i.e. from 1900 to 2005. The model must be able to handle the development of a completely new transport mode, civil air transport, that became available from c. 1920 (Ananthasayanam, 2003). Furthermore, GTTM[dyn] should be able to handle a wide range of policies governing travel cost, time or speed, infrastructure capacity and psychological factors in decision-making processes of tourists (Schäfer, 2012). The long-term and global character of GTTM[dyn] forms a challenging combination to the behavioural (demand) part of the model (Schäfer, 2012).

A common approach in transport modelling is the "four-stage" model (Bates, 2008). The stages are trip generation, trip distribution, modal split and assignment to the grid or infrastructure. In GTTM[dyn] we need the trip generation, distribution and mode-choice stages, but not the grid assignment stage, as detailed global networks are not defined in the model for the main transport modes. In most transport models *trip generation* is a function of population characteristics including income, age, household and trip properties such as motive. Generalised cost (a combination of cost and monetised travel time and sometimes discomfort) is ideally taken into account, but often ignored (Bates, 2008). Trip distribution and modal choice generally are modelled as (multinomial) logit models (Bates, 2008). Multinomial logit models are used in many studies for tourism demand (Huybers, 2003; Lyons *et al.*, 2009; Nicolau, 2008) and tourism transport demand (Bieger *et al.*, 2007; Pettebone *et al.*, 2011). Such models determine the probability of choice for each alternative using an exponential function of utility (Morley, 1994; Papatheodorou, 2006).

Another line of modelling is based on the use of constant elasticity for travel cost and travel time (Schäfer, 2012). Schäfer (2012) shows that most large-scale transport models use a constant elasticity of substitution (CES) or price elasticity as the basic demand function, and in some cases, additionally, a logit type of model to govern distribution of trips over transport modes. Distances are generally determined from distances between (world) regions as given by Schäfer (2012, p. 31). The problem with elasticity-based models is that elasticities are more a statistical artefact than a factor that represents any specific "psychological" behaviour. Elasticities differ when taken over different time periods and general validity is low which is shown by the very wide range of values obtained from different studies for the same kind of behaviour, e.g. choice between air and car transport (Oum *et al.*, 2008).

The kinds of modelling described above are founded in the standard economic model (SEM). The main axioms of SEM are that economic agents make rational decisions, are motivated by utility maximisation, are purely selfish, ignore the impact on others' utility, are Bayesian probability operators, have consistent time preferences (i.e. the discount rate is constant over time) and consider all income and assets to be completely fungible or freely interchangeable (Wilkinson, 2008, p. 5). Mounting criticism of SEM claims that almost none of the above axioms seems to be valid in the real world and result in different strands of thinking like behavioural economics (Wilkinson, 2008) including prospect theory (Kahneman and Tversky, 1979), evolutionary economics (Dopfer, 2005) and ecological economics (Daly and Farley, 2004). It seems risky, specifically in the context of a systems model for a long-term analysis, to ignore known discrepancies in human economic behaviour. Therefore, the behavioural model of GTTM^{dyn} has been founded on insights from prospect theory (Kahneman, 2011; Kahneman and Tversky, 1979) as will be further elaborated in the next section.

The model

An overview of GTTM^{dyn}

The GTTM^{dyn} is a system dynamics simulation model programmed in Powersim Studio 9.2. System dynamic models (SDM) are based on stock and flow structures, e.g. the number of adopters of a certain product is the result of the flow from potential adopters to adopters (Forrester, 1971; Sterman, 2000). Another basic characteristic of SDM is the ability to easily add feedback loops like the effect that adopters may have on the awareness of potential adopters of the existence of the product. This makes it suitable to construct models beyond normal economic equilibrium modelling. SDM is often used where there is a lack of basic theory, lack of detailed data, complexity and flawed cognitive maps (e.g. ignoring >80 per cent of trips – those that are domestic – in most tourism studies), all of which play a role in the assessment of long-term tourism behaviour (Sterman, 2000).

From the introduction, the requirements for the GTTM^{dyn} demand model are that it is:

1   able to handle both international and domestic tourism trips independent of
    geographical regions;
2   able to handle completely new choice options (like the emergence of avi-
    ation in the 1920s);
3   based on psychological mechanisms governing travel behaviour rather than
    pure econometric/statistical relations;
4   able to show effects of long-term policies changing travel cost, travel time,
    infrastructure capacity and psychological factors in choice behaviour; and
5   able to deliver long-term policy analysis from 1900 to 2100 (Lempert *et al.*,
    2003).

The first requirement is fulfilled by defining the model in trips per transport
mode and distance class rather than trying to model all flows between and within
all countries in the world. Global trip generation is based on global income distribution and population size (Peeters and Landré, 2012). The second requirement is solved by choosing a product diffusion model as the core of the
behavioural model, as proposed by Bass (1969). The "Bass diffusion model"
assumes that new product diffusion starts with commercial adoption caused by
commercial activities and is then gradually taken over by a social adoption
(word-to-mouth) mechanism until the market is saturated. In GTTM^{dyn} the
number of potential adopters, the "reservoir" that commercial adoption acts
upon, is a function of global income distribution. The third requirement is implemented by using the psychological value (PV) from prospect theory rather than
linear utilities. In GTTM^{dyn} the PV of generalised travel cost (cost plus weighted
time) and distance (as attractor) is used. Choice probabilities are estimated using
PV in the exponential form of multinomial logit models (Bates, 2008; Ortúzar
and Willumsen, 2011; Papatheodorou, 2006). These probabilities and PVs
provide the Bass model with growth factors for, respectively, commercial and
social adoption. In this way the Bass model is made sensitive to changes in
income, population size, travel cost and time and, thus, all social–economic scenarios and policies affecting these parameters (fourth requirement). As we have
made both the Bass models more dynamic, accounting for long-term changes in
income and population, and have founded the behaviour model more in psychological theory, we feel GTTM^{dyn} is well equipped to fulfil requirement 5 as well.

    Figure 11.1 shows the general layout of the behavioural model in GTTM^{dyn}
(thus, not the whole GTTM^{dyn} which is much more complex). The main (sub-)
models are:

1   A global *Trip generation model* which calculates global number of trips as a
    function of GDP/capita distribution (Peeters and Landré, 2012). The GDP
    and population data are exogenous from several historic databases and
    future scenarios (see supplementary data file 1 to the web-based version of
    this chapter at www.tandfonline.com/toc/rsus20/21/).
2   Three *Bass models*, one per transport mode. The diffusion of transport
    modes is based on a Bass model approach (Bass, 1969, 2004; Bass *et al.*,

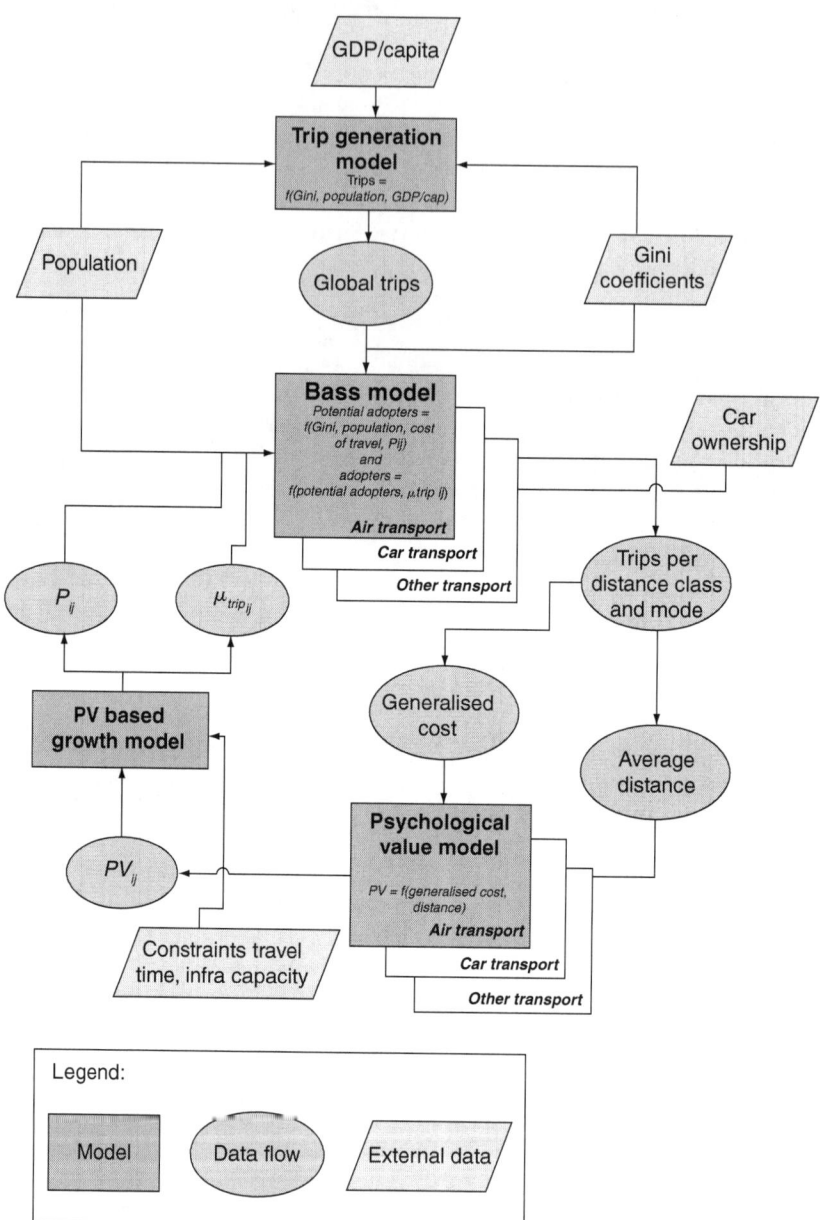

Figure 11.1 Overview of the GTTM^{dyn} behavioural model.

Note
The complete GTTM^{dyn} will be far more complex including infrastructure, car ownership, environmental impacts, and economic and other modules. In this chapter we describe only the behavioural part of the GTTM^{dyn}. PV_{ij} is the psychological value, P_{ij} the probability of choice and $\mu_{trip_{ij}}$ the trip growth rate all for mode i and distance class j.

1994). Each Bass model delivers the distribution of trips over distance class for its transport mode.

3   Based on the Bass model output, three *PV models*, one for each transport mode, calculate the PV per mode and distance class law (Kahneman and Tversky, 1979; Timmermans, 2010) based on generalised cost (weighted money plus weighted time) and distance within some constraints (e.g. infrastructure and fleet capacity and travel time constraints).

4   The PV values are fed to the *PV-based growth model*. This model calculates growth factors proportional to the PV values of each market i, j, and the probability of choice of each of the 60 markets (Morley, 1994; Nijkamp *et al.*, 2004; Papatheodorou, 2006).

The model has a typical feedback structure: the Bass model delivers the trips distribution, which delivers input for the PV model, that again feeds into the PV-based growth model and back to the Bass model. System dynamic modelling is pre-eminently suitable for such modelling problems.

The GTTM^{dyn} behavioural model has the following main characteristics:

- Inputs: global GDP/capita, income distribution, global population, travel cost per transport mode and distance class, and travel time per transport mode and distance class. Indirectly inputs for instance for infrastructure and transport technology investments will affect both cost and travel time and thus, tourism travel behaviour.
- The main outputs are trips per mode i and distance class j. The distribution over distance classes delivers an estimate of the average and total distance travelled.
- Transport modes are air, car and other (mainly rail and coach and including high-speed rail).
- The distance classes are defined in a way that the average distance per class increases according to an exponential function to accommodate a much higher resolution at short distances and still keep the total number of distance classes low: 50–100, 100–125, 125–175, 175–225, 225–300, 300–400, 400–525, 525–675, 675–900, 900–1175, 1175–1550, 1550–2025, 2025–2650, 2650–3500, 3500–4600, 4600–6025, 6025–7925, 7925–10425, 10425–13700, 13700–∞. The second distance class may look a little narrow, but the main parameter used in GTTM^{dyn} is the average distance per class; the class limits do not play a role in the model, except in calibrating to historic data.
- All monetary data have been set to US 1990$ using conversion rates from Pele (2012) and Sahr (2011).

The following section describes all models. A summary of equations and list of symbols can be found in the supplementary data file 1 to the web-based version of this chapter at www.tandfonline.com/toc/rsus20/21/.

The trip generation model

The trip generation model provides the total number of tourist trips for every simulation year. Several authors show that the decision to travel is more or less independently taken, i.e. not weighted extensively to spending on other goods or services (Papatheodorou, 2006; van Raaij, 1986). This means that trip generation is more or less exclusively determined by income (with GDP/capita as proxy) and is independent of the cost of tourism trips. The latter is obvious as the tourism product has a very wide range of costs, allowing people always to adapt. Bigano *et al.* (2004), Peeters and Dubois (2010) and Peeters and Landré (2012) show the existence of a single linear relation between GDP/capita and the number of trips per capita. Furthermore, some evidence from the Netherlands (Mulder *et al.*, 2007) shows the existence of a maximum number of trips per capita, where other than financial constraints become limiting (most likely time constraints). The general equation for trips per capita τ_T in a particular year and for a specific economy (GDP/capita) is:

$$\tau_T = \max(\tau_{T_{max}}, C_{cy} + \alpha_{cy} \times GDP_{cap})$$

(1)

with C_{cy} and α_{cy} constants fitted from data, GDP_{cap} is GDP/capita and $\tau_{T_{max}}$ the maximum number of trips per capita. The coefficients are given in Table 11.1.

The income above which the maximum number of trips per capita occurs is derived from Equation (1),

$$GDP_{cap\,thr} = \frac{\tau_{T_{max}} - C_{cy}}{\alpha_{cy}}$$

(2)

and is $80,780/capita in US 1990$, the currency used for all data in GTTM^{dyn}.

The overall average number of trips per capita now is found from the average GDP/capita below $GDP_{cap_{thr}}$ and the share of the population above $GDP_{cap_{thr}}$ times the maximum number of trips.

Product diffusion: Bass models

In contemporary economic models growth is generally defined in terms of a growth factor per year times the existing volume. However, this creates a

Table 11.1 Baseline values for the parameters determining trip generation

Tourism market	C_{cy}	α_{cy}	$\tau_{T_{max}}$
Total trips	0.2888	0.00005832	5.0

Source: Peeters and Landré, 2012.

Note
The value of α_{cy} is recalibrated to global tourism data for 2005 given by (UNWTO-UNEP-WMO, 2008) because of inclusion of a maximum number of trips for a share of the population above the limit – values for income in US 1990$.

problem when a new product is introduced to the market meaning the growth factor is multiplied with a zero market, preventing the market from emerging. To handle such new products diffusion Bass (1969) introduced the "Bass model" by defining potential adopters, adopters, innovators and imitators. The innovators are potential adopters that acquire the product independently of the number of existing adopters, while imitators do so because of existing adopters. Bass models assume commercial growth to be driven by advertising and marketing and social adoption by word-of-mouth mechanisms from adopters to potential adopters. The growth rate of adoptions is defined as

$$n_{a_{t+1}} = c_c \times N_{p_t} + c_s \times \frac{N_t \times N_{p_t}}{N_t + N_{p_t}} \tag{3}$$

with $n_{a_{t+1}}$ as the growth rate of adoptions at time $t+1$, N_{p_t} the number of potential adopters, N_t the number of adoptions at time t, c_c the commercial adoption coefficient and c_s the social adoption coefficient (Maier, 1998). According to Rich (2008), c_s represents both a contact rate between adopters and potential adopters and a success coefficient of such contacts. Generally, these are considered to be constants and are taken into one coefficient c_s.

In many Bass model applications, the two coefficients and the sum of potential adopters and adopters are constants. In GTTM^{dyn} this is not the case, because on the long time scales that GTTM^{dyn} runs, the idea of a constant population of potential adopters and adopters is not valid: almost all people living at the start of a simulation run will have died by the end of the run a century later. Furthermore, the properties of the product – a certain transport-mode–distance-class combination – is certainly not constant over such long time spans. Therefore, in GTTM^{dyn} the coefficient c_s is not taken as a constant, but as the growth factor derived from the development of the PV for all modes i and distance classes j, an approach also proposed by Maier (1998). Second, the number of potential adopters is restricted by financial or other constraints (Rich, 2008). In GTTM^{dyn} we assume the number of potential adopters per market (the 60 transport-mode–distance class combinations) to be a function of population with sufficient income to acquire the travel. For this we defined "sufficient income" as a fraction of income of the average ticket cost for a certain market. These coefficients are determined by calibrating the model (see section discussing calibration). A common measure for income distribution is the Gini coefficient (Gini, 1912). Gini measures the deviation from a fully equal income distribution as given by the Lorenz curve (Koo *et al.*, 1981). With a Gini coefficient of 0 all incomes are equal, while a coefficient of 1.0 means only one person of the population earns all income. Several authors published historic Gini coefficient time series (Atkinson and Brandolini, 2010; Bourguignon and Morrisson, 2002; Dowrick and Akmal, 2003; Korzeniewicz and Moran, 1996; Milanovic, 2002; O'Rourke, 2001; Pinkovskiy and Sala-i-Martin, 2010). An important novelty in the GTTM^{dyn} is an algorithm that calculates the share of the population above a certain limiting income for a given Gini coefficient. A description of this method can be found in the supplementary data file 1 to the web-based version of this chapter at www.tandfonline.com/toc/rsus20/21/.

There are three Bass models in GTTM^{dyn}, one for each transport mode. In these Bass models the commercial and social "constants" are linked to the psychological value PV for each mode i and distance class j combination.

$$n_{a_{t+1ij}} = c_{c_i} \times P_{ij} + c_{s_i} v_{ij} \times \frac{N_{t_{ij}} \times N_{p_{tij}}}{N_{t_{ij}} + N_{p_{tij}}} \tag{4}$$

with Pij as the probability of choice of alternative ij (see Equation (10)) and vij the PV of alternative ij (see Equation (9)).

The psychological value model

Introduction

The SEM often uses expected utility as its base for modelling discrete choices, even though it has been known since the 1950s that there are problems with the axioms (al-Nowaihi *et al.*, 2008). Expected utility is calculated as a linear weighted summation of all attributes of each choice alternative (see, e.g. Nijkamp *et al.*, 2004; Ortúzar and Willumsen, 2011). Therefore, Kahneman and Tversky (1979) modified "expected utility" into "PV". Van de Kaa (2010) gives an overview of the differences between prospect theory and expected utility theory of which the most relevant for our study are:

1   Framing: people base choices on a reference point and value increases of a utility as gains and a decrease of utility as losses.
2   Change-oriented framing: choices are not made referring to the current state, but based on marginal changes – gains or losses – to the current state.
3   Loss aversion: a certain loss is valued higher than a gain of the same magnitude.
4   Diminishing sensitivity: "The marginal value of both gains and losses generally decreases with their magnitude" (Kahneman and Tversky, 1979, p. 278).

Based on the above assumptions, Figure 11.2 shows the differences regarding loss aversion and diminishing sensitivity between expected utility and prospect theory. The sharp change in slope at the origin represents loss aversion. The deviation of the expected utility line represents the diminishing sensitivity.

The PV function follows a power law (Kahneman and Tversky, 1979; Timmermans, 2010):

$$v_{in_{gain}} = x_{in}^{\alpha}$$

and

$$v_{in_{loss}} = -\lambda \times x_{in}^{\beta} \tag{5}$$

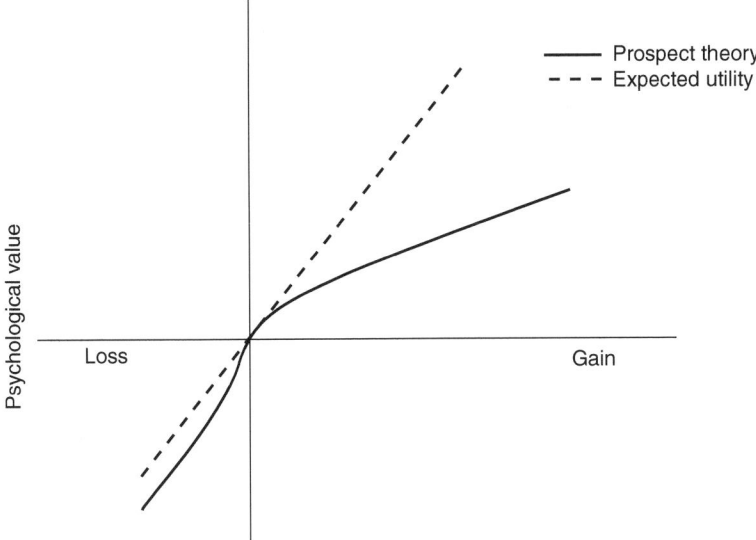

Marginal change of goods with respect to a reference point

Figure 11.2 The reference point is at the origin of the axes, gains are valued less than losses, gains are positive and losses negative, and both values diminish with deviation from the reference point (based on Kahneman and Tversky, 1979, p. 279).

with *vin* as the PV of attribute *xin* with *i* indicating the alternative, *n* the attribute and λ the loss aversion factor. The gain equation accounts for values of $xin \geq 0$ and the loss equation for $xin < 0$ (therefore the minus sign before λ). The power law coefficients α and β have a value between 0.0 and 1.0 and λ larger than 1.0. Van de Kaa (2010) found a value of 2.0 for λ based on 20 experiments, which is slightly lower than the range 2.0–2.5 given by Kahneman (2003). Furthermore, al-Nowaihi *et al.* (2008) show formally the validity of the equation, but also that α and β should be equal.

Often it is assumed that the current situation of a person, e.g. in the current travel from the Netherlands to the south of France at 1100 km, can be considered to be the reference state, i.e. this 1100 km. However, there is evidence that the reference point is better framed as an aspiration level (Kahneman and Tversky, 1979; Van de Kaa, 2010, p. 307). For GTTM^dyn we have chosen to take the reference point for each attribute as a mix of the PV average values per transport mode (a proxy for own reference) and for all 60 markets (a proxy for the general aspiration level). A coefficient between 0 and 1 governs how much of the perceived reference is determined by the own transport mode (air in the case of the PV calculated for air transport) and by the average over all transport modes (the latter when the factor is 0). This coefficient is one of the calibration variables.

The attributes

The most frequently used independent parameters in tourism economic modelling are income, relative prices and transport cost (Lim, 1997). In transport modelling the main independents are generalised cost (the weighted sum of travel cost and time and often some other transport resistances like discomfort) and attractors like the sizes of the population, work forces and economies at the origin and destination (Bates, 2008). In GTTM^{dyn} we only use travel cost and travel time as generalised costs. GTTM^{dyn} does not define origins and destinations, thus, it is not possible to define attraction in this way. Our hypothesis is that in tourism travel, physical distance can be used as a proxy for attraction, meaning that a further away destination, all other (perceived) properties equal, is more attractive than a closer destination. Frändberg and Vilhelmson (2011, p. 1236) find that individuals try to increase "spatial reach, presenting people with new opportunities", which is valid for "the transnational and global levels", thus pointing to a positive value of distance. The desire to travel further away is acknowledged indirectly by Mitchell (1984, p. 11) as increased travel is explained as the result of reduced constraints "as more and more people in the urban-industrial economies of the world have the time and financial ability to engage in long distance travel". The perception of distance as a factor enhancing the tourism experience has been shown to – conceptually – be a driver towards travelling longer distances (Ram *et al.*, 2013).

The generalised cost is calculated using the value of travel time (VoTT) for the three main transport modes based on the values given by Roman *et al.* (2007). Table 11.2 gives the values used in GTTM^{dyn}. However, VoTT is not constant over time (Gunn, 2008) as VoTT decreases at half the rate of growth of income. This has been implemented in GTTM^{dyn} assuming an elasticity of –0.5 for VoTT with respect to income. Again all VoTT values are converted to US 1990\$ using conversion rates from Pele (2012) and Sahr (2011).

For each attribute n (1=generalised cost, 2=distance), the normalised attribute value $xijn$ is calculated using the following equation:

$$x_{ijn} = c_{\text{sign}} \times \frac{(v_{ijn} - v_{\text{ref}_{n_i}})}{v_{\text{ref}_{n_i}}} \qquad (6)$$

$$v_{\text{ref}_{n_i}} = \zeta_i \times \bar{v}_i + (1 - \zeta_i) \times \bar{v}_{\text{all}} \qquad (7)$$

where v_{ijn} is the distance ($n=1$) or generalised cost ($n=2$), $v_{\text{ref}_{n_i}}$ the reference value which is a mix of average $\bar{v}_{\text{all}}$ for all transport-mode–market combinations and $\bar{v}_i$ is governed by the weight of mode i's only reference factor ζ_i; c_{sign} is a factor that determines the sign of the attribute value: it is +1 for $n=1$ and –1 for $n=2$.

Table 11.2 Value of travel time as based on data given by Roman *et al.* (2007)

Transport mode	€2004	1990$ (Geary–Khamis)	Assumptions
Air	12.69	10.92	Assumed the average of one-third is business and of two-thirds is economy class
Car	12.05	10.38	On average three persons per car (for tourism)
Other (ex-HST)	11.45	9.86	Average conventional rail and bus
HST	14	12.05	As in Roman *et al.* (2007)
Other	$\text{VoTT}_{2004}\$ = \alpha_{\text{HST}} \times 14.00 + (\alpha_{\text{HST}} - 1) \times 11.45$ and $\text{VoTT}_{1990}\$ = \alpha_{\text{HST}} \times 12.05 + (\alpha_{\text{HST}} - 1) \times 9.86$		Weighted sum of HST (share is α_{HST}) and other (non-HST)

Notes
1 The conversion from € to $ is taken from Pele (2012), the conversion from 2004 to 1990$ is based on Sahr (2011).
2 HST refers to high-speed train.

The PV-based growth model

The growth model

The PV-based growth model delivers the growth rates for social adoption and the choice probabilities used in the commercial adoption parts of the Bass model. The psychological value v_{ij} is calculated for each transport mode i and distance class j using the power law given by Fishburn and Kochenberger (1979) and Timmermans (2010), and making use of Equation (5).

$$v_{ij} = \text{if}\left(x_{ijn} \geq 0, \sum_{n=1}^{2} \omega_{\text{PV}_{in}} \times x_{ijn}^{\alpha}, -\lambda \times \sum_{n=1}^{2} \omega_{\text{PV}_{in}} \times x_{ijn}^{\alpha} \right) \tag{8}$$

in which $\omega_{\text{PV}_{in}}$ is a weighting factor between the generalised cost (always 1.0) and distance (the distance weights are calibrated per mode i between 0.2 and 5.0).

The probability of each alternative market ij is calculated using the following multinomial logit model (Nijkamp *et al.*, 2004):

$$P_{ij} = \frac{e^{v_{ij}}}{\sum\limits_{i=1, j=1}^{1, J} e^{v_{ij}}} \tag{9}$$

where P_{ij} is the probability of choosing an alternative with transport mode i and distance class j, v_{ij} the direct utility associated with option i and j is the normal exponent. The probability of each alternative is used as the base for the commercial adoption factor c_c of Equation (4). The growth $\mu_{trips_{ij}}$ per market ij is calculated as follows:

$$\mu_{trips_{ij}} = \tau_{ij} \times (C_{fit_{ij}} \times v_{ij} + \Delta\mu_{trips} + \Delta\mu_{trips_{dom}}$$

(10)

with τ_{ij} as travel time constraints (values between 0 and 1, see further down), $C_{fit_{ij}}$ a calibration factor that fits the PVs to "normal" growth rates, μ_{trips} a factor that equals the error of the average trip growth rate from Equation (10) and the calculated growth rate with Equation (10), both of the previous year, and $\mu_{trips_{dom}}$ the dominance growth factor (see further down) from Equations (11) and (12).

Market dominance and compromise

As the behavioural model of GTTM^{dyn} must handle large changes in transport-mode choice, there is a particular effect that might become important, which is "market dominance and market compromise". The attraction of certain markets appears to be not only a function of its direct attributes but also of its position within choices and the size of the market (Simonson, 1989). The first effect is coined as the "compromise" effect, in which a product with "middle" attributes has more attraction at the cost of product with more extreme attributes. The latter effect is known as the "market dominance attraction". Part of this effect is caused by a reduction of abandon rates, because that entails "extremely large switching costs that deter consumers from adopting new alternatives even if they are superior" (Lee and O'Connor, 2003). But dominant products also have a higher attraction as a choice for such a dominant product is more easily justified towards one's peers (Simonson, 1989). This effect is modelled by formulating an additional growth to the dominant transport mode per distance market at the cost of the growth of the smallest market, where a reduction growth is applied. The latter keeps the overall growth at the level given by the trip generation model. The dominance growth/decline is calculated as follows:
    For the mode with highest share,

$$\Delta\mu_{trips_{dom}} = c_{fit_{dom}} \times \mu_{trips} \times \sigma_{tm_{max}}$$

(11)

For the lowest mode share,

$$\Delta\mu_{trips_{dom}} = -c_{fit_{dom}} \times \mu_{trips} \times \sigma_{tm_{min}}$$

(12)

where $\mu_{trips_{dom}}$ is the growth added to overall trip growth, $c_{fit_{dom}}$ a constant "dominance fit factor", μ_{trips} the average growth of all trips, and $\sigma_{tm_{max}}$ the highest and $\sigma_{tm_{min}}$ the lowest share of the three transport modes (per distance class). The calibrated $c_{fit_{dom}}$ is 0.1797. This additional rate of growth is added to the social adoption part of the Bass model as it is a part of social mechanisms.

Constraints

Many economic and behavioural models, including discrete choice models, assume the choice to be limited by "physical" constraints of time and money (Papatheodorou, 2006). The money constraint is accounted for by the Bass model's limiting income assumption. So the main constraint is travel time. Very long distances with surface transport modes are blocked by travel time constraints. From the Dutch continuous holiday survey (CVO, NBTC-NIPO, 2011) we have calculated the occurrence of return travel times and found different constraints per transport mode. The lower end of the last bin with significant numbers of trips was chosen as the start of limiting travel times. This appeared to be 52 hours (return trip time) for both air and car and 42 hours for rail and bus/coach. We assume that a restriction of the growth factors will start at this limiting travel time with factor 1.0 (no restriction), linearly going down to 0.0 at 25 per cent above the limiting time (these 52 or 42 hours). These time constraints are applied directly to the growth rates and probabilities. So the following constraints are applied:

- Air transport shortest distance class (50–100 km) is always set to zero.
- Air transport maximum return travel time constraint is set between 52 at 1.0 and linearly down to 0.0 at 65 hours.
- Car transport maximum return travel time constraint is set between 52 at 1.0 and linearly down to 0.0 at 65 hours.
- Other transport maximum return travel time constraint is set between 42 at 1.0 and linearly down to 0.0 at 52.5 hours.

Calibration

The behavioural model of GTTM^{dyn} as described above has been calibrated to historical data over the period 1900–2005 (see supplementary data file 1 to the web-based version of this chapter at www.tandfonline.com/toc/rsus20/21/).

For calibration we used the evolutionary optimisation module of Powersim Studio 9 (see Hansen, 2006 for background information). This Powersim procedure requires the definition of objectives (the values of model output desired) and decisions, the variables that the module may change to reach the objective. For calibration we defined the objective to reduce the error between number of trips and distances per transport mode as calculated by the model and the historical data below a certain (low) value. The errors are calculated in a cumulative way by summing the square of the error fractions for each year between 1900 and 2005. Because of the mounting uncertainty going back into history and to achieve a good fit for the year 2005, the starting year for future simulations, we have weighted recent errors higher than errors further back in history. This is achieved by calculating the square of the fraction of the error with respect to the final value in 2005 and not the historic value for each year. As all three transport modes show considerable growth between 1900 and 2005, this effectively weighs the errors of the final years above the errors of the early years. One problem occurred with the cumulative error for the distance for other modes. After extensive testing with cumulative errors only, it appeared either the other final distance was far out of the 2005 historic value or the calibration found no solution with reasonably low errors set for all trip and distance errors within the maximum of 1000 optimisation cycles allowed (involving some 13,000 model runs). Therefore, the distance error for other transport modes has been based on the 2005 value only and not the cumulative error. Table 11.3 shows the six objective variables (the squared errors), the limit value (maximum calibration value) and the final calibration value. The calibration requires all error values to be below the limit value. The final calibration was found after some 300 cycles and 4000 model runs.

The decisions are the calibration coefficients in the model governing its behaviour. Each calibration coefficient needs a minimum and maximum value between which the optimisation module tries to find a solution that fulfils the objective. A too narrow search range of values will cause some of the decisions to get "stuck" at the minimum or maximum value and thus prevent a most optimal solution. A too wide range may cause the model to crash during the calibration. As can be seen in Table 1 (see the supplementary file 2 in the web-based version of this chapter at www.tandfonline.com/toc/rsus20/21/), none of the decisions is limited by the search range. Each transport mode is defined by six decision variables. The dominance fit factor governs the "dominance fit" for all transport modes in one single number (see Table 1 in the supplementary file 2 in the web-based version of this chapter at www.tandfonline.com/toc/rsus20/21/).

Results and discussion

Fit to historic data

The main result of the calibration is given in Figure 11.3. The fit is generally good, as shown by the lines following the shaded areas rather well. Only the sharp distance reduction for air transport after 2001 is not well represented by the model. This may have been caused mainly due to the 9/11 attacks in the USA; such international conflicts are not taken into account by GTTM^{dyn}, though

Table 11.3 The limiting and final values for the objective variables

Error (objective)	Limit	Calibrated value
Air cumulative distance error	0.15	0.15
Air cumulative trip error	0.15	0.14
Car cumulative distance error	0.25	0.18
Car cumulative trip error	0.10	0.10
Other distance error in 2005	0.05	0.05
Other cumulative trip error	0.50	0.49

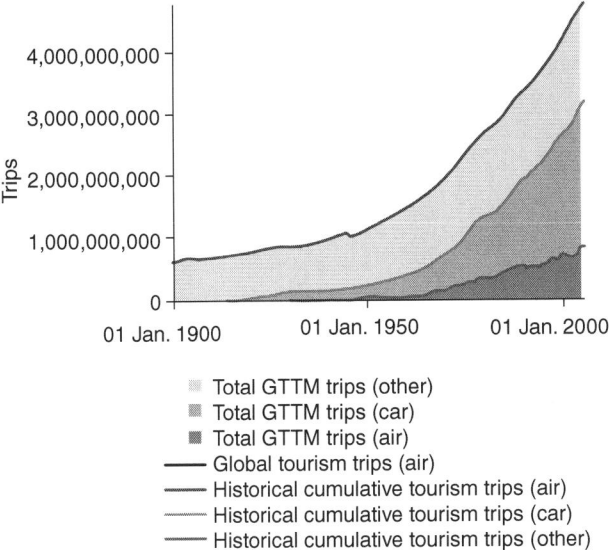

Total GTTM trips (other)
Total GTTM trips (car)
Total GTTM trips (air)
Global tourism trips (air)
Historical cumulative tourism trips (air)
Historical cumulative tourism trips (car)
Historical cumulative tourism trips (other)

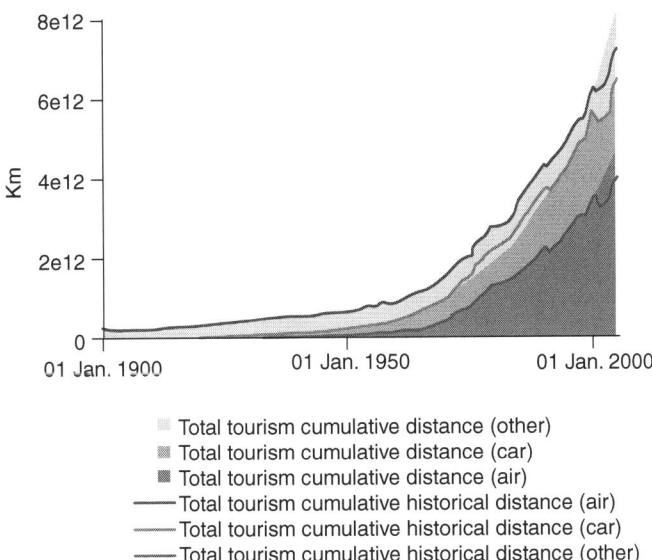

Total tourism cumulative distance (other)
Total tourism cumulative distance (car)
Total tourism cumulative distance (air)
Total tourism cumulative historical distance (air)
Total tourism cumulative historical distance (car)
Total tourism cumulative historical distance (other)

Figure 11.3 Main results after calibration: left figure gives number of trips per transport mode, right figure the total distance travelled in tourism. The shaded areas give the model output, the solid lines give the historic data.

Note
The word "cumulative" in the variable names has no meaning in the context of these figures.

a special input variable (a so-called X-factor per mode i) is available to temper or boost growth in certain years.

The fit to the total number of trips is automatically achieved as the sum of the trips for all three Bass models is each simulation step corrected towards the total trips calculated by the trip generation model. So it is the fit of the distribution of both trips and distances and the total distance that tells us something about the quality of the model, not the total number of trips.

Figure I (see supplementary file 2 in the web-based version of this chapter at http://www.tandfonline.com/toc/rsus20/21/) shows more detailed results of the calibration for each transport mode both for number of trips and distances. The fit is good for trips for both aviation and car. Model results for other transport modes follow some of the relatively strong changes in, for instance, the 1940s, though overall fit is less impressive. For distance the fit is relatively good again for car, a bit less for air, ending relatively high and relatively low for other transport. The latter may be caused by our choice to set the error objective to the error in 2005 only, not the cumulative error.

Model scenarios behaviour

Figure II (see supplementary file 2 in the web-based version of this chapter at http://www.tandfonline.com/toc/rsus20/21/) shows how the calibrated $GTTM^{dyn}$ behaves for scenarios up to the year 2100. The scenarios are based on GDP per capita from the Intergovernmental Panel on Climate Change Special Report: Emissions Scenarios (IPCC SRES) scenarios A1 and A2 (IMAGE-team, 2006), and low, medium and high population growth as projected by the United Nations (2011) and assumptions about the development of Gini fitting the description of the A1 (less equal) and A2 (more equal) scenarios (IPCC, 2000). The first scenario (A1/medium population), a kind of baseline scenario, shows the number of trips to increase to almost 21 billion by 2100, four times the current 5.0 billion. Distances travelled will increase to 74 trillion passenger kilometres (pkm) from the current level of 8 trillion, almost nine times as much. The lowest tourism growth will occur in A2 (low economic growth/low population growth) and just double the total number of trips in 2100 to 9.4 billion, while the distances will reach 29 trillion pkm, 3.5 times the current levels. The A1/high population growth sets the highest tourism growth at 32.5 billion trips (almost seven times the current volume) and 116 trillion pkm (15 times the current transport volume). So the "playing field" in business-as-usual scenarios is growth of trips by two to seven times, with a much stronger increase of distance travelled by a factor of 3.5 to 15.

The $GTTM^{bas}$ and $GTTM^{adv}$ projected 26 trillion pkm in 2035 based on simple exponential growth (Dubois *et al.*, 2011; Peeters and Dubois, 2010), even a little higher than the $GTTM^{dyn}$ figure of 24 trillion pkm for A1/high population scenario. In 2050 $GTTM^{dyn}$ projects, based on the A1/medium population scenario, 36.2 trillion pkm, 65 per cent of which is air transport – which seems a little low compared to the c.37 trillion for high speed modes only (air and rail, mainly tourism) given by Schäfer and Victor (2000). The A1/high population scenario comes in at *c.*41 trillion pkm, of which 26.7 trillion pkm is air transport and thus, again lower than the scenario given by Schäfer and Victor (2000). GTTM^{dyn} compares better with the aviation industry's projections. Boeing (2012) projects a total of 13.7 trillion pkm by air in 2031 and Airbus (2012) expects air transport to be at 12.7 trillion pkm by 2031, which positions GTTM^{dyn} in the middle with 13.3 trillion pkm. Tourism-dedicated long-term scenarios are scarce in contemporary literature and generally incomplete. For instance, Mayor and Tol (2010) suggest that transport will increase by a factor of 16 by 2100, but this is for international aviation only. GTTM^{dyn} projects aviation pkm will increase by 2100 by a factor of almost 17 for all air transport (international plus domestic) for the A1/medium population scenario.

Concluding, the projections of GTTM^{dyn} behavioural demand model are within the order of magnitude of the few other long-term projections now available. Furthermore, the model behaves as expected with different economic and demographic projections. Also, it is interesting to see that low economic growth increases the share of other transport modes and apparently affects mainly air and car. Another observation is that air transport will dominate pkm volumes in all scenarios, but mostly in the high economic-growth scenarios.

Discussion

GTTM^{dyn} was successfully calibrated to time-series data from 1900 to 2005 and its projections are comparable to the few existing projections found in the literature. Still, the GTTM^{dyn} has several weaknesses. For instance, the trip generation model is relatively simple and straightforward and only driven by population growth, income growth and distribution. The latter is an important innovation in GTTM^{dyn}, which has not been implemented before in the "simple" way based on the Gini coefficient. Current literature gives some support for our hypotheses that trip generation is mainly income-driven and not cost-driven, but more research is necessary here.

The Bass model adopters are corrected for the death rate of people (assuming adopters also die and thus reduce the adopter population), but not for birth rate. The latter could be argued to be necessary because new-borns automatically become a part of the adopters group as tourism travel is often a family-based activity. But this is only valid until the children start living on their own, developing their own habits, lifestyles and limited by their own income and time constraints. As this would result in a rather small growth factor, we decided to ignore it. Tests revealed the difference to be very small, which is caused by the trip generation model that governs the total number of trips.

The cost of tourism trips consists of travel cost along with cost for lodging, etc. To be able to travel, at least some accommodation costs are involved. In GTTM^{dyn} the limiting income is solely based on the cost of travel. So there is an argument to include accommodation costs when calculating the potential adopters population's lower-limit income. GTTM^{dyn} ignores accommodation costs because they differ over a very wide range, from almost zero (private homes of friends and relatives and wild camping) to the very high cost of luxury hotels. So accommodation cost is probably not the limiting factor for people to start travelling; it is mainly transport cost.

A final problem with this kind of model is the long time span (over two centuries) they cover. Of course there is no way to "predict" the future with any model and certainly not over such long time spans and for travel behaviour that varies within less than a year at the individual level. On the other hand, a model with a time horizon of 15–20 years would reduce uncertainties but not be very helpful to analyse the impacts of climate mitigation policies on tourism's sustainability as the response times of the tourism–climate system are too long to be assessed on such a relatively short time scale. Therefore, system dynamics as a modelling environment has been chosen for GTTM^{dyn}. The objective of system dynamics is not to project the future but to learn how complex systems based on cause–effect relations and including feedback loops behave under different assumptions (Sterman, 2000). A system dynamics model that best avoids purely statistical relationships uses "real" mechanisms in cause–effect relationships. Therefore, in the GTTM^{dyn} behavioural model we have chosen not to use the commonly applied SEM but to use prospect theory which is more based on known psychological mechanisms for the distribution of trips over the 60 tourism travel markets. The latter is important as these psychological, even neurological mechanisms can be assumed to be more or less constant for humans, or at least change only slowly, while the parameters for economic models, like price elasticities, are shown to be neither constant over time, nor a property of the "brains" of humans. Finally, the diffusion of products has been split into a commercial and a social effect behaving in different ways and thus better representing "real" mechanisms.

Conclusion

This chapter seeks to describe a tourism travel behaviour model able to describe long-term developments. To do so the behavioural model of the GTTM^{dyn} has been founded partly in psychological mechanisms (prospect theory, Kahneman and Tversky, 1979) rather than standard economic theory. Only global trip generation is a statistical model assuming basically a linear relation between GDP/capita and trips/capita/year. The model distinguishes three transport modes and twenty distance classes: psychological mechanisms have been used to distribute trips over these modes and classes. This method has delivered global tourism travel distances. The model is able to reproduce the large increase in trip distances over the period 1900–2005 just by the growth differences occurring from the differences in PVs between different distance classes and transport modes. A novel idea to accomplish this is to assume attraction in the utility-function-based PV to be distance itself: the longer the distance to a destination, all other attributes equal, the more attractive this destination becomes. The successful calibration shows this method works as expected.

Another challenge was to have the model successfully introduce a new transport mode (aviation) during its calibration runs between 1900 and 2005. This was accomplished by using the Bass product diffusion model (Bass, 1969) that makes a distinction between commercial adoption and social adoption, the former allowing a market to develop from zero. The main conclusion is that, using the classical economic ideas of utility maximisation as the driver for tourism travel choice behaviour, but modified with ideas from psychological economics, specifically prospect theory, and applied in a system dynamics model environment, is fit to reproduce the development of revealed tourism travel behaviour between 1900 and 2005. Also the model behaved in a stable and reasonable way for scenarios combining different economic growth, population growth and income distribution developments. Most of these projections fall in line with the few projections of global tourism transport found in the literature. Of course the validity of the projections cannot be proven and the reproduction of the past only shows that the model can be calibrated to these past developments, which does not necessarily validate the model. The model was calibrated through 19 coefficients that define the model's behaviour. Of course each simulation is also informed by input variables like population growth, GDP/capita, income distribution and cost and speed of three transport modes and differentiated to distance class for speed and in the case of aviation also for cost.

A couple of long-term scenario runs reveal that GTTM^{dyn} compares reasonably with projections found in the literature. The same scenario runs also show that tourism is not likely to be able to reduce its carbon dioxide emissions without very strong policy interventions. Even in the unlikely lowest growth scenario, total transport increases by a factor of 3.6, requiring technological solutions to reduce emission factors for transport of more than 70 per cent to just keep emissions at current levels. This might be possible, but the most recent population estimates point rather to a high population growth than a low one. In the highest growth scenario tourism transport increases by a factor of 14.5, requiring emission factors to be reduced by 93 per cent to keep emissions level. This is most likely prohibitive (Peeters, 2010), emphasising the need for policies changing demand for tourism and tourism transport, specifically the development of distances travelled. GTTM^{dyn} promises to become an instrument to explore such policies in more detail.

Acknowledgements

I would like to thank Bert van Wee, Els van Daalen, Wil Thissen, Jaap Lengkeek and two anonymous reviewers for their very valuable comments on various drafts of this chapter.

References

Airbus. (2012). *Navigating the future. Global market forecast 2012–2031*. Paris: Airbus S.A.S.

al-Nowaihi, A., Bradley, I. and Dhami, S. (2008). A note on the utility function under prospect theory. *Economics Letters*, *99*(2), 337–339.

Ananthasayanam, M.R. (2003). *Pattern of progress of civil transport airplanes during the twentieth century.* Paper presented at the AIAA Atmospheric Flight Mechanics Conference and Exhibit, Austin, Texas, USA.

Atkinson, A.B. and Brandolini, A. (2010). On analyzing the world distribution of income. *The World Bank Economic Review, 24*(1), 1–37.

Bass, F.M. (1969). A new product growth for model consumer durables. *Management Science, 15*(5), 215–227.

Bass, F.M. (2004). Comments on "A new product growth for model consumer durables": The Bass model. *Management Science, 50*(12), 1833–1840.

Bass, F.M., Krishnan, T.V. and Jain, D.C. (1994). Why the Bass model fits without decision variables. *Marketing Science, 13*(3), 203–223.

Bates, J. (2008). History of demand modelling. In D.A. Hensher and K.J. Button (eds), *Handbook of transport modelling* (Vol. 1, pp. 11–34). Amsterdam: Elsevier.

Bieger, T., Wittmer, A. and Laesser, C. (2007). What is driving the continued growth in demand for air travel? Customer value of air transport. *Journal of Air Transport Management, 13*, 31–36.

Bigano, A., Hamilton, J.M., Lau, M., Tol, R.S.J. and Zhou, Y. (2004). *A global database of domestic and international tourist numbers at national and subnational level* (Working Paper No. FNU-54). Hamburg: Research Unit Sustainability and Global Change, Hamburg University and Centre for Marine and Atmospheric Science.

Boeing. (2012). *Current market outlook 2012–2031.* Seattle, WA: Boeing Commercial Airplanes.

Bourguignon, F. and Morrisson, C. (2002). Inequality among world citizens: 1820–1992. *The American Economic Review, 92*(4), 727–744.

Ceron, J.P. and Dubois, G. (2007). Limits to tourism? A backcasting scenario for sustainable tourism mobility in 2050. *Tourism and Hospitality Planning & Development, 4*(3), 191–209.

Ch'eze, B., Gastineau, P. and Chevallier, J. (2011). Forecasting world and regional aviation jet fuel demands to the mid-term (2025). *Energy Policy, 39*, 5147–5158.

Daly, H.E. and Farley, J. (2004). *Ecological economics. Principles and Applications.* Washington, DC: Island Press.

Dopfer, K. (ed.). (2005). *The evolutionary foundations of economics.* Cambridge: Cambridge University Press.

Dowrick, S. and Akmal, M. (2003). Contradictory trends in global income inequality: A tale of two biases. Paper presented at the Conference on Inequality, Poverty and Human Well-being, Helsinki, Finland.

Dubois, G., Ceron, J.P., Peeters, P. and Gössling, S. (2011). The future tourism mobility of the world population: Emission growth versus climate policy. *Transportation Research – A, 45*(10), 1031–1042.

Fishburn, P.C. and Kochenberger, G.A. (1979). Two-piece von Neumann-Morgenstern utility functions. *Decision Sciences, 10*(4), 503–518.

Forrester, J.W. (1971). *World dynamics.* Cambridge, MA: Wright-Allen Press.

Forum for the Future. (2009). *Tourism 2023 – creating a sustainable tourism industry. Four scenarios, a vision and a strategy for UK outbound tourism and travel.* London: Forum for the Future.

Frändberg, L. and Vilhelmson, B. (2011). More or less travel: Personal mobility trends in the Swedish population focusing gender and cohort. *Journal of Transport Geography, 19*(6), 1235–1244.

Gini, C. (1912). Variabilit'a e mutabilit'a [Variability and mutability]. In E. Pizetti and T. Salvemini (eds), Rome: Libreria Eredi Virgilio Veschi, as repr. in *Memorie di metodologica statistica* [*Memoirs of statistical methodology*], p. 1.

Girod, B., van Vuuren, D.P. and Deetman, S. (2012). Global travel within the 2°C climate target. *Energy Policy*, *45*, 152–166.

Girod, B., Wiek, A., Mieg, H. and Hulme, M. (2009). The evolution of the IPCC's emissions scenarios. *Environmental Science & Policy*, *12*(2), 103–118.

Gunn, H.F. (2008). An introduction to the valuation of travel time-savings and losses. In D.A. Hensher and K.J. Button (eds), *Handbook of transport modelling* (Vol. 1, pp. 503–517). Amsterdam: Elsevier.

Hansen, N. (2006). The CMA evolution strategy: A comparing review. In J.A. Lozano, P. Larrañaga, I. Inza and E. Bengoetxea (eds), *Studies in fuzziness and soft computing: Towards a New Evolutionary Computation* (Vol. 192, pp. 75–102). Berlin: Springer-Verlag.

Huybers, T. (2003). Domestic tourism destination choices – a choice modelling analysis. *International Journal of Tourism Research*, *5*(6), 445–459.

IMAGE-team. (2006). *The IMAGE 2.2 implementation of the SRES scenarios. A comprehensive analysis of emissions, climate change and impacts in the 21st century* (No. CD-ROM 500110001 (former 481508018)). Bilthoven: National Institute for Public Health and the Environment.

IPCC. (2000). Special report on emission scenarios (No. ISBN-10: 0521800811 (web version)): International Panel on Climate Change.

Kahneman, D. (2003). Maps of bounded rationality: Psychology for behavioral economics. *American Economic Review*, *93*(5), 1449–1475.

Kahneman, D. (2011). *Thinking, fast and slow*. London: Allen Lane.

Kahneman, D. and Tversky, A. (1979). Prospect theory: An analysis of decision under risk. *Econometrica: Journal of the Econometric Society*, *47*(2), 263–291.

Koo, A.Y.C., Quan, N.T. and Rasche, R. (1981). Identification of the Lorenz curve by Lorenz coefficient. *Review of World Economics*, *117*(1), 125–135.

Korzeniewicz, R.P. and Moran, T.P. (1996). World-economic trends in the distribution of income, 1965–1992. *American Journal of Sociology*, *102*(3), 1000–1039.

Lee, Y. and O'Connor, G. (2003). New product launch strategy for network effects products. *Journal of the Academy of Marketing Science*, *31*(3), 241–255.

Lempert, R.J., Popper, S.W. and Bankes, S.C. (2003). *Shaping the next one hundred years. New methods for quantitative, long-term policy analysis*. Santa Monica, CA: RAND.

Lim, C. (1997). Review of international tourism demand models. *Annals of Tourism Research*, *24*(4), 853–849.

Lyons, S., Mayor, K. and Tol, R.S.J. (2009). Holiday destinations: Understanding the travel choices of Irish tourists. *Tourism Management*, *30*(5), 683–692.

Maier, F.H. (1998). New product diffusion models in innovation management – a system dynamics perspective. *System Dynamics Review*, *14*(4), 285–308.

Mayor, K. and Tol, R.S.J. (2010). Scenarios of carbon dioxide emissions from aviation. *Global Environmental Change*, *20*(1), 65–73.

Milanovic, B. (2002). True world income distribution, 1988 and 1993: First calculation based on household surveys alone. *The Economic Journal*, *112*(476), 51–92.

Mitchell, L.S. (1984). Tourism research in the United States: A geographic perspective. *GeoJournal*, *9*(1), 5–15.

Morley, C.L. (1994). Experimental destination choice analysis. *Annals of Tourism Research*, *21*(4), 780–791.

Moss, R., Edmonds, J., Hibbard, K., Manning, M., Rose, S., van Vuuren, D., ... Wilbanks, T.J. (2010). The next generation of scenarios for climate change research and assessment. *Nature, 463*(7282), 747–756.

Mulder, S., Schalekamp, A., Sikkel, D., Zengerink, E., van der Horst, T. and van Velzen, J. (2007). Trendanalyse van het Nederlandse vakantiegedrag van 1969 tot 2040. *Vakantiekilometers en hun milieu-effecten zullen spectaculair blijven stijgen.* (No. E 4922 18–02–2007). Amsterdam: TNS NIPO.

Müller, H. and Weber, F. (2007). Climate change and tourism – Scenarios for the Bernese Oberland in 2030. Paper presented at the 42nd TRC Meeting, Bolzano, Italy.

NBTC-NIPO. (2011). Continu Vakantie Onderzoek (continuous holiday survey). Retrieved from www.nbtcniporesearch.nl/nl/Home/Producten-en-diensten/cvo.htm.

Nicolau, J.L. (2008). Characterizing tourist sensitivity to distance. *Journal of Travel Research, 47*(1), 43–52.

Nijkamp, P., Reggiani, A. and Tsang, W.F. (2004). Comparative modelling of inter-regional transport flows: Applications to multimodal European freight transport. *European Journal of Operational Research, 155*(3), 584–602.

O'Rourke, K.H. (2001). *Globalization and inequality: Historical trends* (NBER Working Paper No. 8339). Cambridge, MA: National Bureau of Economic Research.

Ortúzar, J.D. and Willumsen, L.G. (2011). *Modelling transport* (4th edn). Chichester: John Wiley & Sons.

Oum, T.H., Waters, W.G. and Fu, X. (2008). Transport demand elasticities. In D.A. Hensher and K.J. Button (eds), *Handbook of transport modelling* (Vol. 1, pp. 239–255). Amsterdam: Elsevier.

Owen, B., Lee, D.S. and Lim, L. (2010). Flying into the future: Aviation emissions scenarios to 2050. *Environmental Science & Technology, 44*(7), 2255–2260.

Papatheodorou, A. (2006). Microfoundations of tourist choice. In L. Dwyer and P. Forsyth (eds), *International handbook on the economics of tourism* (pp. 73–88). Cheltenham: Edward Elgar.

Peeters, P. (2010). Tourism transport, technology, and carbon dioxide emissions. In C. Schott (ed.), *Tourism and the implications of climate change: Issues and actions* (Vol. 3, pp. 67–90). Bingley: Emerald.

Peeters, P.M. and Dubois, G. (2010). Tourism travel under climate change mitigation constraints. *Journal of Transport Geography, 18*, 447–457.

Peeters, P. and Landré, M. (2012). The emerging global tourism geography – an environmental sustainability perspective. *Sustainability, 4*(1), 42–71.

Pele, L. (2012). Historical rates. Retrieved from http://fxtop.com/en/historates.php?MA=1.

Pettebone, D., Newman, P., Lawson, S.R., Hunt, L., Monz, C. and Zwiefka, J. (2011). Estimating visitors' travel mode choices along the Bear Lake Road In Rocky Mountain National Park. *Journal of Transport Geography, 19*(6), 1210–1221.

Pinkovskiy, M. and Sala-i-Martin, X. (2010). Parametric estimations of the world distribution of income. Retrieved from http://voxeu.org/index.php?q=node/4508.

Ram, Y., Nawijn, J. and Peeters, P. (2013). Happiness and limits to sustainable tourism mobility: A new conceptual model. *Journal of Sustainable Tourism, 21*(7).

Rich, E. (2008). Management fads and information delays: An exploratory simulation study. *Journal of Business Research, 61*(11), 1143–1151.

Rogelj, J., Hare, W., Lowe, J., van Vuuren, D. P., Riahi, K., Matthews, B., ... Meinshausen, M. (2011). Emission pathways consistent with a 2°C global temperature limit. *Nature Climate Change, 1*(8), 413–418.

Roman, C., Espino, R. and Martin, J.C. (2007). Competition of high-speed train with air transport: The case of Madrid-Barcelona. *Journal of Air Transport Management*, *13*(5), 277–284.

Sahr, R. (2011). Inflation conversion factors for dollars 1774 to estimated 2021. Retrieved from http://oregonstate.edu/cla/polisci/faculty-research/sahr/sahr.htm.

Schäfer, A. (2012). *Introducing behavioral change in transportation into energy/ economy/environment models* (Policy Research Working Paper No. 6234). Washington, DC: The World Bank.

Schäfer, A. and Victor, D.G. (2000). The future mobility of the world population. *Transportation Research – A*, *34*, 171–205.

Schwaninger, M. (1984). Forecasting leisure and tourism – scenario projections for 2000–2010. *Tourism Management*, *5*(4), 250–257.

Scott, D., Peeters, P. and Gössling, S. (2010). Can tourism deliver its "aspirational" greenhouse gas emission reduction targets? *Journal of Sustainable Tourism*, *18*(3), 393–408.

Sgouridis, S., Bonnefoy, P.A. and Hansman, R.J. (2010). Air transportation in a carbon constrained world: Long-term dynamics of policies and strategies for mitigating the carbon footprint of commercial aviation. *Transportation Research – A*, *45*(10), 1077–1091.

Simonson, I. (1989). Choice based on reasons: The case of attraction and compromise effects. *Journal of Consumer Research*, *16*(2), 158–174.

Sterman, J.D. (2000). Business dynamics. *Systems theory and modeling for a complex world*. Boston: Irwin McGraw-Hill.

Timmermans, H. (2010). On the (ir)relevance of prospect theory in modelling uncertainty in travel decisions. *EJTIR*, *4*(10).

United Nations. (2011). World population prospects: The 2010 revision. Retrieved from http://esa.un.org/unpd/wpp/unpp/panel_indicators.htm.

UNWTO. (2011, October). Tourism towards 2030. Global overview: Advance edition presented at UNWTO 19th General Assembly, Madrid, Spain.

UNWTO-UNEP-WMO. (2008). *Climate change and tourism: Responding to global challenges*. Madrid: UNWTO.

Van de Kaa, E.J. (2010). Prospect theory and choice behaviour strategies: Review and synthesis of concepts from social and transport sciences. *EJTIR*, *10*(4).

van Raaij, W.F. (1986). Consumer research on tourism mental and behavioral constructs. *Annals of Tourism Research*, *13*(1), 1–9.

Wilkinson, N. (2008). *An introduction to behavioral economics*. New York: Palgrave Macmillan.

WTO. (1998). *Tourism 2020 vision. Executive summary* (2nd edn). Madrid: WTO.

12 Promoting public transport use in tourism

Diem-Trinh Le-Klähn, C. Michael Hall and Regine Gerike

Introduction

It has long been recognised in the academic literature that developing tourism in a sustainable manner is vital for all destinations. Sustainable development in general and sustainable tourism development in particular have received growing attention in government and business planning and policies. Some companies also started to be more responsible with the environment. There have been growing numbers of supporting services from other sectors including green hotels, bio-products and low emissions rental cars. It is of course best to combine efforts from all product suppliers and consumers to achieve sustainability in tourism. However, one very important area in sustainable tourism development is transport services for tourists. Tourism contributes at least 4.4 per cent of global CO_2 emissions and a large part of it (75 per cent) comes from transport, although this figure may be considerably larger if radiative forcing is also considered (Dubois *et al.*, 2011; Hall *et al.*, 2013; Scott *et al.*, 2012a). Consequently, reducing transport emissions is a priority task in mitigating the emissions from tourism (Gössling *et al.*, 2013). This has become even more important due to the continuing problems of growing population, increasing traffic congestion and pollution, and the impacts of climate change. Of the several initiatives to reduce emissions in tourism transport, shifting to more eco-friendly transport modes (e.g. public transport) is one promising direction (Dickinson and Dickinson, 2006; Scott *et al.*, 2012b).

This chapter discusses the importance of public transport in sustainable tourism development. It explains why public transport plays a key role at an urban tourism destination. Using a case study of visitor use of public transport in Munich, it explains how visitors use public transport in the city. The chapter identifies the most important attributes of public transport services for visitors, which reflect significant differences from the resident users and are therefore worthy of attention with respect to public transport management and marketing. Marketing strategies targeting the visitor groups are then presented, which shows how public transport could be promoted to visitors at an urban destination.

Public transport and tourism development

The importance of promoting public transport use for sustainable tourism development

Public transport (also referred to as public transportation, mass transit and/or public transit) is a shared transport service that is available for public use with set fares. According to the International Association of Public Transport (UITP), public transport includes rail, bus, scheduled ferries, taxicab and other systems that transport the public members (UITP, 2013). Compared to other engine-powered modes of transport, public transport has certain advantages such as lower cost for passengers, less demands on space for the cities and being more energy-efficient and safer (APTA, 2013; UITP, 2013). Travelling by train is also much safer than by car. Road accidents claim more than 90 per cent the total number of deaths resulting from transport accidents in Europe. In 2011, for example, road accidents caused 30,300 fatalities, much higher than the 2,325 from rail accidents, which has experienced a declining trend since 2004 (Eurostat, 2011, 2012). Hence, train or public transport in general is often viewed as a more sustainable mode of transport as opposed to private car (Holmgren, 2007).

Since its arrival in the 1980s, sustainable tourism has become the focus of many destinations' planning and development policies (e.g. Connell *et al.*, 2009; Risteskia *et al.*, 2012; Waligo *et al.*, 2013). Although the concept is contested, sustainable tourism can be broadly understood as a form of tourism that meets the needs of stakeholders (visitors, industry, environment and local communities) while respecting the needs of present and future generations (Hall, 2011). Tourism brings multiple economic and social benefits, but it is also responsible for some negative impacts, especially on the environment (Hall and Lew, 2009). However, as most tourism emissions are from transport (Dubois *et al.*, 2011), for sustainability in tourism to be achieved, sustainable transport development is a necessity (Høyer, 2000; Scott *et al.*, 2012b).

Emission reductions in tourism are believed to only be feasible by combining a substantial modal shift together with shortened travel distances and improved low-carbon technologies (Dickinson and Dickinson, 2006; Dickinson *et al.*, 2009; Scott *et al.*, 2012b). Guiver *et al.* (2008) suggested reducing the number of trips and length of trips, switching to alternative transport modes such as walking or cycling, and reducing the number of vehicles used (by car-sharing or public transport use). However, the first option may affect the tourism industry at the destination. Cycling and walking are environmentally friendly but are restricted under many circumstances such as long distance, unsuitable road conditions and limited by personal endurance. A modal shift to public transport is therefore a promising strategy to reduce CO_2 emissions (Dubois *et al.*, 2011; Martín-Cejas and Sánchez, 2010).

Transport offers access and facilitates tourist mobility to, from and within a destination (Duval, 2007). Tourists appreciate a destination that is accessible but being able to reach the attractions at the destination is equally important (Sorupia, 2005). Therefore, a reliable and effective public transport system is vital to provide accessibility and enhance mobility for tourists. More importantly, promoting greater use of public transport could lead to less use of private vehicles and thus fewer emissions to the environment.

The role of public transport in urban tourism

According to the UITP (2013), public transport at urban, suburban and regional level carried 60 billion passenger trips in 2008 in the EU-27 region; equivalent to 120 trips per inhabitant per year. With ridership increasing continuously since 2000 (UITP, 2013), public transport is expected to grow by 1.6–4.4 per cent per year up to 2050 (Dubois *et al.*, 2011). At the upper end, this rate of growth is greater than that expected for international tourism up to 2030 (Hall *et al.*, 2013).

Many urban areas are now facing the issues of a growing population and high level of dependence on car transport. The pressure to reduce traffic congestion and private vehicle use becomes even more urgent for tourist destinations where there is also an increasing number of tourists sharing space and infrastructure with residents. Tourism is a major income generator and significant contributor to the economic growth and destination image of many cities (Page and Hall, 2003). Income from tourists could provide funds for public transport yet this does not often lead to an increase in supply (Albalate and Bel, 2010). The results are crowded vehicles, congestion and limited services, which both residents and visitors have to endure.

While tourism could be a driver for the development of a public transport system, public transport in return plays a key role in urban tourism development (Albalate and Bel, 2010; Mandeno, 2011). Tourists visit a destination for many reasons and are attracted by its urban features, including transport infrastructure (Ashworth and Page, 2011; Khadaroo and Seetanah, 2008). Public transport contributes to a tourist experience and may even influence his or her satisfaction with a destination (Thompson and Schofield, 2007). An efficient public transport system can attract visitors to the city, facilitate tourist mobility, help to extend their stay and consequently benefit the tourism economy (Mandeno, 2011; Yang, 2010). For example, an improved public transport network is believed to be of vital importance for Dublin to be more attractive to tourists (Kinsella and Caulfield, 2011). In other words, public transport is an essential element and plays a key role in the sustainable development of an urban destination. Tourists should be encouraged to use more public transport for travel. However, in order to promote public transport use, it is important to understand how tourists use public transport and what is particularly required by them. The next section examines a case study of public transport use by visitors in the city of Munich, Germany.

Public transport use by visitors: the case of Munich, Germany

Munich as a major urban tourism destination

As the capital of Bavaria, Germany's wealthiest state, Munich has over 1.4 million inhabitants, making it the country's third largest city. Munich is the economic centre of Southern Germany and serves as a home to some of the world's largest corporations such as BMW, Siemens AG and Allianz. In addition to its strong industrial and financial services, tourism is one of the city's important revenue generators. With 5.4 million overnight stays in 2011, Munich is the second most visited city in Germany (after Berlin) (German National Tourist Board, 2012). Tourists visit Munich for its multiple remarkable art museums, historical sites and well-known events. However, as with most other cities, Munich experiences high and low seasons. The city often strains for two weeks in September/October, when millions of tourist come to celebrate Oktoberfest, the world's largest beer festival. In 2012 for example, 6.4 million visitors were recorded in the period from 22 September to 7 October (City of Munich, 2013), causing enormous stress for transport in the city. To accommodate the large and increasing inflow of tourists, having an efficient public transport system is therefore of high importance to the city.

The public transport systems in Munich

The public transport network in Munich is one of the most well-developed and extensive systems in Europe. The network includes 442 km of S-Bahn (suburban trains), 95 km of U-Bahn (underground trains), 79 km tram and 454 km of local bus routes. All are under the supervision of the Munich Transport and Tariff Association (MVV – Münchner Verkehrs- und Tarfiverbund). In 2009, public transport systems in Munich carried 500 million passengers. Of the residents of Munich, 66 per cent use the underground, bus and tram several times per week and 35 per cent of them are daily users of the systems. Public transport therefore has taken over the car (31 per cent daily users) to be the most common means of travel in Munich (Münchner Verkehrsgesellshaft mbH, 2010).

Munich has a relatively long history in urban planning and transport management, which dates back to the early twentieth century. Several transport projects and development plans have been undertaken in Munich including the *Perspective Munich* which was initiated in 1998 aiming at better urban expansion management (City of Munich, 2005a, 2005b). With the motto "compact, urban, green", *Perspective Munich* is a flexible guide founded on two principles, namely sustainability and urbanism. The city also invested €1 million per year to implement the mobility management concept "*München – Gscheid mobil*", targeting increased (sustainable) mobility for four groups: new citizens, children and young people, companies and other important target groups including the elderly (Schreiner, 2007). Committed to becoming a sustainable city, mobility has been an important focus in the city urban planning and development. Since October 2008, the city centre of Munich has been designated as a so-called "low emission zone"; high emissions vehicles are not allowed to drive in. Sustainable mobility is also a primary component of its "Green City" project (Green City Projekt GmbH, 2013).

Given the increasing number of tourists to the city, some initiatives have been made to promote the use of public transport by the visitors. The Munich

CityTourCard was first introduced in 2007, targeting the visitor group. The Card offers tourists unlimited travel on public transport plus discounts at several attractions. It comes in several varieties and prices, and can be purchased at tourist centres, ticket vending machines, Munich Transport and Tariff Association (MVV) and Munich Transport Company (MVG) customer centres, and through MVG/MVV partners.

Visitors' use of public transport in Munich

The survey

To examine how visitors use public transport in Munich, data were collected from a self-completed questionnaire-based survey, a standard method in transport behaviour research (see, for example, Bansal and Eiselt, 2004; Fellesson and Friman, 2008). Respondents were filtered by the question "Have you used public transport in Munich during this visit?" Users of public transport were then asked to rate their perception of public transport services using a five-point Likert scale from 1 (lowest) to 5 (highest). Popular tourist sites in Munich including the English Garden, the Residenz and the Pinakothek Museums were chosen at study sites. Following pilot testing, the survey was conducted in April and May 2012. Overall, 2,481 people were approached and about 500 questionnaires were distributed. Of the 483 questionnaires collected, 466 were usable, 17 were rejected because the questionnaire was not properly completed, with most of the important questions being skipped or the respondents were not considered as visitors.

The respondent profile

Among 466 respondents, 82% (380 visitors) used public transport in Munich during their visit. As simplified in Table 12.1, the visitor users of public transport in Munich tended to be of a younger age group (40% aged 18–29). Most reported high education (62% are graduates and postgraduates) and indicated no health restrictions (87%). Interestingly, almost half of the respondents were repeat visitors in Munich, mostly for holiday (54%) and VFR purposes (22%).

Visitors mainly used public transport for tourism related purposes such as to get to attractions (77%) or to travel around Munich for an overview of the city (54%). Public transport was also used for other activities including shopping (47%), visiting friends and relatives (21%) and business related purposes (13%). For the majority of the visitors (51%), public transport was the main transport mode in Munich. Among different types of public transport, the U-Bahn (underground train) was the most popular (used by 88% of respondents), followed by S-Bahn (suburban train) (67%). Other types (tram and bus) are relatively less common (43% and 39% respectively). There are a variety of tickets for public transport in Munich, however, the visitors in the survey reported highest use of the partner day ticket (29%), three-day ticket (27%), single-day ticket (20%) and

Table 12.1 Profile of visitor user of public transport in Munich

---

Who am I?
- Of younger age
- Well-educated
- In good physical condition
- On holiday trip

Use of public transport
- *Why*: for tourist activities and shopping
- *What*: underground train and suburban train
- *Which tickets*: partner day ticket, three-day ticket, single-day ticket
- *How*: main mode of transport

---

single-trip ticket (18%). Other types of tickets such as a weekly ticket, monthly ticket, Bavarian ticket, etc. were only used by less than 10% of the respondents. Interestingly, the CityTourCard was only used by around 5% of the respondents.

Important attributes in public transport services for visitors

Most urban public transport systems are built essentially to accommodate the needs of the residents. Attributes such as reliability, frequency, pricing, speed and comfort are often appreciated by public transport passengers (Redman *et al.*, 2013). Visitors as users of public transport however are expected to have distinctive requirements and perceptions from the locals. The survey findings highlight four most important attributes of public transport services, namely: service quality, ease of use, information and price. These items were resulted from a Discriminant Function Analysis to determine the service attributes which have strongest influence on visitors' overall satisfaction with public transport in Munich.

Service quality

Quality should always be the focus of all product and service providers including transport. Similar to resident users (Budiono, 2009; Tyrinopoulos and Antoniou, 2008), visitors value a transport system that offers good quality. Public transport in Munich is highly rated by visitors yet there is still room for improvement. In particular, respondents stressed the importance of service frequency, cleanliness and space on the vehicles. Munich has an extensive public transport network nonetheless trains and buses do not run very frequently. During off-peak hours, the underground trains (U-Bahn) run every ten minutes, whereas most suburban trains (S-Bahn), tram and buses have intervals of 20 minutes. This of course leads to long waits at the stations/bus stops and also more crowds on vehicles. The common suggestions from visitors are to have high frequency during the day and services the entire night. Additionally, more seating facilities at stations and bus stops is also recommended.

Ease of use

Ease of use is "the degree to which travellers spend affective and cognitive effort on a journey by public transport" (Dziekan, 2008, p. 12). Less effort spent means that the passenger experiences better convenience and feels more secure and comfortable in learning to use the transport system. Having a user friendly system is important for public transport operators to increase passenger penetration. This is especially so in the case of visitors, who are new to the place and probably even the language. Apparently most tourists are unfamiliar with and may even be intimidated by the public transport systems at the destinations (Lew and McKercher, 2006). Therefore, complex public transport systems may discourage visitors from using them. As Thompson and Schofield (2007) pointed out, ease of use is more important for visitors than the efficiency and safety of a network.

Visitors in Munich were relatively satisfied with the public transport ease of use (mean=3.87). Nevertheless, their perception is slightly influenced by whether or not they had been in Munich before. As expected, returning visitors found public transport easier to use than first-time visitors. Similarly, the number of previous trips has a positive effect on visitor perception: the more trips he/she has made to Munich, the easier it is for him/her to use the transport system. Making public transport easy to use is essential as visitors tended to recommend others to use the system if public transport was perceived as "easy".

Information

Information has been emphasised as highly important for public transport users (Friman *et al.*, 2001; Friman and Gärling, 2001). For the visitors in Munich, getting public transport information was also essential (mean=3.98). The majority of visitors (90%) collected some kind of information before their journey on public transport. However, while most looked for information during their time in Munich or just before getting on public transport (58%), other tourists (32%) started searching for information on public transport before coming to the city. The fact that tourists collected information early in their trip stage confirms the importance of information for them and also shows that tourists require a lot of information to assist in their decision-making. As Thompson (2004) asserted, tourists need more information than residents. This could be because much transport information is linked to local knowledge (e.g. train station location, departure and arrival points), whereas tourists are unfamiliar with the place and the systems. Second, there are differences in terms of the information sources that are referred to. Real-time information was considered most important by local public transport users (Molin and Timmermans, 2006). Tourists, however, prefer information sources such as tourist information centres, word-of-mouth, attraction leaflets, the Internet and hotel reception (Thompson, 2004). Similarly, the visitors in Munich tended to use a variety of information sources. Train stations and bus stops were most used (52%), followed by the Internet (36%). Visitors also relied on the local people, accommodation reception and tourist centres for information (Table 12.2).

Table 12.2 Public transport information sources for visitors

Information sources	% responses	% cases
Train stations and bus stops	31.0	52.0
Internet	21.4	35.9
Local people	15.4	25.7
Accommodation reception	10.2	17.2
Tourist information centres	10.1	16.9
Mobile phone applications	4.8	8.0
Other tourists	2.9	4.8
Other	4.2	7.0

Price

Pricing of the service is an important factor that determines the attractiveness of public transport (Budiono, 2009; Redman *et al.*, 2013). Visitors tended to be least satisfied with ticket prices of public transport services in the city (mean=2.93). In April and May 2012 (when the survey was conducted), a single-ride ticket in the inner city, which is valid for three hours and does not allow a return trip, costs €2.5. A single day ticket is priced at €5.60. The prices, according to some visitors, are "high" and "expensive". However, visitors travelling in groups of up to five people have a better option of a partner day ticket (€10.2). There are many ticket options available, which on the one hand offers tourists variety, yet on the other hand causes confusion. This is especially so in the case of foreign visitors, as one respondent indicated "I don't speak German well so it was hard for me to buy the right kind of ticket…" The visitor suggestions could be summarised as to have "fewer ticket options" and "lower prices".

Marketing public transport in tourism

Based on the most important attributes of public transport for visitors, this section discusses the marketing strategies to encourage the use of public transport in tourism. It is suggested that in order to make public transport more attractive to tourists, transport planners and operators should place more attention in service improvement, promotion and pricing, which respectively reflects the attributes discussed earlier: service quality and ease of use, information and price.

Service improvement

The importance of service quality is well recognised in the tourism industry (Chen and Chang, 2005; Erdil and Yıldız, 2011). In the case of public transport, service quality is the attribute that is most valued by passengers (Budiono, 2009; Tyrinopoulos and Antoniou, 2008). Redman *et al.* (2013) suggested that an improvement in public transport services can attract more private car users.

However, most public transport systems were built according to the residents' needs. Public transport services should be modified and customised to better accommodate the visitor users.

Tourists are different from the local users and so are their needs for public transport. Passengers look for an effective and efficient transport system with high punctuality and reliability, frequent services, convenient schedule and good network connection. However, while local residents were concerned with aspects such as quality and safety of the vehicles, visitors to a city emphasised service reliability, frequency and punctuality (Kinsella and Caulfield, 2011), or as in this study: service frequency and clean and spacious vehicles. An efficient and reliable public transport system is thus critical to attract more visitor users. Nonetheless, it is most important to make the system easy to use. This is because public transport's ease of use has a stronger influence on destination satisfaction than efficiency and safety (Thompson and Schofield, 2007). A complicated system would put off visitors from trying to use it. Munich public transport has a high number of visitor users (82 per cent of the respondents in the survey used public transport) which is partially due to the fact that it is perceived as relatively easy to use. However, visitors have most problems figuring out which tickets to buy or how to find their ways through large stations. Language is of course the first problem, and therefore more English signage showing different exits is recommended, especially at interchange stations or stations close to tourist attractions. Including pictures of the tourist sites on the network map to help visitor find their route easily should also be considered. With respect to ticket issues, there should be clearer explanations of the benefits for each ticket type, which include examples of target group so that visitors can easily identify the most suitable one for them. The use of electronic smartcard tickets would greatly simplify the ticketing system. However, despite having been in planning for many years, it has still not been implemented in Munich.

Furthermore, similar to other studies (Lew and McKercher, 2006; Yang, 2010), buses appeared to be least favoured by visitors in Munich. It could be because bus networks are often too problematic for visitors to be able to use effectively (Lew and McKercher, 2006). Providing visitors with bus route maps and incorporating bus information in the sign board at train stations could be useful. Bus stations should also be within walking distance to major sites and indicated with clear signposting in both German and English. Moreover as Lumsdon (2006) recommended, friendly bus drivers with good local knowledge are also desirable.

Promoting public transport

Even though the public transport system could be excellent, visitors may not use it simply because they are not aware of its existence. Lack of information is one of the most common reasons for visitors for not using public transport (Edwards and Griffin, 2013). Having information about public transport at the destination is highly important for tourists (Kinsella and Caulfield, 2011). Therefore, an important part of promoting public transport to visitors is to provide them with sufficient information. In line with findings from Thompson (2004), the Munich case study suggests that tourists require more information than local people and they often refer to traditional information sources. Public transport operators should cooperate with tourist centres, tourist attractions and hotels in order to give tourists accurate and updated information in a form that is easily understandable by visitors. Providing booklets and brochures at train stations, bus stops, accommodations or tourist centres will enable tourists to conveniently reach the information needed. Even though mobile phone applications were not used by many visitors in the case study, online information plays an important role in tourism marketing (Buhalis and Law, 2008; Doolin *et al.*, 2002; Wang *et al.*, 2002). Passengers of public transport value real-time information (Dziekan and Kottenhoff, 2007) and mobile phone applications could be a good way to deliver the information effectively to the visitors. Furthermore, providing an additional English version and/or other major languages such as French, Spanish or Chinese is necessary for the visitors as many of them may not speak the local language. Similarly, transport and ticketing signage in several languages at purchase points can be extremely helpful.

Pricing

Pricing often has significant influence on customer use of public transport (Budiono, 2009; Redman *et al.*, 2013). Fare promotion and special ticket schemes are often effective in the case of local residents and could also be applied for visitors. The ticket prices for public transport in Munich were perceived to be relatively high. While pricing of public transport is a decision involving multiple stakeholders and depends on several factors, there should perhaps be considerations for a discount ticket for special groups of visitors such as students or seniors. Moreover, promotion could be applied to the use of transport services outside peak hours. Big cities often have problems with large crowds on public transport during peak hours, whereas demand during off-peak periods has high elasticity (Paulley *et al.*, 2006). Tourists, on the other hand, are more flexible in terms of timing use. Applying a different pricing scheme for high/low periods is a common revenue management practice to boost demand in tourism (Talluri and Van Ryzin, 2005). Similarly, there should be discounted tickets to motivate tourists to use public transport during off-peak hours. The IsarCard9Uhr, the ticket valid only after 9 a.m., should be offered in daily and weekly forms, in addition to its present monthly option, which most tourists cannot benefit from.

The regional public transport operators in Germany, Switzerland and Austria have been successful in attracting passengers by offering a wide variety of tickets and by combining public transport fares with entrance fees (Pucher and Kurth, 1995). However, this may not be very effective for visitors. As mentioned earlier, visitors in Munich were confused by the many types of tickets. Together with the language problem, visitors would have difficulty in choosing the best option for them. Therefore, offering a wide selection is always good but, most importantly, transport operators should provide clear instructions and explanations for visitors.

Figure 12.1 summarises the marketing strategies addressing the four most important attributes of public transport valued by visitors: service quality, ease of use, price and information. Visitors prefer a system that runs frequently and offers clean and spacious vehicles. Public transport should also be reliable and efficient; constant quality monitoring and customer survey are thus necessary. Most importantly, the system should be easy to use, even for newcomers to the city. One essential part of this is the provision of sufficient and accessible information in multiple languages. To reach out to the potential visitor users, information should be provided in different forms including brochures/booklets at tourist traditional information sources such as train stations, accommodations and tourist centres, as well as real-time information made available on the Internet, mobile phone applications and ticketing vending machines. In addition to ticket counters and vending machines, tickets should be made available online for tourists to buy in advance. This is especially useful for those who fly in. A discount offer in combination with the flight ticket for example would possibly contribute to encouraging more use of public transport at the destination. Planning for pricing strategies should give consideration to students and seniors. Group discount, off-peak hour tickets and a Tourist Card that provides multiple access to attractions and public transport could also be adopted to motivate more use of public transport.

Conclusion

Transport infrastructure plays a major role in determining the attractiveness of a destination (Khadaroo and Seetanah, 2008). Public transport also plays an important role in sustainable urban tourism development and sustainable mobility.

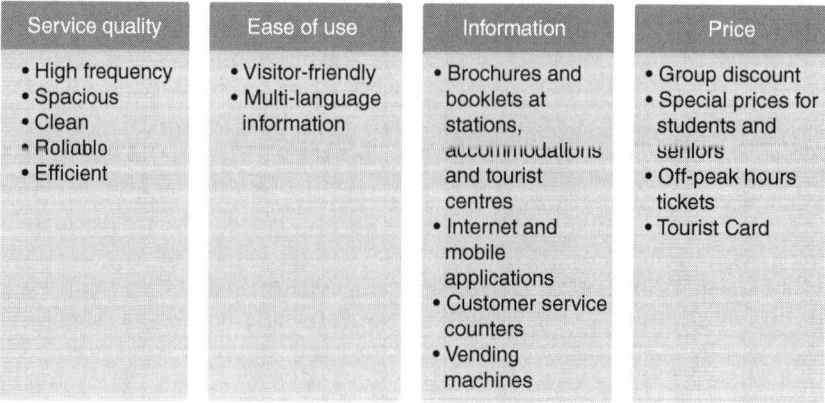

Service quality	Ease of use	Information	Price
• High frequency • Spacious • Clean • Reliable • Efficient	• Visitor-friendly • Multi-language information	• Brochures and booklets at stations, accommodations and tourist centres • Internet and mobile applications • Customer service counters • Vending machines	• Group discount • Special prices for students and seniors • Off-peak hours tickets • Tourist Card

Figure 12.1 Marketing strategies to promote public transport in tourism.

Therefore, adaptive public transport planning is important for the success of urban tourism (Kwan, 2012), reducing traffic congestion, as well as lowering the per trip emissions of visitors. Cities with excellent public transport systems can also potentially attract more visitors (Mandeno, 2011; Yang, 2010). However, to promote the use of public transport in tourism, transport operators should start from the visitors' needs (Gronau and Kagermeier, 2007) and engage in appropriate marketing programmes (Hall, 2014). The initial target visitor users of public transport are younger and well-educated visitors travelling for holiday purposes. As indicated from the findings in the case study, this group of visitors is more likely to use public transport than others. However, over time other visitor market segments can be engaged, especially as marketing campaigns are developed that engage visitors and travel intermediaries with appropriate pre-trip information. As users of public transport, visitors greatly emphasise service quality, ease of use, information and price. Service improvement therefore should be the core of any marketing plan yet it is most important to make the public transport system visitor-friendly. Transport providers should also make information widely available for visitors, especially at their preferred information channels (e.g. train stations, accommodation, tourist centres). A public transport system may be simple for the residents to use but visitors may find it problematic if they cannot read the information on stops or trains. Finally, visitors should be motivated by different ticket categories that offer more discounts or other benefits.

References

Albalate, D. and Bel, G. (2010). Tourism and urban public transport: Holding demand pressure under supply constraints. *Tourism Management*, *31*(3), 425–433.

American Association of Public Transport (APTA). (2013). *Facts at a glance*. Retrieved from www.publictransportation.org/news/facts/Pages/default.aspx.

Ashworth, G. and Page, S. J. (2011). Urban tourism research: Recent progress and current paradoxes. *Tourism Management*, *32*(1), 1–15.

Bansal, H. and Eiselt, H. A. (2004). Exploratory research of tourist motivations and planning. *Tourism Management*, *25*(3), 387–396.

Budiono, O. A. (2009). *Customer satisfaction in public bus transportation: A study of travelers' perception in Indonesia*. (Master thesis), Karlstad University.

Buhalis, D. and Law, R. (2008). Progress in information technology and tourism management: 20 years on and 10 years after the Internet – The state of eTourism research. *Tourism Management*, *29*(4), 609–623.

Chen, F.-Y. and Chang, Y.-H. (2005). Examining airline service quality from a process perspective. *Journal of Air Transport Management*, *11*(2), 79–87.

City of Munich. (2005a). Shaping the future of Munich, Perspective Munich – Strategies, Principles, Projects *Development Report 2005*. Munich: Department of Urban Planning and Building Regulation.

City of Munich. (2005b). Transport Development Plan, Perspective Munich *Development Report 2005*. Munich: Department of Urban Planning and Building Regulation.

City of Munich. (2013). *Das Oktoberfest in Zahlen*. Retrieved from www.muenchen.de/rathaus/Stadtverwaltung/Referat-fuer-Arbeit-und-Wirtschaft/Tourismusamt/Presse-und-Medienservice/Pressetexte/Oktoberfest.html.

Connell, J., Page, S. J. and Bentley, T. (2009). Towards sustainable tourism planning in New Zealand: Monitoring local government planning under the Resource Management Act. *Tourism Management, 30*(6), 867–877.

Dickinson, J. E. and Dickinson, J. A. (2006). Local transport and social representations: Challenging the assumptions for sustainable tourism. *Journal of Sustainable Tourism, 14*(2), 192–208.

Dickinson, J. E., Robbins, D. and Fletcher, J. (2009). Representation of transport: A rural destination analysis. *Annals of Tourism Research, 36*(1), 103–123.

Doolin, B., Burgess, L. and Cooper, J. (2002). Evaluating the use of the Web for tourism marketing: A case study from New Zealand. *Tourism Management, 23*(5), 557–561.

Dubois, G., Peeters, P., Ceron, J.-P. and Gössling, S. (2011). The future tourism mobility of the world population: Emission growth versus climate policy. *Transportation Research Part A: Policy and Practice, 45*(10), 1031–1042.

Duval, D. T. (2007). *Tourism and transport: Modes, networks and flow.* Clevedon, Buffalo: Channel View Publications.

Dziekan, K. (2008). *Ease-of-use in public transportation.* Stockholm: Department of Transport and Economics, Royal Institute of Technology.

Dziekan, K. and Kottenhoff, K. (2007). Dynamic at-stop real-time information displays for public transport: Effects on customers. *Transportation Research Part A: Policy and Practice, 41*(6), 489–501.

Edwards, D. and Griffin, T. (2013). Understanding tourists' spatial behaviour: GPS tracking as an aid to sustainable destination management. *Journal of Sustainable Tourism, 21*(4), 580–595.

Erdil, S. T. and Yıldız, O. (2011). Measuring service quality and a comparative analysis in the passenger carriage of airline industry. *Procedia – Social and Behavioral Sciences, 24*, 1232–1242.

Eurostat. (2011). *Transport accident statistics.* Retrieved from http://epp.eurostat.ec.europa.eu/statistics_explained/index.php/Transport_accident_statistics.

Eurostat. (2012). *Railway safety statistics.* Retrieved from http://epp.eurostat.ec.europa.eu/statistics_explained/index.php/Railway_safety_statistics.

Fellesson, M. and Friman, M. (2008). Perceived satisfaction with public transport service in nine European cities. *Journal of the Transportation Research Forum, 47*(3), 93–104.

Friman, M. and Gärling, T. (2001). Frequency of negative critical incidents and satisfaction with public transport services. II. *Journal of Retailing and Consumer Services, 8*(2), 105–114.

Friman, M., Edvardsson, B. and Gärling, T. (2001). Frequency of negative critical incidents and satisfaction with public transport services. I. *Journal of Retailing and Consumer Services, 8*(2), 95–104.

German National Tourist Board. (2012). *Incoming Tourism Germany.* Retrieved from www.germany.travel/media/en/DZT_Incoming_GTM11_web.pdf.

Gössling, S., Scott, D. and Hall, C. M. (2013). Challenges of tourism in a low-carbon economy. *Wiley Interdisciplinary Reviews: Climate Change, 4*(6), 525–538.

Green City Projekt GmbH. (2013). Retrieved from www.greencity.de/.

Gronau, W. and Kagermeier, A. (2007). Key factors for successful leisure and tourism public transport provision. *Journal of Transport Geography, 15*(2), 127–135.

Guiver, J., Lumsdon, L. and Weston, R. (2008). Traffic reduction at visitor attractions: The case of Hadrian's Wall. *Journal of Transport Geography, 16*(2), 142–150.

Hall, C. M. (2011). Policy learning and policy failure in sustainable tourism governance: From first and second to third order change? *Journal of Sustainable Tourism, 19*(4–5), 649–671.

Hall, C. M. (2014). *Tourism and social marketing.* Abingdon: Routledge.

Hall, C. M. and Lew, A. (2009). *Understanding and managing tourism impacts: An integrated approach.* London: Routledge.

Hall, C. M., Scott, D. and Gössling, S. (2013). The primacy of climate change for sustainable international tourism. *Sustainable Development, 21*(2), 112–121.

Holmgren, J. (2007). Meta-analysis of public transport demand. *Transportation Research Part A: Policy and Practice, 41*(10), 1021–1035.

Høyer, K. G. (2000). Sustainable tourism or sustainable mobility? The Norwegian case. *Journal of Sustainable Tourism, 8*(2), 147–160.

International Association of Public Transport (UITP). (2013). *What is public transport?* Retrieved from www.uitp.org/Public-Transport/why-public-transport/index.cfm.

Khadaroo, J. and Seetanah, B. (2008). The role of transport infrastructure in international tourism development: A gravity model approach. *Tourism Management, 29*(5), 831–840.

Kinsella, J. and Caulfield, B. (2011). An examination of the quality and ease-of-use of public transport in Dublin from a newcomer's perspective. *Journal of Public Transportation, 14*(1), 69–81.

Kwan, C. Y. J. (2012). *To investigate slow mode transport for urban tourism in Hong Kong.* (Master of Science in Urban Planning), The University of Hong Kong.

Lew, A. and McKercher, B. (2006). Modeling tourist movements: A local destination analysis. *Annals of Tourism Research, 33*(2), 403–423.

Lumsdon, L. M. (2006). Factors affecting the design of tourism bus services. *Annals of Tourism Research, 33*(3), 748–766.

Mandeno, T. G. (2011). *Is tourism a driver for public transport investment?* (Master of Planning), University of Otago, Dunedin, New Zealand.

Martín-Cejas, R. R. and Sánchez, P. P. R. (2010). Ecological footprint analysis of road transport related to tourism activity: The case for Lanzarote Island. *Tourism Management, 31*(1), 98–103.

Molin, E. J. E. and Timmermans, H. J. P. (2006). Traveler expectations and willingness-to-pay for Web-enabled public transport information services. *Transportation Research Part C: Emerging Technologies, 14*(2), 57–67.

Münchner Verkehrsgesellschaft mbH (MVG). (2010). Sustainable mobility for Munich (2010). Retrieved from www.mvg-mobil.de/en/images/mvg_nachhaltigkeitsbericht_02052011_eng.pdf.

Page, S. and Hall, C. M. (2003). *Managing urban tourism.* Harlow: Prentice Hall.

Paulley, N., Balcombe, R., Mackett, R., Titheridge, H., Preston, J., Wardman, M., Shires, J. and White, P. (2006). The demand for public transport: The effects of fares, quality of service, income and car ownership. *Transport Policy, 13*(4), 295–306.

Pucher, J. and Kurth, S. (1995). Verkehrsverbund: The success of regional public transport in Germany, Austria and Switzerland. *Transport Policy, 2*(4), 279–291.

Redman, L., Friman, M., Gärling, T. and Hartig, T. (2013). Quality attributes of public transport that attract car users: A research review. *Transport Policy, 25*, 119–127.

Risteskia, M., Kocevskia, J. and Arnaudov, K. (2012). Spatial planning and sustainable tourism as basis for developing competitive tourist destinations. *Procedia – Social and Behavioral Sciences, 44*, 375–386.

Schreiner, M. (2007). München – gscheid mobil: Munich invests 1 million €/a for realising new mobility management concept. Retrieved from www.kpvv.nl/files_content/schreiner.pdf.

Scott, D., Gössling, S. and Hall, C. M. (2012a). International tourism and climate change. *WIRES Climate Change*, 3(3) (published online: 22 March 2012). doi: 10.1002/ wcc.165.

Scott, D., Gössling, S. and Hall, C. M. (2012b). *Tourism and climate change: Impacts, adaptation and mitigation*, 213–232. London: Routledge.

Sorupia, E. (2005). Rethinking the role of transportation in tourism. *Proceedings of the Eastern Asia Society for Transportation Studies*, 5, 1767–1777.

Talluri, K. T. and Van Ryzin, G. J. (2005). *The theory and practice of revenue management*. New York: Springer-Verlag US.

Thompson, K. (2004, 04/10/2004–06/10/2004). *Tourists' use of public transportation information: What they need and what they get*. Paper presented at the Association for European Transport Strasbourg, France.

Thompson, K. and Schofield, P. (2007). An investigation of the relationship between public transport performance and destination satisfaction. *Journal of Transport Geography*, 15(2), 136–144.

Tyrinopoulos, Y. and Antoniou, C. (2008). Public transit user satisfaction: Variability and policy implications. *Transport Policy*, 15(4), 260–272.

Waligo, V. M., Clarke, J. and Hawkins, R. (2013). Implementing sustainable tourism: A multi-stakeholder involvement management framework. *Tourism Management*, 36, 342–353.

Wang, Y., Yu, Q. and Fesenmaier, D. R. (2002). Defining the virtual tourist community: implications for tourism marketing. *Tourism Management*, 23(4), 407–417.

Yang, Y. (2010). *Analysis of public transport for urban tourism in China*. (Master of Arts in Transport Policy and Planning), The University of Hong Kong.

13 Understanding tourists' perceptions of distance

A key to reducing the environmental impacts of tourism mobility

Gunvor Riber Larsen and Jo W. Guiver

Introduction

The average distances travelled by tourists are increasing, especially since flying has become widely used for vacation travel (Peeters *et al.*, 2007). The availability of low-cost air fares has brought many distant destinations within reach and inside acceptable time periods and cost constraints (Knowles, 2006). Lower fares and provision of many new direct flights have together resulted in a higher proportion of the tourist population visiting distant places, with future projections of yet more growth and more distant travel (Peeters, 2007). Unfortunately longer travel generates more greenhouse gas emissions. Although tourism is currently only responsible for about 5 per cent of global greenhouse gas emissions, both its absolute and relative contributions are growing while other industries are reducing their emissions (UNWTO–UNEP–WMO, 2008).

Transport accounts for 75 per cent of tourism's emissions, largely from flying (UNWTO–UNEP–WMO, 2008), and reducing the distance travelled, particularly by air, is therefore a high priority, if tourism's emissions are to be reduced (Peeters, 2007). In a report on climate change and tourism's global challenges UNWTO–UNEP–WMO (2008) argue for a shift towards surface modes to reduce tourism transport's emissions.

While the increasing distances being travelled have been measured and aggregate movements of tourists analysed (see Ankomah *et al.*, 1996; Lin and Morais, 2008; McKercher and Lew, 2003; Nicolau and Mas, 2006), there are no accounts of how tourists *themselves* view distance and the role it plays in their travel behaviour. This research identifies different ways in which Danish tourists talk about distance and uses the insights gained to suggest ways to encourage tourists to travel shorter distances.

After more detailed evidence of the increase in tourist travel, the chapter explores thinking about distance in several disciplines, including geography, tourism and mobility. A methodology section is followed by a summary of the findings and a discussion about how these results might inform efforts to decrease the distances tourists travel.

Increasing tourism travel

It has been estimated that tourism contributed between 3.9 per cent and 6.0 per cent of global CO_2 emissions in 2005 of which 75 per cent were from travel and 40 per cent were from air travel (UNWTO–UNEP–WMO, 2008).

Although air travel still accounts for a small proportion of tourist trips (9 per cent in 2005 (UNWTO–UNEP–WMO, 2008)), it has a high environmental impact because of the distances travelled and its high emissions (Gössling and Upham, 2009), making it an ideal target for reduction of tourism's emissions (Dubois *et al.*, 2011; Scott and Becken, 2010).

Further, Peeters (2007) and Peeters *et al.* (2007) project that total passenger-kilometres will increase for European outbound tourism from 2021 billion kilometres in 2000 to 4480 billion kilometres in 2022, representing a rise of 122 per cent, whereas the number of trips will "only" increase by 57 per cent because of the growth of the long-haul air market (Peeters *et al.*, 2007).

These trends towards taking more frequent holidays in more distant destinations with an increasing proportion of travel by air will result in increasing emissions from tourism at a time when other industries, through technological change and economic downturn, seem to be moving towards lower emissions. Tourism, if emission reduction aspirations fail, risks becoming a major greenhouse gas source (Scott *et al.*, 2010). Scott *et al.* (2010) argue that the only scenario that would reduce total tourism emissions would be high energy efficiency gains through technological developments, combined with considerable modal shifts and tourists choosing closer destinations and staying in these destinations for longer periods of time. Reduction of transport volumes is important to make mitigation strategies of emissions successful. It means uncoupling the growth of tourism and the growth of passenger-kilometres by changing current mobility trends towards shorter and more frequent trips to longer and less frequent ones and to shift destination choices away from long haul (Peeters, 2007, p. 23). This chapter concentrates on the issues involved in reducing the distances travelled, thus reducing fuel burn through distance reduction, and further, by making non-flying modes more attractive.

Travel and distance

Travelling across a distance is obviously an essential element of tourism; tourists being defined as people "traveling to and staying in places outside their usual environment" (UNWTO, 1995, p. 10). Fridgen (1984) identified five phases of tourism experience: planning and anticipation, travel to the destination, being at the destination, the return journey and recollection, and distance features in all five phases. In planning a holiday, distance is an element of destination choice. The journey to the destination and back again represents very tangible engagements with the distance through travel, and most holidays also involve travel within the destination. Lastly, distance becomes an element in holiday recollection through being the spatial separator between the tourist's home and their holiday space.

Much literature has focused on the importance travel has for tourism: Moscardo and Pearce (2004) identified five different roles of travel, ranging from where travel is not undertaken even if it is desired to travel that dominates the experience and is enjoyed and desired. Lumsdon and Page (2004) outline how tourism transport can be understood as a continuum, from the position where tourism transport is viewed purely as a utility with low intrinsic value as a tourism experience to the opposite position where transport is viewed *as tourism* with high intrinsic value for tourism experience. Specific modes of transport are linked to this continuum, with fast modes (such as flying) generally associated with the transport that yields low intrinsic experience values and slow modes of transport (such as walking and cycling) associated with high intrinsic experience values.

Viewing travel as a part of the tourist experience differs from the more traditional view of travel to and from the destination being a practical problem (Haldrup, 2004). It opens the potential of enhancing the experience, rather than crossing the distance as quickly and cheaply as possible, focusing on the values travel can have as a tourist experience (see Dickinson and Lumsdon, 2010; German Molz, 2009; Larsen, 2001; Page, 2005). The journey to and from the holiday destinations holds more importance than just being a practicality to be dealt with, and illustrates how distance has value beyond being a spatial separation to be transcended by tourists en route to their chosen destinations.

Tourists' understanding of distance in relation to tourism mobility remains largely unexplored. Previous studies have shown that distance can be both a positive and negative factor in influencing travel behaviour (Nicolau, 2008) and distance has been viewed by researchers as a proxy for other variables which impact holiday mobility (McKercher *et al.*, 2008). Aggregate studies of tourism often utilise measured distance to explain quantitative relationships between origins and destinations (e.g. Duval, 2007; Mazanec *et al.*, 2007), but they do not attempt to understand the subjective perceptions of distance and how these might underpin tourists' travel behaviour.

Distance

As well as understanding distance as a measured separation between places, most people are also aware that not all kilometres, miles, etc. are the same. While such understandings of distance are sufficient in most everyday situations, for research focused on perceptions of distance, a more nuanced understanding of distance is needed.

Distance is a central concept in both geography and mobilities studies which offer insights into how distance can be understood theoretically. Distance emerges as a concept with both a physical and numerous relative dimensions (Pirie, 2009) and it also denotes a relationship between places (Gatrell, 1983). Distance can be as simple as "close or far away" or express degrees of closeness or separation such as "nearby, further away or a long way away".

Distance's *physical* dimensions are measured in units derived from the physical world and its mappings, such as kilometres or miles (Pirie, 2009). These measurements allow comparisons between distances in different areas and are as such decontextualised, but often form a reference baseline for discussions about distance. Physical distance representations are familiar to most people, and "make sense" in many different spatial situations.

The *relative* dimensions of distance refer to distance using units that are not directly linked to physical spatiality (Chapman, 1983), denoting distance in terms of time, cost, accessibility, travel experience or familiarity (with routes and modes, or culture at destination) (Pirie, 2009). These factors can change the relative distances between places (Cooper and Hall, 2008; Gatrell, 1983) but involve a certain degree of subjectivity because of the different capability of and access to resources to cross the same physical distance. Further, distance may also be perceived differently in the estimation of physical distance (Ankomah *et al.*, 1996), the resources needed to overcome it (Hall, 2005) or the inclination to traverse it (Pirie, 2009). The different degrees of familiarity with the route, mode, journey or destination, various motivations to travel and attitudes towards all of these, alongside different affective and symbolic meanings (Stradling *et al.*, 2001) attributed to these and any alternatives to travelling, all influence choices to travel. Distances between places may therefore hold multiple significances for potential travellers.

The concept of relative distance theoretically challenges the view that distance to tourists is just miles and kilometres. It also offers new possibilities about how the distances travelled by tourists might be decreased. Accepting that perceptions of physical and relative distances exist simultaneously opens the understanding of how tourist travel reaches beyond the need for spatial movement to reach the holiday destination.

Methodology

An abductive (Reichertz, 2007) and qualitative research approach was adopted because of lack of previous studies and knowledge about how tourists perceive the distance they travel. The research aimed to identify different ways in which tourists understand distance and to develop a theoretical understanding of their perceptions rather than seeking a representative sample of tourists' view on distance. Interviews were conducted and recorded with 30 Danish tourists, aged between 26 and 67 years and all with experience of international travel. A wide range of socio-economic backgrounds was deliberately chosen to capture a wide variety of understandings and discourses about distance. A table summarising the socio-economic backgrounds of the interviewees can be found as a supplementary file on the web-based version of this chapter.

The respondents were chosen through theoretical sampling (Corbin and Strauss, 2008); each respondent was chosen based on an assessment of whether they could contribute to expanding the knowledge on tourists' understandings of distance. Snowballing was used as the method of identifying potential respondents, whereby interviewees were asked to suggest other respondents. Each interview lasted between one and two hours and was conducted in a place chosen by the respondent. All were recorded and transcribed and the interview guidelines can be found as a supplementary file on the web-based version of this chapter.

Using semi-structured interviews allowed the respondents to expand on the topics introduced by the researcher and introduce new ones (Kvale, 1997). The 30 interviews were conducted in three stages. Initially, 11 interviews were conducted, transcribed and analysed. These were analysed using discourse analysis in order to understand the language and concepts the respondents used to talk about distance. Discourse analysis focuses on ways of talking about the world (Hall, 1997) and sets of meanings and representations, which together produce a version of events (Burr, 1995). The analysis identifies and examines discourses or ways of talking about a subject. The discourse is the unit of analysis, rather than the respondent. This means contradictory attitudes and actions, even from the same respondent, provide examples of discourses used about the same entity. This was relevant for the identification and understanding of how tourists view distance in a variety of ways.

Following the data collection the interview transcripts were coded using Atlas.ti qualitative data analysis software. First, open coding was used to identify and label all the data relevant for further analysis into various concepts, while "at the same time qualifying those concepts in terms of their properties and dimensions" (Corbin and Strauss, 2008, p. 195). Thereafter axial coding was used in order to explore in more detail the various concepts identified through open coding and to group concepts together into themes. This "bring[s] the data back together again to a coherent whole after the researcher has fractured them through line-by-line [open] coding" (Bryant and Charmaz, 2007, p. 603), and focuses on the concepts that have emerged through open coding, identifying the relationships one concept has to other concepts originating from the same data.

The findings from the data analysis based on open and axial coding were recorded in memos, the written records of analysis (Corbin and Strauss, 2008), which were used to focus the researcher's engagement with the data, while the concepts in the data are developed into more abstract themes. "Writing successive memos keeps the researchers involved in the analysis and helps them to increase the level of abstraction of their ideas" (Bryant and Charmaz, 2007, p. 608). The findings are the discourses identified through open and axial coding and thematically developed in successive memos about the concepts identified in the data.

The next seven respondents were chosen, and the questions put to them were informed by the analysis of the first set of interviews. Likewise the final 12 interviews explored and expanded themes and concepts which had been identified in the previous interviews. The number of respondents in each round of interviews was determined by the point at which empirical saturation was reached and little new knowledge emerged from the analysis. Conducting the interviews in stages therefore allowed knowledge emerging from the data analysis to influence subsequent data collection, creating an interplay between data and theory (Blaikie, 2007) which developed the understanding of how respondents viewed and talked about distance. Examples of topics that received further attention in the subsequent interviews based on the analysis of earlier interviews are: a more detailed discussion of the links between distance and holiday mobility, the relationships between distance perceptions and experiences, and transport modes, and a deeper inquiry of how distance can become a motivational factor and an object for desire for the tourist. The supplementary file on the web-based version of this chapter shows how the interview schedule developed through this three-stage process.

The aim was to explore the range of distance understandings among tourists, rather than to assess their statistical representativeness, and although the discourses identified originate from a small sample of tourists they are likely to represent discourses used more widely by tourists. In the presentation of the findings, translated quotes are used to illustrate the themes and serve as examples of the discourses employed by the respondents.

Findings

The analysis focused on how the interviewees spoke about distance in the context of their holiday mobility. Three categories emerged which are relevant for a discussion of how the distances travelled by tourists can be reduced:

- distance as a use of resources;
- distance as an experience;
- distance as ordinal and zonal.

When talking about distance, the interviewees rarely speak about physical measurements of distance. Distance is predominantly understood and spoken about in relative dimensions. This does not mean that physical distance is not relevant, and most interviewees reflect that their relative understandings of distance are supplemented by their knowledge of the physical distance to a given place. However, the analysis showed that, generally, the interviewees' knowledge of physical distance measured in kilometres to various countries and continents is sketchy. Most interviewees were unable to say in kilometres how far it is from Denmark to some of the favourite Mediterranean holiday destinations and some appeared to not be aware of the relative location of different countries, as exemplified by one interviewee:

> Actually I don't know if it (Egypt) is far away in terms of kilometres in comparison to some of the other places I have mentioned [USA, Asia].
>
> (Female, 26)

The analysis showed that distance is a factor for the interviewees in relation to their holiday mobility, both the choices of holiday destination and travel mode and when they reflect on the holiday experiences, but typically it is the relative dimensions of distance that are relevant and not physical distance.

Distance as a use of resources

The distance travelled by the interviewees is predominantly determined by the availability of time and money: how long a time period do they have for travelling in and how much a journey will cost them. Limited or designated time and/or money budgets can be constraints within which a holiday is planned and which limit the distance travelled:

> It [that, which determines how far to travel] is how long I can get off from my work and how much money I have decided to spend. So it is time and economy.
>
> (Male, 30)

> It [that, which determines how far to travel] is a combination of the cost and then how much time I have at my disposal.
>
> (Male, 37)

Distance becomes measured in time or money. The temporal measurement of distance relates to two issues: the time spent on a specific holiday in relation to the total annual leave and the holiday journey time in relation to the total duration of a specific holiday. Especially, the latter understanding of distance is used to compare the relative distances to different destinations or explain why certain holiday destinations were chosen. Time distance emerged in the interviews as the single most used way of measuring distance.

Cost was also mentioned as an important factor, because cost distance has a strong determining influence on destination choice and transport mode. The price of a holiday is important for most people as they have a budget to work within and the interviewees correlated price with distance. However, it appears that the *total* price of the holiday is important rather than the price of the journey:

> We had discussed, because usually every other year we travel a bit further, so this year we were to stay just in Europe, that would be better. But then we found out that there were cheap tickets to Singapore. So we decided to go down there, and it is cheaper to stay in the East, so that was why [they travelled to Asia instead of Europe, as previously planned].
>
> (Female, 29)

Where time is viewed as a strong signifier of distance, cost-distance is less rigid because all the holiday costs are included, and not only the journey or ticket costs, but the interviews make it clear that cost is perceived as the factor which most constrains the distances travelled on holidays.

Several respondents also pointed out that accessibility rather than physical distance influenced their choice of destination, because inaccessible destinations require more time or money to reach them and often have a hassle factor when the journey is not straightforward:

> Distance is one thing, and accessibility another. Because the distance is still the same. But it is obviously easier to get to Phuket in Thailand than to Nuuk in Greenland.
>
> (Male, 30)

> You can travel far by plane, and there are some places that have very good flight connections, I mentioned Stavanger in Norway before, that is not an easy place to get to, but it is closer than for example Tenerife would be ... but I think it would be easier to get to Tenerife.... It does something for the accessibility [a direct flight route].
>
> (Female, 26)

This understanding of distance in the form of the resources necessary to overcome distance is also reported in the literature (e.g. Hall, 2005) and is an intuitive and relevant representation in everyday life, recognised by many in relation to daily and holiday mobility.

Distance as experience

After distance understood as resources, the way in which the interviewees most often refer to distance is in the form of distance as experience, i.e. the experiences associated with travelling across distance. Distance as experience mostly relates to the journey itself, when the journey assumes intrinsic qualities as well as the instrumental aim of just reaching a destination. The experience value may be derived from the mode, the areas travelled through or the travel companions:

> To move on a bicycle, that is very enjoyable. It is not fast, but the journey becomes part of the holiday.... You are closer to the surroundings, and nothing just flies past you, you see more of the scenery and experience the area in a different way.
>
> (Male, 29)

> I often enjoy the journey very much, even when it is by car, and you have ten hours driving with other people to a ski resort or somewhere, I find that part of the journey quite fun. It is often very nice, you are in the car telling stories, listening to music, drinking coffee. It becomes part of it, it is part of the concept. And I like the feeling of moving.
>
> (Female, 32)

There is also an element of savouring the arrival and enjoying the transition from home to holiday, especially when it involves changes in climate, landscape or culture:

> It is nice [travelling from Denmark to Austria by train]. You arrive in a foreign country peacefully and quietly. And then you sleep on the way, that

is you travel at night and that is actually a good way, you start in the late afternoon from Denmark and arrive the next morning, and the night is spent sleeping and travelling. When it goes to plan you wake up when you have come to Southern Germany and continue into Austria and see more and more mountains, hopefully with snow on them, and that is really a very good way of arriving.

(Male, 63)

For example when I travelled to Spain after high school I deliberately chose the coach, which was 54 hours. I could have taken the plane which would have been two hours, but it was important that I experienced the journey, to be able to adjust to something new.

(Female, 32)

Another aspect of the experience of distance relates not to the journey but to an appreciation of the differences in cultures encountered by travelling further. For some of the respondents experiencing foreign places and cultures is a driver for their holidays and often they make a strong correlation between physical distance and the possibility of encountering that which is different. When distance is spoken about by the interviewed tourists as experience, their behaviour and reflections can often be interpreted as a "desire for distance" and it is clear that distance in itself sometimes becomes an attraction for tourists and a driver for tourism mobility, through the desire to encounter somewhere culturally different. For the interviewed tourists, distance signifies these attributes that are sometimes, though not always, desired in a holiday, and through this association, distance itself becomes desired and an attraction. The attraction of physical distance to tourists has been suggested elsewhere (see Nicolau, 2008; Peeters and Eijgelaar, 2012) but the findings of this research also indicate that distance in relative dimensions can be an attraction to tourists and can motivate travel. The manifestation of a distance desire into actual travel behaviour is through the level of engagement with distance as a part of the journey, which, sometimes, becomes the entire holiday, and through the deliberate choice of travelling to destinations that are culturally different from the usual context of the tourist. Thus, a desire for distance can be identified in two different forms of travel behaviour: distance can be desired when a tourist wants to travel distance in order to achieve difference and associates physical distance and cultural difference. Distance can also be desired when the aim of a holiday is the actual journey from one place to another (or a circular trip), where it is the movement itself that is the holiday experience.

This knowledge that relative distances are desired by some tourists gives an insight into potential reasons why people travel and especially the roles distance has the potential to play in the motivation of holiday travel. Where previously travelling across distance has often been viewed as mostly an instrumental part of tourism mobility, the knowledge about distance desires gives weight to the arguments that the journey can be the most important element of a holiday and shows how engagement with distances (in various relative dimensions) must also be viewed as an intrinsic factor of travelling.

Ordinal and zonal distance

The analysis of the interviews also showed that tourists understand distance as either ordinal or zonal. The expression "ordinal distance" was used by Tobler (2004) to capture the understanding of distance expressed through something or somewhere being near, close or closer, or far, further or furthest away, without necessarily specifically measuring the distance in any physical or relative dimensions:

> It takes longer to fly to New York than it does to Egypt, so in that sense New York is further away. Even though culturally it is more similar, and in that sense New York is closer than Egypt.
>
> (Female, 29)

> I will rather travel from Copenhagen airport, even though it is the airport furthest away, than from Billund or Aarhus, because those are difficult to access if you don't have a car, and therefore they feel further away.
>
> (Female, 59)

That such an understanding of places being at different distances is present in the tourist interviews was not unexpected, as this is an obvious component of a tourist destination choice, theorised by Stouffer (1940) and Hall (2005) in the intervening opportunities model. This model projects that of two destinations offering the same to the traveller, the closest will be chosen. However, the analysis showed that ordinal distance and the destination choice model described by Stouffer (1940) and by Hall (2005) do not necessarily result in the choice of the destination that is closest in terms of physical distance.

> Between Mallorca and Thailand as charter destinations you might as well go to Mallorca, but only if it is the cheapest.
>
> (Male, 33)

> The price matters more than the actual distance [in the destination choice].
>
> (Male, 34)

Often the destination choice is based on an effort to reduce distance in one of its relative dimensions, such as time or cost, so ordinal distance can, just as any other distance, be measured in relative dimensions. The respondents' judgement of which places they perceive to be close or far away then rests on an assessment of relative distance rather than physical distance.

The zonal distance widely expressed by the interviewees is when distance is understood as a zone, where it signifies an unspecified location such as "here" or "there" or "not here" or "not there". Distance becomes relevant just as the spatial separation it signifies and not in terms of any quantitative measure of that distance. In the interviews it is typically seen when the tourists talk about just wanting to get away from home and distance comes to signify somewhere (often anywhere) else but home. Holidays to sun or ski destinations also appear to include a zonal understanding of distance:

> I go away every year on a summer holiday, a week or fourteen days to some almost unimportant place, just to get some sun and summer.... In essence it is just to get away for a while and as long as it is warm, that is the important criterion.
>
> (Female, 29)

From the analysis it is clear that tourists can use both ordinal and zonal understandings of distance: for some holidays an ordinal understanding of distance is predominant, while other holidays are influenced more strongly by a zonal understanding of distance.

Discussion

One of the major challenges in reducing emissions from tourist transportation is reversing the trend to ever-longer travel distances (Bows *et al.*, 2009). Any acceptable reduction in emissions from tourism is likely to require a combination (Scott *et al.*, 2010) of technological and infrastructural improvements (although perhaps significant technological improvement in aviation is limited (Peeters *et al.*, 2009)), regulatory and price-based policies (Daley and Preston, 2009) and changes in travel behaviours. Peeters (2007) argues that emissions could be cut by reducing the number of passenger-kilometres, i.e. if tourists choose closer destinations. UNWTO–UNEP–WMO (2008) highlights the necessity of a modal shift to increase rail and coach travel and decrease air travel. Voluntary change in tourists' travel behaviour on the grounds of environmental concerns appears unlikely to achieve the necessary reduction in emissions (Gössling and Upham, 2009; McKercher *et al.*, 2010). The finding that distance can be a direct source of valued holiday experiences also reduces the potential of voluntary decreases in the distance travelled and emissions.

Time and cost distance

That the tourists interviewed primarily understand distance in relation to temporal and financial resources expenditure is not unexpected. Such an understanding of distance is also reported both within and outside tourism studies (e.g. Hall, 2005; Pirie, 2009). This way of representing distance poses significant challenges to reducing the average distances travelled by tourists as well as reducing high-emission air travel. Holiday destination choice is, according to the interviewed tourists, highly influenced by minimising journey time and journey cost.

This leaves air travel the most preferred mode of holiday transport (although not the favourite one). The tourists interviewed chose flying as a default holiday transport mode after assessing that it is the cheapest and fastest way to reach a destination (a finding echoed by Hares *et al.*, 2010).

This suggests that changes will be needed to the temporal and financial contexts of holidays to effect change. Changing the time budget for tourists is a potential approach, where by design of annual leave budgets could favour more sustainable travel patterns, by for example giving more time off work to people who then spend that extra time travelling a more sustainable way than by air. Scott and Becken (2010) advocate that people travelling long distances should go less frequently and stay longer, currently very difficult with annual leave allowances. More flexible arrangements allowing leave to be accumulated over a few years might allow this. Employers wishing to improve their carbon ratings might give more annual leave to employees using slower modes, but currently holiday carbon spending is not viewed as of any relevance to employers and differential leave allowances might prove contentious. This is also a top-down regulation that could prove very difficult to achieve, as it would involve significant changes in labour regulations and introduce differentiation between workers according to choices regarding their leisure time. Schemes that try to influence people's travel behaviour have, however, been used in relation to work-related mobility, such as the cycle to work schemes in the UK and Denmark, and these might provide inspiration about levers to change holiday mobility.

Another approach to encouraging a change in travel behaviour away from aviation is an increase in air fares, thus making tourists pay for the environmental externalities of their air travel. This could probably reduce aeromobility because of the known power of price differentials. However, Hares *et al.* (2010) found considerable loyalty to low-cost airlines because they had opened up travel and far destinations "to the masses" (Hares *et al.*, 2010, p. 470), making such price rises politically difficult to implement. It would necessitate negotiations with and regulation of the airline industry. Many attempts at regulating air travel have been countered by the argument of reduced competitiveness in the global market and would probably meet resistance from airlines, their customers and long-haul destinations. The latter are especially important in this discussion because, as shown earlier in this chapter, the perceived overall cost of long-haul travel is a summation of at-destination costs and transport costs.

Distance as experience and attraction

The analysis indicates that tourists' experience of distance increases when using a mode which facilitates a closer engagement with the distance being travelled. The slower the tourist travels, the more they engage with the distance, so slower travel modes have the potential to not only make holiday travel more environmentally friendly, but also to satisfy some tourists' desire for distance. Hence, it could be possible to accommodate the environmental aim of fewer passenger-kilometres without tourists necessarily experiencing less distance through a change in the transport mode. Lumsdon and McGrath (2011) argue that slow travel is emerging as a new form of tourism mobility, with an emphasis on travel using slow transport modes (i.e. avoiding planes, cars and possibly high-speed trains), that allows for a deeper engagement with the space and places the tourist travels through and to. Slow travellers incorporate their travel time into the holiday experience (Dickinson and Lumsdon, 2010), so there is potential for a tourist's engagement with distance to be enhanced through deliberately choosing slower surface modes of travel, as opposed to longer distances travelled by air. If tourists could be encouraged to travel by modes that in themselves offer experience and thereby add to the overall holiday experiences, it is likely, from the interview analysis, that they would travel shorter physical distances. So, not only would they not fly, but also use more environmentally sustainable modes for shorter journeys.

The desire for distance is also a result of the link tourists make between physical distance and cultural dissimilarity, i.e. the further one travels, the more likely an encounter with a different culture. Tourists search for novelty (Urry, 2002) and when novel cultures become associated with physical distance, distance appears attractive. The analysis shows that the holidays undertaken to experience different cultures are generally to destinations that are further away than, for example, the annual sun-holiday. This insight suggests that if tourists' desire for experiencing unfamiliar cultural contexts can be disassociated with physical distance, it would be possible to facilitate a reduction in the distance travelled. Supporting this argument are comments made in the interviews about how little tourists know about relatively close destinations, with one interviewee commenting that Ukraine is close compared to some of the other destinations he discussed, but seems very different. Others reflected that many destinations that are close in physical distance actually "feel" far away because of the experience of cultural dissimilarity. If this "feeling" of far away could be harnessed to satisfy a distance desire, physical travel distance could be reduced.

Holiday travel has often been framed as just a practical problem, with few intrinsic or positive emotional values (Haldrup, 2004), but this view of travel as a disutility, a cost of reaching the destination, has been challenged (see Baxter, 1980; Cao *et al.*, 2008; Jain and Lyons, 2008; Moscardo and Pearce, 2004). Travelling across distance does sometimes hold intrinsic values for the traveller, and the analysis of the interviews showed that the journey element of a holiday is often embraced as both a physical and mental transition from one place to another, with surface travel modes having a high intrinsic value, while air travel was never attributed any intrinsic value. Instead of forcing tourists to travel on slower modes through regulation of time and financial resources available for the tourists, policies that encourage a modal shift could be introduced, perhaps in the form of incentives for travelling using surface modes which could satisfy distance desires. Further, a more targeted marketing of nearby destinations could also inform tourists in their destination choice of less distant holiday destinations that have the potential to fulfil their holiday desires.

Zonal and ordinal distance

Scott *et al.* (2010, p. 9) point out that "many tourists do not seek a holiday at a specific destination but seek a specific holiday experience that can be had at several destinations that may be at a range of distances". This corroborates the research findings that many interviewees wanted a certain type of holiday, either "sun and sea" or winter sports, rather than a specific destination. If distance only matters to the tourist as a signifier of absolute separation, in the form of here or there, home or away, the choice of closer destinations would not devalue the holiday experience, as long as the destination fulfils the holiday expectations. This suggests a potential of a choice of nearer destinations within the desired zone, perhaps entailing short-haul, rather than long-haul, flights which could reduce total emissions (Scott *et al.*, 2010). Closer destinations within a desired zone may even meet the threshold at which surface travel offers a viable alternative to air travel.

However, closer destinations within the desired zone will only be attractive if they are closer in "ordinal" terms of money and travel time than further destinations. Although the intervening opportunities model (Hall, 2005) suggests that closer destinations will be favoured, the interviewees' use of "close" to mean cheaper or quicker suggests that physical distance is less important than relative distance, mostly in terms of money and time.

Conclusion

Conceptualising how tourists understand distance provides insights into how possible reductions in travel distances and modal shifts could be encouraged. The research found that tourists understand distance as both physical and relative, measured in kilometres and non-spatial entities such as time, cost and cultural difference. Tourists' understanding of distance was shown to be either ordinal or zonal, where the spatial separation that distance signifies is viewed as either scalar or absolute. Tourists' understanding of distance in relation to their holiday mobility is therefore far from only spatial, and distance to tourists is often represented in dimensions that are not direct attributes of the physical world but rather a result of how tourists manage to engage with distance. Distance was also identified as being an attraction for tourists, especially for holidays that are perceived as free of temporal and financial constraints, where long-distance travel is embraced as a positive and desired holiday element. This seems a logical extension of predicted holiday trends, towards increasing future numbers of trips and the distances travelled (Peeters *et al.*, 2007).

It has previously been established that the scope for significant voluntary changes in travel behaviour towards shorter (in distance) holidays is minimal. This can be explained by understanding how tourists perceive distance. Most seek to minimise the time and cost of their holidays and flying currently offers the best way of doing this. Coupled with the knowledge from the analysis that many tourists desire distance in the form of experience and meeting that which is different, which they associate with long physical distances, this strongly suggests that voluntary travel behaviour change is unlikely.

However, the research also showed that the tourists' desire for distance can be satisfied through destination choices and travel modes that can facilitate tourism mobility to become more environmentally sustainable than it currently is. Distance desires can be satisfied through the choice of culturally different destinations, but not necessarily far away, because what matters is being in a culturally different zone, rather than being distant in physical terms. This also opens the possibility of visitation using surface transport modes. Indeed, land-based transportation satisfies some distance desires as the engagement with distance increases significantly. The physical distances might be shorter but the experiences of distance are better and thus the "need" for distance to satisfy the distance desire decreases.

The major challenges in encouraging tourists to satisfy their desire for distance are who should instigate this change and how. Change towards more sustainable tourism mobility is unlikely to happen through tourists becoming more aware of the damage caused by their current travel behaviour. Change could be encouraged if tourists were more aware of good and valued holiday experiences at closer destinations with more sustainable transportation choices, combined with policy changes that will mitigate constraints felt on tourists' time and financial budgets. Unless changes are made to how these two factors influence the potential for holiday mobility, significant changes in tourism mobility are difficult to envisage, as time and money are the factors that determine how far tourists travel.

References

Ankomah, P., Crompton, J. and Baker, D. (1996). Influence of cognitive distance in vacation choice. *Annals of Tourism Research, 23*(1), 138–150.

Baxter, M. (1980). The interpretation of the distance and attractiveness components in models of recreational trips. *Geographical Analysis, 11*(3), 311–315.

Blaikie, N. (2007). *Approaches to social enquiry* (2nd edn). Cambridge: Polity Press.

Bows, A., Anderson, K. and Peeters, P. (2009). Air transport, climate change and tourism. *Tourism and Hospitality Planning & Development, 6,* 7–20.

Bryant, A. and Charmaz, K. (eds). (2007). *The SAGE handbook of grounded theory.* London: Sage.

Burr, V. (1995). *An introduction to social constructionism.* London: Routledge.

Cao, X., Mokhtarian, P. and Handy, S. (2008). No particular place to go: An empirical analysis of travel for the sake of travel. *Environment and Behavior, 41*(2), 233–257.

Chapman, K. (1983). *People, pattern and process: An introduction to human geography.* London: Edward Arnold.

Cooper, C. and Hall, C.M. (2008). *Contemporary tourism.* Oxford: Butterworth Heinemann.

Corbin, J. and Strauss, A. (2008). *Basics of qualitative research* (3rd edn). London: Sage.

Daley, B. and Preston, H. (2009). Aviation and climate change: Assessment of policy options. In S. Gössling and J. Upham (eds), *Climate change and aviation – Issues, challenges and solutions* (pp. 347–372). London: Earthscan.

Dickinson, J. and Lumsdon, L. (2010). *Slow travel and tourism.* London: Earthscan.

Dubois, G., Peeters, P., Ceron, J.-P. and Gössling, S. (2011). The future tourism mobility of the world population: Emission growth versus climate policy. *Transportation Research Part A*, *45*, 1031–1042.

Duval, D.T. (2007). *Tourism and transport: Modes, networks and flows*. Clevedon: Channel View Publications.

Fridgen, J. (1984). Environmental psychology and tourism. *Annals of Tourism Research*, *11*, 19–39.

Gatrell, A. (1983). *Distance and space: A geographical perspective*. Oxford: Clarendon.

German Molz, J. (2009). Representing pace in tourism mobilities: Staycations, slow travel and the amazing race. *Journal of Tourism and Cultural Change*, *7*(4), 270–286.

Gössling, S. and Upham, J. (2009). *Climate change and aviation – Issues, challenges and solutions*. London: Earthscan.

Haldrup, M. (2004). Laid-back mobilities: Second-home holidays in time and space. *Tourism Geographies*, *6*(4), 434–454.

Hall, C.M. (2005). *Tourism – Rethinking the social science of mobility*. Harlow: Prentice Hall.

Hall, S. (1997). Introduction. In S. Hall (ed.), *Representation: Cultural representation and signifying practices* (pp. 1–11). London: Sage.

Hares, A., Dickinson, J. and Wilkes, K. (2010). Climate change and the air travel decisions of UK tourists. *Journal of Transport Geography*, *18*, 466–473.

Knowles, R. (2006). Transport shaping space: Differential collapse in time-space. *Journal of Transport Geography*, *14*, 407–425.

Jain, J. and Lyons, G. (2008). The gift of travel time. *Journal of Transport Geography*, *16*, 81–89.

Kvale, S. (1997). *Interview – Introduktion til det kvalitative forskningsinterview* [Interview – Introduction to the qualitative research interview]. Copenhagen: Hans Reitzels Forlag.

Larsen, J. (2001). Tourism mobilities and the travel glance: Experiences of being on the move. *Scandinavian Journal of Hospitality and Tourism*, *1*(2), 82–98.

Lin, C. and Morais, D. (2008). The spatial clustering effect of destination distribution on cognitive distance estimates and its impact on tourists' destination choices. *Journal of Travel and Tourism Marketing*, *25*(3–4), 382–397.

Lumsdon, L. and McGrath, P. (2011). Developing a conceptual framework for slow travel: A grounded theory approach. *Journal of Sustainable Tourism*, *19*(3), 265–279.

Lumsdon, L. and Page, S. (2004). Progress in transport and tourism research: Reformulating the transport–tourism interface and future research agendas. In L. Lumsdon and S. Page (eds), *Tourism and transport: Issues and agenda for the new millennium*. London: Elsevier.

Mazanec, J., Wöber, K. and Zins, A. (2007). Tourism destination competitiveness: From definition to explanation? *Journal of Travel Research*, *46*, 86–95.

McKercher, B. and Lew, A. (2003). Distance decay and the impact of effective tourism exclusion zones. *Journal of Travel Research*, *42*(2), 159–165.

McKercher, B., Chan, A. and Lam, C. (2008). The impact of distance on international tourist movements. *Journal of Travel Research*, *47*(2), 208–224.

McKercher, B., Prideaux, B., Cheung, C. and Law, B. (2010). Achieving voluntary reductions in the carbon footprint of tourism and climate change. *Journal of Sustainable Tourism*, *18*(3), 297–317.

Moscardo, G. and Pearce, P. (2004). Life cycle, tourist motivation and transport: Some consequences of the tourist experience. In L. Lumsdon and S. Page (eds), *Tourism and transport: Issues and agenda for the new millennium*. London: Elsevier.

Nicolau, J. (2008). Characterising tourist sensitivity to distance. *Journal of Travel Research*, *47*, 43–52.

Nicolau, J. and Mas, F. (2006). The influence of distance and prices on the choice of tourist destinations: The moderating role of motivations. *Tourism Management*, *27*, 982–996.

Page, S. (2005). *Tourism and transport: Global perspectives*. Harlow: Pearson.

Peeters, P. (2007). Mitigating tourism's contribution to climate change – An introduction. In P. Peeters (ed.), *Tourism and climate change mitigation – Methods, greenhouse gas reductions and policies*. Breda: NHTV Academic Studies No. 6.

Peeters, P. and Eijgelaar, E. (2012, July). *Modelling tourist travel behaviour for a global tourism flow model*. Paper presented at the Conference of Psychological and Behavioural Approaches to Understanding and Governing Sustainable Tourism Mobility, Freiburg, Germany.

Peeters, P., Szimba, E. and Duijnisveld, M. (2007). Major environmental impacts of European tourist transport. *Journal of Transport Geography*, *15*, 83–93.

Peeters, P., Williams, V. and de Haan, A. (2009). Technical and management reduction potentials. In S. Gössling and J. Upham (eds), *Climate change and aviation – Issues, challenges and solutions*. London: Earthscan.

Pirie, G. (2009). Distance. In R. Kitchin and N. Thrift (eds), *International encyclopedia of human geography* (Vol. 1). Oxford: Elsevier.

Reichertz, J. (2007). Abduction: The logic of discovery of grounded theory. In A. Bryant and K. Charmaz (eds), *The SAGE handbook of grounded theory*. London: Sage.

Scott, D. and Becken, S. (2010). Editorial introduction. Adapting to climate change and climate policy: Progress, problems and potentials. *Journal of Sustainable Tourism*, *18*(3), 283–295.

Scott, D., Peeters, P. and Gössling, S. (2010). Can tourism deliver its "aspirational" greenhouse gas emission reduction targets? *Journal of Sustainable Tourism*, *18*(3), 393–408.

Stouffer, S. (1940). Intervening opportunities: A theory relating mobility and distance. *American Sociological Review*, *5*(6), 845–867.

Stradling, S., Hine, J. and Wardman, M. (2001). *Physical, cognitive and affective effort in travel mode choice*. Paper presented at the International Conference on traffic and transport psychology – ICTIP (2000, September), Berne, Switzerland.

Tobler, W. (2004). On the first law of geography: A reply. *Annals of the Association of American Geographers*, *94*(2), 304–310.

UNWTO (1995). *Collection of tourism expenditure statistics. Technical manual no. 2*. Madrid: UNWTO.

UNWTO–UNEP–WMO (2008). *Climate change and tourism: Responding to global challenges*. Madrid: UNWTO–UNEP–WMO.

Urry, J. (2002). *The tourist gaze* (2nd edn). London: Sage.

Part III

Governance and policies based upon psychological, behavioural and social mechanisms

14 Towards a new model for communicating climate change

Sander van der Linden

Introduction

It has been well documented that for most people, the media is a prominent and integral source for acquiring information about climate change (e.g. Boykoff and Rajan, 2007; Ungar, 2000). Moreover, the way that information about climate change is framed and communicated can significantly influence the public's knowledge, attitude and perception (e.g. Sampei and Aoyagi-Usui, 2009; Sharples, 2010; Stamm *et al.*, 2002; Weingart *et al.*, 2000). As a result, a popular strategy for inducing behavioural change has been the deployment of persuasion techniques embedded in communication strategies. To this extent, a major area of concern is the apparent disparity between public communication and the lack of actualised behavioural change observed in the general public (Whitmarsh *et al.*, 2008). While public polls often indicate that people express general awareness and concern (e.g. GlobeScan, 2000, 2006), individuals remain reluctant to take personal action. This has also been dubbed the "value-action" gap (e.g. Kollmuss and Agyeman, 2002), "attitude-behaviour" gap or "intention-behaviour" gap (e.g. Sheeran, 2002) depending on where the focus is applied. Traditionally, most communication campaigns have tried to address this gap by providing people with more information, a strategy that has become better known as the "information-deficit" model of human behaviour. In fact, a content analysis by Devine-Wright (2004) suggests that a deficit model of human behaviour has played a predominant role in past public behavioural change campaigns and continues to do so. On the whole, public interventional campaigns only seem to produce modest behavioural changes (Steg, 2008). For example, a 1999 mass public media campaign in the UK: "Are you doing your bit" only elicited small consequent changes in attitudes and behaviours (O'Neill and Hulme, 2009). Similar disappointing findings have been observed in The Netherlands (e.g. Staats *et al.*, 1996). While new communication strategies have been undertaken in recent years, more substantial analyses of such campaigns often remain elusive (i.e. those that go beyond media hits and broad-opinion brushes), making it hard to identify benefits and limitations (Moser, 2010; Steg and Vlek, 2009). Recent research is increasingly pointing out that communication interventions need to be made more locally relevant and designed in such a way that they meaningfully involve and engage the public with climate change (Moser

2006; O'Neill and Nicholson-Cole, 2009). Moser (2006, p. 3) defines effective communication as: "any form of public engagement that actually facilitates an intended behavioural, organizational, political or other social change consistent with identified mitigation and adaptation goals". While there certainly has been no shortage in the number of publications that offer "practical" shortlists for effective climate change communication (CCCAG, 2010; CRED, 2009; Futerra, 2005; Moser, 2010), there is currently no systematic overview of the theoretical and empirical pathways that explain how to get from merely communicating information to actually changing people's behaviour. Moreover, while behavioural change is of course, to a certain extent, a practical matter, a more theory-driven perspective is generally welcomed by behavioural researchers (Steg and Vlek, 2009). Attaining a more holistic understanding of the link between designing persuasive messages, the communication and processing of that information on one hand and eliciting behavioural change on the other inevitably begs for the integration of insights from all relevant disciplines that deal with the subject matter. Indeed, integrative theoretical research can help synthesise, connect and combine dispersed research findings from various disciplines to advance new insights and improve current knowledge and understanding. Yet, in order to validate the value of a new integrative communication model, it is pivotal to first discuss the theoretical and empirical evidence of past models as well as their limitations. In an attempt to provide a more systematic overview, the current chapter delineates the "evolution" of public climate change campaigns according to the following typology:

1   The "*cognitive-analytical*" type (consistent with the traditional knowledge-attitude-behaviour model);
2   The "*affective-experiential*" type (congruent with the "risk-as-feelings" framework and the use of negative emotional appeals such as fear and guilt messaging); and
3   The "*social-normative*" type (consistent with the "normative" paradigm – which seeks to leverage the persuasive potential of social and moral norms on behaviour).

In addition, three major shortcomings of past and current public communication interventions are identified:

1   Most public interventions ought to be, but are not designed in an integrative manner;
2   Current campaigns do not sufficiently target specific behaviours nor pay sufficient attention to the psychological determinants of the behaviours that they are trying to change; and
3   Public campaigns often fail to make the climate change context explicit.

In the first section of this chapter, the theoretical and empirical evidence for each of these three public communication strategies is critically discussed. In the following section, a more integrated understanding of human behaviour and decision-making is advanced by looking at the combined influence of cognitive, experiential and normative influences on behaviour. In addition, the importance of understanding the psychological determinants of environmental behaviour is outlined for both, the communication process as well as its integral role in eliciting behavioural change. Finally, a new integrative conceptual framework is proposed in an attempt to advance current understanding of how to transition from merely communicating information about climate change to actually changing individual behaviour.

The cognitive-analytical approach

The homo logicus?

> "I know that you believe you understand what you think I said, but I'm not sure you realize that what you heard is not what I meant."
>
> (Robert McCloskey)

Until recently, the tradition has been to communicate information about climate change in a relatively scientific and analytical format (CRED, 2009), operating under the assumption that people process (uncertain) information predominantly in an analytical matter (Marx *et al.*, 2007). As a result, technical terms such as "stratospheric ozone depletion", "anthropogenic climate change" and "significant probability" have often been used in communicating information about the long-term developments in the earth's climate. Yet, whether or not cognitive reasoning abilities in humans are really that well developed is questioned by both comparative neuroanatomical work as well as cognitive psychology. In particular, it has been argued that the "neocortex" (the rational, higher functioning) part of the brain was developed last in the chain of human evolution and is in fact the least developed part of the brain (MacLean, 1990). Similarly, in their heuristics and biases approach, Kahneman *et al.* (1982) have highlighted that, when forming judgements under uncertainty, people employ relatively simple heuristics and cognitive short cuts that may lead to erroneous and biased decision-making strategies. In short, recent research has questioned how proficient individuals are in dealing with abstract, descriptive and analytical information about climate change (Marx *et al.*, 2007).

Because climate change is such a complex and elusive global hazard, the concept is difficult to communicate to various publics (Moser and Dilling, 2004). This process is even further complicated by the fact that people tend to process information so that it is congruent with their pre-existing beliefs. Selectively attending to evidence that confirms pre-existing beliefs and the negligence, re-interpretation as well as distortion of information to the contrary is generally referred to as "confirmation bias" (Lewicka, 1998). In fact, most information that is eventually retained in an individual's memory tends to be information that supports pre-existing thoughts and beliefs. For example, in one US study increased levels of knowledge seemed to increase concern for some people (e.g.

Democrats) but not for those (e.g. Republicans) that were already sceptical about climate change from the outset (Malka *et al.*, 2009). Similarly, in a study where US farmers were asked to recall weather statistics, those who believed that their region was undergoing climate change recalled weather statistics consistent with those beliefs while farmers that believed that their region had a constant climate recalled weather statistics congruent with those beliefs (Weber and Sonka, 2004). In sum, the way in which people process information and structurally organise their knowledge can have significant impacts on their behaviour. Yet, how do people cognitively understand climate change? And to what extent does more knowledge of the climate change problem affect people's behaviour?

Information processing and the structural organisation of knowledge

Cognitive psychologists have often described the way in which individuals process and organise incoming information as an elaborate network of mental structures that represents an individual's understanding of the external world, perhaps better known as "schema theory" (Anderson, 1977). More recently, the study of "mental models" has gained increased attention. Essentially, a mental model is a person's internal, personalised, intuitive and contextual understanding of how something works (Kearney and Kaplan, 1997). What is important to take away from this is that mental models carry three important functions: (1) they serve as a framework into which people fit new information; (2) they define how individuals approach and solve problems; and, perhaps most important, (3) they help formulate actions and behaviour (Carey, 1986; Morgan *et al.*, 2002). The majority of research on mental models has identified several fundamental gaps in the public's knowledge and understanding of climate change.

For example, the American Psychological Association concluded in a recent report that the understanding of climate change, both in its causes and in its likely effects by the average citizen around the world is limited (APA, 2010). Nationally representative surveys in the US point out that climate literacy seems to be low in general (e.g. Leiserowitz *et al.*, 2010). In particular, people do not seem to understand the human causes that contribute to climate change nor the scientific consensus on this matter (Leiserowitz, 2007). In some cases, people even perceive a few degrees increase in global mean temperature as something rather pleasant, not understanding the potentially large harmful geophysical consequences (Meijnders, 1998). Despite widespread media coverage of climate change and related issues, typical mental models of global climate change tend to suffer from several severe fundamental misconceptions (Bostrom *et al.*, 1994). For example, most explanations given of the physical mechanisms underlying climate change are inconsistent and incomplete. Kempton *et al.* (1995) found that Americans assimilated information about climate change into pre-existing mental models of ozone depletion. In particular, people mistakenly believe that ozone depletion is a cause of climate change (Meijnders, 1998). This is not just the case in the US: a survey performed by GlobeScan in 1999 (covering 25 countries) found that people worldwide (mistakenly) identified "depletion of the earth's ozone layer" as a main cause. This development has led to much confusion between the two issues. A likely explanation is that the hole in the ozone layer has been scientifically well documented over the years and is much easier to imagine and remember (Ungar, 2000). It is interesting to observe that this was pointed out by Bostrom *et al.* (1994) and by Kempton *et al.* (1995) and yet again some 11 years later by Lorenzoni *et al.* (2006), suggesting that despite past communication efforts, this popular misconception still persists.

People also tend to have difficulty understanding the difference between climate change and other environmental problems (Heskes, 1998; Read *et al.*, 1994). In particular, it is hard for people to differentiate between good environmental conduct more generally and specific actions that help reduce climate change. Often measures such as not buying aerosol spray cans, recycling and reducing waste are mentioned as effective strategies for mitigating climate change, possibly because such apposite behaviours are generally known to be harmful to the environment (Bostrom *et al.*, 1994; Leiserowitz *et al.*, 2010; Read *et al.*, 1994). In addition, misconceptions about the relative importance of the various causes of climate change are also widespread. Particularly, people tend to overweigh the effects of deforestation and non-recycling and underestimate the effects of fossil fuel consumption (Bostrom *et al.*, 1994; Whitmarsh, 2009). This also becomes evident from the fact that the general public remains mostly unaware of the link between air travel and climate change (Becken, 2007; Cohen and Higham, 2011; Gössling and Peeters, 2007; Gössling *et al.*, 2006) or meat consumption and climate change (de Boer *et al.*, 2012).

Occasionally, some local studies find that people are relatively well informed about basic concepts (e.g. Lorenzoni and Langford, 2001; Truelove, 2009). Although, on net, improvement seems to be slow and while awareness of climate change is relatively high a more sophisticated understanding still appears to be random and inconsistent (Anable, Lane and Kelay, 2006). Furthermore, it would be erroneous to suggest that a flawed understanding of climate change solely exists among the "lay" or general public. In fact, Sterman and Booth Sweeney (2002, 2007) and Sterman (2008) conducted a series of experiments that identified widespread incorrect beliefs about climate change among highly educated MIT science and engineering majors. In particular, the students were unable to correctly describe the process mechanisms that underlie climate change. Sterman and Booth Sweeney (2007) hypothesised that these deep-seated misperceptions arise as a limitation of people's mental model with regard to the relationship between concepts of stock and flow in phenomena of accumulation (Sterman and Booth Sweeney, 2007).

In conclusion, public understanding of climate change still reveals great diversity, confusion and often ignorance. Understanding the way that individuals process, classify and organise new information is important because incorrect mental representations of climate change are likely to contribute to a "wait and see" attitude (Xiang, 2011).

Does knowing make a difference? The relationship between knowledge and behaviour

Based on the evidence reviewed, it would not be unreasonable to conclude that there is still widespread misunderstanding about (1) the process mechanisms of climate change, (2) its underlying causes and (3) effective response behaviours. In the face of addressing these issues, there are varying theoretical assumptions concerning the role of knowledge in behaviour. Knowledge is often believed to be a background factor that influences a person's attitude toward a certain behaviour (Ajzen, 1991) and given the known association between attitude and behaviour (Armitage and Connor, 2001), knowledge is then assumed to influence behaviour through a mediating variable. The idea behind the attitude-behaviour relationship is that the more people know about and understand the connections between their own behaviour and a range of environmental threats, the more likely it is a person will adjust their behaviour accordingly. Such models essentially assume a linear progression from increased knowledge to a favourable change in attitude which in turn is thought to produce a change in behaviour – a framework that has become better known as the Knowledge-Attitude-Behaviour (KAB) model (Kollmuss and Agyeman, 2002).

While there is little doubt about the fact that environmental attitudes correlate significantly with environmental behaviour, the KAB model has received fierce criticism in recent years. For example, Bulkeley (2000, p. 314) states: *"recent research challenges the assumption that public confusion and an apparent gap between stated beliefs and action, arises from a deficit in public knowledge and understanding of environment issues"*. Similarly, Kollmuss and Agyeman (2002) and Moser (2006) criticise information-based campaigns for being too rationalist and outdated. While it is certainly true that knowledge is a necessary but not sufficient condition for behavioural change, it would be erroneous to suggest that the role of knowledge is outdated or not important. In fact, the role of knowledge in environmental behaviour is important but often underestimated (Kaiser and Fuhrer, 2003), particularly because researchers fail to make a distinction between three *converging* types of environmental knowledge, namely; *declarative* knowledge (i.e. factual knowledge), *procedural knowledge* (i.e. knowledge of appropriate courses of action) and *effectiveness* knowledge (i.e. knowledge of how effective each course of action is). To illustrate, information about the causes of climate change (e.g. CO_2 emissions) can help create a better understanding of appropriate response behaviours (e.g. reducing energy consumption) and vice versa. While Kollmuss and Agyeman (2002) state that only a small fraction of environmental behaviour can directly be explained by environmental knowledge, this argument neglects to consider that the effect of knowledge is often overlooked because it is mediated by other important psychological processes (Kaiser *et al.*, 1999).

For example, one of the first studies that systematically reviewed the psychological determinants of environmental behaviour (Hines *et al.*, 1986/87) reported positive and significant correlations between environmental knowledge and environmental behaviour ($r=0.30-0.37$, $p<0.001$). Building on their work, a more recent and extensive meta-review conducted by Bamberg and Moser (2007) found these findings (largely) to be still accurate. Other studies have also corroborated these findings. For example, a study by Meinhold and Malkus (2005) supports the theory that there exists a linear relationship between environmental knowledge, attitudes and behaviour. In particular, environmental knowledge was found to moderate the attitude-behaviour relationship, where strong pro-environmental attitudes and high environmental knowledge predicted more pro-environmental behaviour. Other critics have argued that if individual knowledge about climate change is generally limited then knowledge should not be able to explain much of the variance in response behaviours (Maibach *et al.*, 2008).

It is worth noting here that predicting specific climate change mitigation behaviours (e.g. air travel) based on general environmental knowledge and attitudes may sometimes (unsurprisingly) cause distortion in measurements (Bamberg, 2003; Kaiser *et al.*, 1999; Whitmarsh, 2009) – as someone can hold general pro-environmental knowledge and beliefs but still maintain different attitudes toward specific behaviours. For these reasons, it is perhaps more appropriate to review to what extent knowledge about climate change is able to predict specific climate change mitigation behaviours. To this extent, some evaluative studies have found that general knowledge about climate change is only weakly related to actual self-reported behaviours (e.g. Staats *et al.*, 1996). Yet, research by Bord *et al.* (2000) and O'Connor *et al.* (1999) provides evidence that knowledge is in fact an important predictor. In both studies, knowledge about climate change kept its statistical validity as an independent predictor of behavioural intentions (even after controlling for general environmental attitudes). In fact, knowledge was the strongest relative predictor of intentions, explaining 11 per cent of the variance to take voluntary action and 20 per cent of the variance to support new government policies (Bord *et al.*, 2000). Similarly, Ngo *et al.* (2009) also found that knowledge successfully predicted a range of climate change mitigation behaviours.

While knowledge of climate change impacts has also been implicated in eliciting behavioural change (e.g. Nillson and Kuller, 2000), Truelove (2009) found that knowledge of appropriate response behaviours was the strongest predictor of mitigating intentions. Similarly, research by Semenza *et al.* (2008) and Hounsam (2006) found that the most popular self-reported barrier to behavioural change was simply the fact that people did not know how to change their behaviour to reduce their own contribution to climate change.

In conclusion, knowledge can be considered a necessary condition for behavioural change, given that knowledge about the causes, consequences and solutions to climate change have all been implicated as significant predictors of behavioural outcomes. Yet, in order to maximise the effect of environmental knowledge on behaviour, knowledge must converge (Kaiser and Fuhrer, 2003). Thus, popular recommendations that public campaigns should prioritise one type of knowledge (e.g. response strategies) over another (e.g. understanding process mechanisms) must be exercised with caution as in some cases they may neglect the interdependent relationship that exists among these knowledge structures.

The affective-experiential approach

The homo expertus?

> "A man who carries a cat by the tail learns something he can learn in no other way."

> (Mark Twain)

What Mark Twain jokingly points out is an important fact of life: knowledge and cognition can only do so much for human learning and understanding. Humans have inherited a well-equipped sensory system, and, through interaction with our natural and social environments we are able to learn and understand things in a way that abstract knowledge is unable to provide. While traditional cognitive-knowledge-based approaches assume that the public is not changing their behaviour because they fail to understand the issue (Lorenzoni *et al.*, 2007), scholars are increasingly pointing out that highlighting scientific narratives in the media is unlikely to elicit more engagement (Hargreaves *et al.*, 2003), particularly because of the "yawn factor" that science tends to have on non-experts (Abbasi, 2006). In addition, several recent studies have shown that climate change is a temporally and spatially distant phenomenon for most individuals (Maibach *et al.*, 2008; O'Neill and Nicholson-Cole, 2009; Spence *et al.*, 2012).

One reason for this is that individuals have difficulty visualising future periods (Tonn *et al.*, 2006). For example, a study by O'Neill and Nicholson-Cole (2009) indicated that respondents could not really articulate what climate change might mean for the United Kingdom. Individuals also tend to display an unrealistic sense of optimism (Weinstein, 1980), particularly to the extent that climate change is likely to affect others (e.g. the third world) but not the individual in question (O'Neill and Nicholson-Cole, 2009). Because climate change cannot be experienced directly, it is likely that individuals will continue to distance themselves psychologically (Lorenzoni and Pidgeon, 2005). Yet, a study covering 34 countries found that the majority of people in each country believed that climate change was a somewhat to very serious problem (GlobeScan, 2000). In 2006, GlobeScan repeated the study and found that the percentage of respondents that believed that climate change was a "very serious threat" increased significantly in most countries (GlobesScan, 2006). Similarly, a study performed in the UK also indicated that over 80 per cent of the respondents reported to be concerned about climate change (Poortinga *et al.*, 2006). Yet some researchers have argued that there is an issue with the way this apparent "concern" is conceptualised. Because there is no one coherent method of how an individual's risk perception is assessed, measures vary greatly and the terms "concern", "worry" and "perceived seriousness" are often used interchangeably. Yet, the literature often fails to note that these terms mean slightly different things. To illustrate, it is possible to have general concern for an issue without actively worrying about it. Worry is then considered to be a much more active emotional state and a stronger predictor of behaviour than either concern or perceived seriousness (Leiserowitz, 2007). For example, a survey by the Pew Global Attitudes Project (2006) found that, while varying among countries, personal levels of worry about climate change are generally lower than perceived seriousness or general stated concern. Thus, while general concern about climate change seems to be well established among the general public, the same cannot be said for personal worry.

It is also questionable whether stated concern is related to the perception that the problem of climate change is urgent or of high priority. For example, while many people are concerned about climate change, they rank it as less important than many other social issues such as terrorism, health care and the economy (Krosnick *et al.*, 2006). Similar evidence is provided by Poortinga and Pidgeon (2003) – based on 1,547 face-to-face interviews the researchers found that while there was some moderate concern for all environmental risks mentioned in the study, climate change was ranked among the least important issues. This evidence leads to the conclusion that although general concern is expressed, there is also a dominant belief that climate change is a distant, non-urgent and non-personal threat (Darnton, 2005), possibly hindering proactive behavioural responses (Lorenzoni and Langford, 2001). These findings have lent support for the hypothesis that if general concern can somehow be transformed into personal worry then perhaps people are more likely to change their behaviour accordingly.

Risk as feeling

It has become increasingly apparent that individuals have a hard time relating to technical, descriptive and abstract risk messages. In fact, the public may not act upon simple information about probabilities unless this information is given emotional meaning (Slovic *et al.*, 2004). Accordingly, converging evidence from cognitive, social and clinical psychology has indicated that human perceptions of risk (across domains) are very much influenced by affective and emotion-driven processes (Chaiken and Trope, 1999; Esptein, 1994; Loewenstein *et al.*, 2001; Sloman, 1996; Slovic *et al.*, 2006; Weber, 2006).

At this point it is perhaps warranted to make a conceptual distinction between "emotion" and "affect". Perhaps a definition encompassing the most pivotal characteristics of emotion states that: "emotion is a complex state of feeling that results in psychophysiological changes that influence thought and behaviour" (Myers, 2004, p. 500). Emotions can be regarded as relatively transient and tied to a particular stimulus or event, manifesting in a specific state such as fear or happiness. *Affect* is a more subtle form of emotion defined specifically as a positive (like) or negative (dislike) evaluative feeling towards an external stimulus (Slovic, 1999). For example, affective images can be regarded as a broad construct to which positive and negative feeling states have become attached through learning and experience (Slovic *et al.*, 1998, p. 3). Thus, while emotions are more complex in-depth feelings that cause psycho-physiological changes, affect is a rather fast, specific and automatic evaluation of a stimulus object. In the context of climate change, Leiserowitz (2006) and Smith and Leiserowitz (2012) found that negative affect and imagery toward climate change were the strongest predictors of risk perception. More specifically, the research indicated that people tend to display a strong negative affective feeling towards the term "global warming". Similar findings were reported in a study by O'Neill and Nicholson-Cole (2009): while respondents seemed to have a wide range of imaginations and mental visions related to the concept of global warming, most of them were negative and bleak.

Thus, negative affective imagery towards climate change seems to be widespread. Yet, to what extent do (negative) affect and emotions affect willingness to help reduce climate change? According to Böhm (2003), environmental behaviours are guided by so-called "prospective" and "retrospective" consequence-based emotions, such as fear and worry, which also happen to be the most intense emotions associated with environmental risks. In fact, Böhm and Pfister (2001) theorised that feelings of fear and worry over consequences should lead people to prevent and reduce environmental damage. In line with this train of thought, a large amount of research has been directed towards eliciting "fear" – with the underlying hope that fear will serve as a strong motivator for behavioural change.

The link between personal experience, risk perception and behavioural change

While it is well known that emotions are an important and significant predictor of environmental behaviour in general (Grob, 1995; Maloney *et al.*, 1975), less is known about the specific relationship between experience, emotion, risk perception and behaviour in the context of climate change. Direct experience is thought to influence risk perception and behaviour (Whitmarsh, 2008), in particular, because experiences can invoke strong memorable feelings, possibly making them more dominant in processing (Loewenstein *et al.*, 2001). Consider an individual that encounters an approaching tornado. Such a direct threat can elicit strong instinctive emotions such as fear and anxiety that subsequently guide immediate behaviour, the so-called *fear-flight response*. These instinctive emotions primarily arise in the brain's limbic system, an evolutionary older part of the brain that guides behaviour through fast and automatic responses, especially in reaction to threats (MacLean, 1990).

Yet, it is unclear to what extent this model applies to the context of climate change. For example, a sensible response to flooding would be moving away from the danger zone or perhaps buying flooding insurance (i.e. adaptation measures). In fact, instinctively, the goal of the response behaviour is to mitigate immediate threats (not climate change as a broader concept in itself). It is not at all obvious that whenever a person's house floods, that person is actively going to diminish his or her carbon footprint in response, unless that person explicitly links the flooding event to climate change (Helgeson *et al.*, 2012) – which is not always the case. For example, a study by Whitmarsh (2008) reported no difference in risk perceptions of climate change between respondents that had experienced flooding before and those who had not.

Yet, people who live in low-lying coastal areas do tend to have a heightened sense of personal risk (Brody *et al.*, 2008) and a recent study by Spence *et al.* (2011) did find that past flooding experiences were significantly related to increased preparedness to reduce energy use. In particular, past flooding experiences mediated onto level of concern and perceived local vulnerability, which in turn, increased individual preparedness. Moreover, a significant amount of studies indicate that risk perception is an important predictor of individual willingness to help reduce climate change (e.g. Heath and Gifford, 2006; Hidalgo and Pisano, 2010; Leiserowitz, 2006; Ngo *et al.*, 2009; O'Connor *et al.*, 1999; Semenza *et al.*, 2008).Yet, a second issue revolves around the idea that even if direct experience does matter, there is a disassociation between the cognitive information that informs individuals that there is in fact a risk to be worried about and the inability for many people to observe or experience this risk in their direct environment (Weber, 2006). This lack of personal experience with the potentially negative consequences of climate change is causing a lower level of individual concern than advisable (APA, 2010; Weber, 2006). Thus, although "experience" may indeed raise level of concern to what is considered a more appropriate level of personal "worry", direct experience with the effects of climate change is generally lacking.

A potential solution to this problem has been to try to inflate personal worry among the general public through measures that do not require actual personal experience. One such measure is the use of negative emotional appeals, where the centre of focus revolves around "fear-appeals" (Stiff and Mongeau, 2003). This approach has gained popularity in climate change communication (Moser and Dilling 2004; O'Neill and Nicholson-Cole, 2009) as the advertisement of extreme events is thought to do better than the idea of slow ongoing change (Brönnimann, 2002). To understand why fear appeals are often believed to work (at least in theory) it is helpful to briefly consider the development of various theories in the field of persuasive communication.

To start with, the experience of "fear" is a negatively valenced emotion accompanied by a high level of arousal and is elicited by a threat that is perceived to be significant and personally relevant (Witte, 2000). In particular, fear appeals are a method of communication that attempts to influence attitudes and behaviours through the threat of some danger (Tanner *et al.*, 1989). A large amount of research has been performed over the years concerning the role of negative threat-related emotion in communication, largely inconclusive about its general effectiveness. One of the earliest theories on fear appraisal was put forth by Janis (1967), who proposed that the relation between fear and attitude change is curvilinear (U shaped). In effect, the theory suggests that moderate levels of fear are more persuasive than lower or higher levels. More specifically, Janis's drive theory argues that fear arousal is needed to elicit a motivational drive state (i.e. tension) that individuals seek to resolve. From this point of view, for fear-appealing communication to be persuasive two requirements have to be met: (1) the level of fear induced by communication must be sufficiently high to function as a drive state and (2) recommendations have to be included in the communication as to how to reduce this drive state. Unpleasant emotional tension can be resolved by individuals through either "adaptive responses" (i.e. useful behavioural changes) or through "maladaptive responses". For example, Leventhal (1970) made a distinction between an internal "fear control" process and an external "danger control" process. Maladaptive behaviours are targeted at controlling the fear response (e.g. through denial) but they often leave the actual threat (e.g. climate change) intact.

Rogers' (1975, 1983) protection motivation model (PMM) has arguably been one of the most applied theories to understanding fear appeals (Witte, 1992). The PMM basically states that a threat-related message will only be effective if (1) it convinces the reader that he or she is seriously threatened (threat appraisal) and (2) actually capable of averting the threat (coping response). In particular, Rogers (1983) postulated that people continue to engage in maladaptive behaviours (e.g. "binge flying") if the rewards (convenience) of that behaviour exceed both the perceived severity of the threat (e.g. climate change) and the individual's perceived susceptibility to that particular threat (low). The intention to protect one's self then depends on four factors: (1) a threat's malignancy; (2) its probability of occurrence; (3) the effectiveness of a coping response (i.e. response efficacy); and (4) an individual's ability to perform the response (i.e. self-efficacy).

Scaring people, does it work?

Empirical evidence supporting either Janis (1967) or McGuire's (1969) model has been lacking (Higbee, 1969; Sutton, 1982). Furthermore Leventhal's (1970) and Rogers' (1975) models have been criticised for being imprecise (e.g. Witte, 1992). In general, the empirical evidence for the fear approach is mixed. While some meta-reviews point out that a positive linear relationship is found between the level of fear and a change in behavioural outcomes (Boster and Mongeau, 1984; Sutton, 1982) this does not mean that a curvilinear relationship should be rejected (Meijnders, 1998). More generally, Boster and Mongeau (1984) found that fear appeals are modestly correlated with attitudes and to a lesser extent with intention and behaviour. Yet, overall, it is generally agreed upon that without efficacy messaging (i.e. an individual's perceived capability to avert the threat), fear appeals tend to be rather unsuccessful (O'Neill and Nicholson-Cole, 2009). Indeed, if anything can be learned from 50 years of theory development, it is that *strong fear appeals with high efficacy messaging* produce the highest level of behavioural change whereas *strong fear appeals with low efficacy messaging* produce the most maladaptive responses (Witte, 2000).

However, much of the research on fear appeals has been conducted on health risks, which are both personal and direct (de Hoog *et al.*, 2005) – two important conditions that are perceived to be absent in the context of climate change. Yet, there is some evidence available on the use of fear appeals in the context of climate change. For example, Meijnders *et al.* (2001a, 2001b) found that moderate fear messages related to global warming induced more systematic processing of the perceived risks as well as more favourable attitudes toward energy conservation (when compared to the low-fear condition). Lowe *et al.* (2006) carried out a pre/post-test study after individuals had watched the climate change disaster movie *The Day After Tomorrow* (Emmerich, 2004). Although a majority of the respondents (67 per cent) were in agreement that everyone needs to do something about climate change, this sense of urgency quickly faded in a focus group meeting a month after the screening. However, very different results were presented by Leiserowitz (2004). In a similar study concerning the same movie, the author found that movie-watchers versus non-watchers showed higher levels of both concern and worry, estimated various impacts on the US more likely and significantly increased their intentions (in all stated categories) to engage in personal action to address climate change (Leiserowitz, 2004). Similarly, Jacobson (2011) used a spatial-econometric analysis to measure increases in the purchase of voluntary carbon offsets within a ten mile radius of US movie theatres after the release of Al Gore's *An Inconvenient Truth* (Guggenheim, 2006). Shortly after the release, the purchase of carbon offsets went up as far as 50 per cent, yet no renewals were recorded in subsequent years. O'Neill and Nicholson-Cole (2009) conclude that dramatic, sensational, fearful and shocking representations of climate change (both visual and iconic) can successfully capture people's attention and drive a general sense of urgency to the issue (O'Neill and Nicholson-Cole, 2009). Yet, the same researchers also found that while capturing attention and raising concern, fear messaging disengages people from climate change and renders them feeling hopeless and overwhelmed. The authors further mention that catastrophic and fearful representations of climate change are unlikely to motivate a sense of personal engagement and can possibly trigger psychological barriers such as anxiety, apathy, paralysis and denial (Lorenzoni *et al.*, 2007: Moser and Dilling, 2007). Consistent with these claims, other research has found that fear messaging can have different impacts on different audiences. For example, individuals with strong "just-world" beliefs tend to resort to maladaptive responses (e.g. denial) when faced with fear-messages about climate change, negatively affecting intentions to curb carbon footprints (Feinberg and Willer, 2011). The authors recommend that messages should include sufficient information on potential solutions, which is consistent with the idea that a message is more persuasive when negative emotions about one's vulnerability are coupled with positive thoughts about potential solutions (Das *et al.*, 2003).

All in all, these findings point to one (of several) potential problem(s) that characterise the "fear" approach. To start with, it is likely that fear appeals have a rather short-term effect. In fact, Weber (1997) coined the term "single action bias" to explain the tendency for individuals to only take a single action to reduce a perceived threat and subsequently neglect further steps that would provide incremental protection. For instance, a 2008 poll in the United States indicated that while 28 per cent of Americans thought the environment was getting better, after having elected President Barack Obama in 2009, this number rose to 49 per cent (Silver, 2009). A possible explanation for the single action bias is that the first measure people take often sufficiently reduces the active level of worry/vulnerability.

An additional shortcoming is that fear appeals offer diminishing returns (Hastings et al., 2004). That is, communicators run the risk of desensitising or "emotionally numbing" people to the risks involved, as familiarity with a risk reduces its salience (Fischoff et al., 1978). Furthermore, the "finite pool of worry" hypothesis states that people can only worry about a limited number of problems at any given time. As a result, increased concern for one risk (e.g. economic crisis) might decrease concern for other risks such as climate change (Hansen et al., 2004).

In conclusion, "fear as a motivator" should be used with caution (Futurra, 2005). While fear messaging definitely has its place in the communication strategy mix (capturing attention and breeding concern), it will be difficult to retain such level of interest and arousal as people need a reason to stay engaged and often quickly shift their attention. Furthermore, when using narratives of an impending "catastrophic" and "looming disaster" without promoting actions that can help reduce the threat, fear messages are likely to trigger maladaptive coping responses and leave people feeling disempowered and disengaged. In the words of O'Neill and Nicholson-Cole (2009, p. 376): "depicting a state of crisis does not sit comfortably with the suggestion of individual action".

The social-normative approach

The homo sociologicus?

> "No man is an island, entire of itself; every man is a piece of the continent, a part of the main."
>
> (John Donne)

In addition to cognitive and experiential processing, human behaviour is also shaped by a wide range of normative factors. In fact, social-psychological research that explores the effect of social influences on behaviour is pervasive – as it is through social comparison with referent others that people validate the correctness of their opinions and decisions (Festinger, 1954). People derive descriptive and prescriptive social norms from observing others (Heath and Gifford, 2006), apply a logic of appropriateness in unfamiliar situations (March, 1994) and unsurprisingly, tend to behave as their friends and peers (Cialdini et al., 1999). There are many examples of how social factors influence environmental behaviour. For example, people's energy use tends to decrease when they are told that their neighbours are conserving energy as well. People tend to alter their use of energy more generally to conform to the group-norm (e.g. Schultz *et al.*, 2007). Usually a distinction is made between descriptive and prescriptive social norms. While prescriptive norms contain information about how others think someone ought to behave, descriptive norms merely describe how others are behaving (Cialdini *et al.*, 1991). When communicating social information it is important to understand the relation between these two concepts. For example, merely describing that CO_2 emissions are increasing because a lot of people are choosing to fly short distances instead of using alternative modes of transportation (a descriptive norm) may have unintended effects if it is not clearly mentioned that this behaviour is in fact undesirable (prescriptive norm). In other words, it should be made clear that people *ought* to avoid flying short distances, otherwise the message is easily misread as: "it's okay because everyone's doing it".

Another important question is how moral norms are theoretically distinct from social norms. Moral norms refer to the idea that some behaviours are just inherently right or wrong regardless of their personal or social consequences (Manstead, 2000). While there certainly is a strong link between social and moral norms, it is nevertheless possible that a person's moral convictions do not coincide with the expectations that exist in that person's social environment. One way to think of the relationship between these two concepts is that both cultural and social learning play an important role in acquiring moral beliefs (Krebs and Janicki, 2002), as social reference groups deliver standards for what is viewed as right or wrong. It is over time, when people have internalised social norms that they become a (personal) moral norm. Moral norms are then considered to be the link between internalised (general) values and more specific opinions and expectations about how to behave in a tangible situation (Schwartz, 1977). Thus, even though moral norms may originate from social group norms, once they have become internalised, they exercise influence over an individual's behaviour independently from any immediate social context (Manstead, 2000). Similarly, Bicchieri (2006) highlights that while social norms are followed conditionally upon the satisfaction of expectations of others, moral norms are followed unconditionally based on internal (emotional) processes.

Moral norms have always played a central role in explaining pro-environmental behaviour. A particularly influential framework is Stern *et al.*'s (1999) *Value-Belief-Norm* (VBN) theory. According to the VBN, people's motivation for caring about the environment can be traced back to a specific set of personal values. For example, someone could be aware of the potentially negative consequences of climate change because they wonder how it might affect them personally (i.e. egoistic values), how it will affect other humans (altruistic values) or how it will affect the earth more generally (biospheric values). These values are then thought to influence more specific belief structures about human-environment interactions (a person's ecological worldview) which in turn determines the extent to which people are *aware of consequences* (AC) and *ascribe responsibility to their own actions* (AR) – eventually leading to the activation of an individual's moral norm, which according to the VBN, is thought to be the main driver of pro-environmental behaviour.

Evidence for the role of normative influences on environmental behaviour

Empirical evidence for the persuasive power of normative influences on behaviour is growing. A frequently quoted study concerns a conservation experiment on hotel towel reuse. In the experiment, a simple "normative" prompt (i.e. "75 per cent of guests in this room reuse their towel when asked") significantly increased the reuse of towels (Goldstein et al., 2008) – illustrating the potential of behavioural change through the communication of social information. A range of other studies that have used social norm manipulations in the context of energy conservation have showed similar positive results (e.g. Dolan and Metcalfe, 2012; Schultz et al., 2007). In addition to social pressure, moral norms are an equally (if not more) powerful tool for encouraging pro-environmental behaviour (Markowitz and Shariff, 2012). To this extent, a recent field experiment by Bolderdijk et al. (2013) set out to explore what message frame is most successful when asking drivers to pull over to get their tyre pressure checked. Results strongly indicated that a moral message frame was most effective. Similarly, other recent research found that both moral and social norms are significant predictors of consumer decisions to purchase carbon offsets (Blasch and Farsi, 2012).

Yet, there are a number of identifiable problems inherent to the normative approach. First, observed effect sizes are typically small and short-lived (John et al., 2011). The latter is particularly true for social norms, since they are conditional upon the existence of exogenous social pressure (extrinsic motivation) while moral norms elicit motivation through internal processes (i.e. intrinsic motivation). A second problem concerns the use of "guilt appeals". Guilt usually arises as a result of violating some moral or social norm (Baumeister, 1998). In theory, guilt is thought to be a motivator of pro-environmental behaviour because guilt often leads to a moral obligation to compensate for any caused damages (Bamberg and Moser, 2007). Studies show that under certain conditions guilt can be effective in changing behaviour (O'Keefe, 2002) and some support is found in the context of climate change, for example, Ferguson and Branscomble (2010) illustrate that feelings of guilt mediated beliefs about global warming and willingness to engage in mitigation behaviours. Truelove (2009) comments, however, that similar to the case of "fear appeals", the associated drawbacks are significant. A particular drawback is the risk of eliciting negative emotions (such as anger toward the guilt-inducer) as this can potentially undermine the technique's overall effectiveness.

It is also interesting to note that in survey studies, social norms are often identified as one of the weakest predictors of behaviour (e.g. Armitage and Connor, 2001). A potential explanation for this is that it may very well be the case that the effect of social norms on behaviour is systematically underdetected because people display a strong tendency to underestimate the extent to which they are subject to social influences (Griskevicius *et al.*, 2008). Yet, in a similar manner it can be argued that the positive results found in many experimental studies are simply the result of artificially inflated social pressure. In fact, it is crucial to understand that in order for social norms to affect behaviour, they must first be activated and made salient (Cialdini *et al.*, 1999). Unfortunately, strong social norms are generally absent for most pro-social behaviours (van der Linden, 2011). In fact, negative social identities associated with performing environmentally unfriendly behaviours are currently not very well articulated. Thus, throwing a few numbers at people for social comparison purposes is not going to have much effect when there is no negative social identity to leverage in the first place (Corner, 2011). As a result, while social and moral norms may affect behaviour, in order to leverage their full potential, a strong pro-environmental norm must be established first. Several governmental advisory bodies have recently advised the UK government to use more "deep-frames" in their communication. This entails a community-based approach where the discourse is shifted from "you" to "we" and from "I" to "us", encouraging the elicitation of moral values, collectivism and social identity (CCCAG, 2010). For example, the UK government is currently actively trying to harness the persuasive potential of social norms in its design of large-scale behavioural change campaigns (Cabinet Office, 2011). Yet, empirical evaluations remain elusive.

Towards a new framework for communicating climate change

Building a more integrated understanding: theories of dual-processing in the brain

"Information's pretty thin stuff unless mixed with experience."

(Clarence Day)

So far, all three major approaches to public climate change campaigns have been considered in isolation. Yet, cognitive, experiential and normative influences do not affect human behaviour independently of each other – on the contrary, most behaviour is the result of carefully integrated neurological processes. The ancient Greek philosophers Plato and Aristotle long debated the intricacies of the fine line between passion and reason and, ever since, a substantial amount of research in social, cognitive and neuropsychology has lent its support for a theory of "dual-processing" in the brain. Either a distinction is made between cognitive and affective processing (e.g. Damasio, 1994; Epstein, 1994; LeDoux, 1996; Zajonc, 1998) or between controlled and automated processes (e.g. Kahneman, 2003; Sloman, 1996). It is important to realise that these processing systems do not function independently from each other. Instead, they operate in parallel and continuously interact with each other, where higher analytical reasoning may evoke strong (basic) emotions and simple reflexes can be triggered by higher functioning neocortical processes (Marx *et al.* 2007; Weber, 2006). In fact, rational decision-making cannot be effective unless it is guided by emotion and affect (Damasio, 1994, 1999), moreover, without emotions, humans are not able to learn effectively at all (Baumeister and Bushman, 2008). While dual-process theories aid conceptual understanding, they often provide an overly simplified understanding of neurological functioning. Camerer *et al.* (2003) present a more useful categorisation of human neural functioning (Table 14.1).

To illustrate how these four quadrants (Table 14.1) operate in relation to everyday consumer behaviour, consider the following example: let's assume that a customer walks into a travel agency and wants to book a well-deserved exotic vacation. Upon entering the store, the customer's attention is immediately drawn to a big fancy flyer displaying the ultimate vacation, including a sunny location, palm trees, white sandy beaches and a breath-taking turquoise sea. The brain's motor cortex will guide that person's arm to reach for the flyer drawing on two processes, namely the *cognitive and automatic* quadrant III (reaching) and the *affective and automatic* quadrant IV (pleasure and enjoyment). However, at the same time higher level processing might occur in the brain. For example, it could be that this par-ticular person has recently been exposed to a documentary on sustainable tourism and anticipates that going on this holiday would perhaps disappoint important family members who recommended watching the documentary. These processes (explicit memory) and anticipation (planning) draw on two areas of the brain; the hippocampus and the prefrontal cortex, which are involved in controlled cognitive (quadrant I) and controlled affective (quadrant 2) processing.

To keep things relatively simple, no social context was made explicit. Yet, it should already become clear from this hypothetical example that in most realis-tic decision-environments, all four neurological quadrants can potentially be activated (and interact) in a matter of seconds. A clear implication of a more advanced understanding of human behaviour is that in order to make communi-cation efforts more effective, substantial efforts should be directed towards integrating *cognitive, experiential* and *normative* aspects of climate change com-munication. Particularly, increased cognitive understanding can help make beha-vioural change more sustainable in the long-term while experiential approaches can help elicit affective associations and facilitate learning and understanding through visualisation of the information presented. For example, research by Marx *et al.* (2006) indicated that people retain more factual information about climate change when that information is presented in an experiential format. In addition, the overall message should be designed and framed in a context that

Table 14.1 Categorisation of human neural functioning (adopted from Camerer *et al.*, 2003)

	Cognitive processes	Affective processes
Controlled Processes (e.g. effortful, evoked deliberately, serial)	I	II
Automated Processes (e.g. effortless, reflexive, parallel)	III	IV

illustrates that other people are also acting sustainably and that a strong pro-environmental norm is both expected, desired and rewarded. In conclusion, knowing how information is processed and integrated in the brain and subsequently affects behaviour should lead to the understanding that communication designs that take into account positive interactive feedback loops between, cognitive, experiential and normative processes are likely to be more effective than "either/or" type strategies.

The pivotal role of determinants of behaviour in communicating change

While the previous section has established that cognitive, experiential and normative processes often operate simultaneously and affect environmental behaviour in an integrative manner, the relative importance (or contribution) of each factor is not always known. To this extent, it is useful to introduce a distinction between theories of change and models of behaviour. While models of behaviour aid in understanding specific behaviours by identifying the underlying psychological factors that influence them, theories of change show how behaviours can be changed and/or change over time (Darnton, 2008). Thus, while theories of change generally describe more generic processes, models of behaviour are diagnostic and help illuminate the psychological determinants that explain and predict a given behaviour (van der Linden, 2012). Psychological determinants refer to the behavioural factors and processes that explain and predict a certain behaviour. For example, both the role and relative importance of cognitive (knowledge), experiential (affect) and normative (moral norms) factors in explaining transport behaviours is currently an active area of research (see Bamberg and Schmidt, 2001, 2003; Bamberg *et al.*, 2007; Steg, 2005).

It should be noted that while theories of change and models of behaviour have distinct purposes, they are also highly complementary. In fact, it is argued here that the ineffectiveness of climate change campaigns can, in part, be attributed to the fact that most public climate change interventions pay little to no attention to the psychological determinants of the behaviours that they are trying to change. For example, public campaigns that promote sustainable lifestyles and "good environmental conduct" across the board (e.g. Doyle, 2011) do not take into account the determinants of different environmental behaviours. When campaigns do get specific, for example, in the case of meat consumption (Meat Free Monday, 2010; Peta2, 2008), little attention is paid to the social-psychological determinants of the behaviour. In fact, a report by the Government Communication Network (2009) points out that attaining a better understanding of how relevant behaviours are determined and influenced should be considered a prerequisite for the design of effective communication campaigns (GCN, 2009). It is important for evaluators to not only look at behavioural outcomes, as it is from studying the psychological determinants of behaviour that we gain understanding of why certain interventions were successful or not (Steg and Vlek, 2009). In short, successfully trying to change any given behaviour involves a thorough understanding of all the factors that determine and influence the behaviour under investigation.

A new framework for communicating climate change

A conceptual framework to help guide the design of public climate change campaigns is presented in Figure 14.1. The central argument behind the framework is that persuasive communication is only persuasive (i.e. likely to elicit behavioural change) if it is based on an integrated understanding of the psychological processes that underlie and influence pro-environmental behaviour. In order to achieve this, three criteria need to be met: (1) interventions should design *integrative* communication messages that appeal to cognitive, experiential as well as normative dimensions of human behaviour; (2) the *context* and *relevance* of climate change needs to be made explicit; and (3) *specific* behaviours should be targeted, paying close attention to the psychological *determinants* of the behaviours that need to be changed.

To illustrate that few of these criteria are typically met in practice, consider the following three illustrative cases. One of the earlier large-scale information-based campaigns on climate change was conducted in The Netherlands in 1996 and evaluated by Staats *et al.* (1996). The campaign employed a wide range of media tools, including billboards, posters, television commercials and information pamphlets. The aim of the campaign was to raise knowledge and awareness about the causes, consequences and solutions to climate change. Two strong

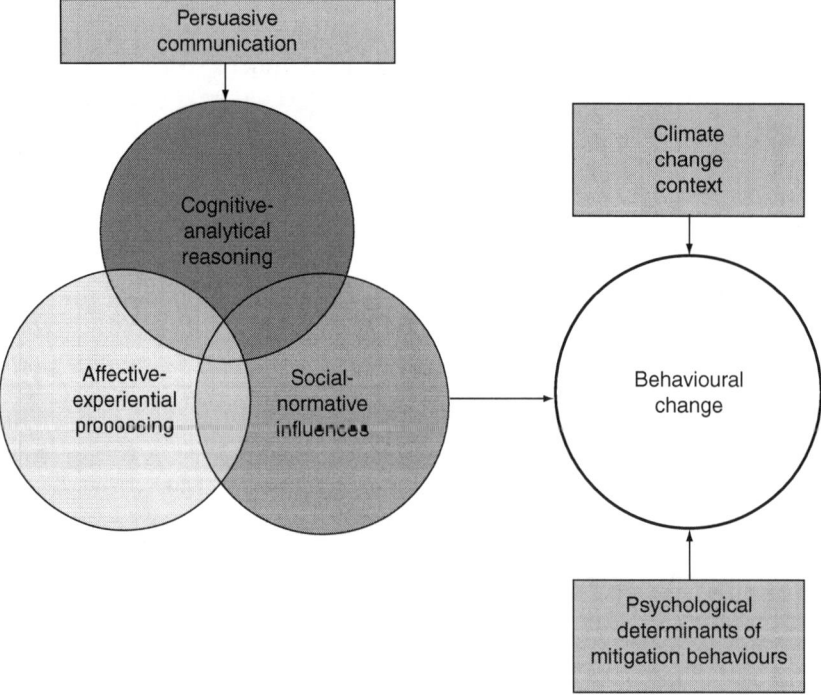

Figure 14.1 An integrated framework for public communication interventions.

points of the campaign were the strong explicit focus on climate change and a link was made between the greenhouse effect, climate change and relevant behaviours. Nevertheless, being one of the earlier campaigns, the intervention was nearly entirely focused on cognitive-information- and knowledge-based factors (although some imagery was used to symbolically illustrate the greenhouse effect). Yet, only marginal focus was applied to "affective and experiential" processes and no attention was paid to "normative" influences. In fact, in their evaluation, Staats *et al.* (1996) highlight that the disappointing results of the campaign can be attributed to the fact that little attention was paid to social-normative factors. In addition, no effort was made to research the psychological determinants of the target behaviours.

More than ten years later, the "Act on CO_2" (2009) campaign in the UK featured an advertisement where a little girl is read a scary bedtime story by her father about the potentially horrible consequences of climate change. While the commercial did feature some knowledge/information, the content was predominantly controlled by a "fear frame" – focusing on the negative, scary and threatening consequences of climate change in the form of a cartoon that depicted "climate monsters" and drowning people. The cartoon implied that a "happy ending" is uncertain (hinting that a happy ending is dependent on people changing their behaviour). Falling prey to all the common pitfalls associated with the use of guilt and fear appeals, the advertisement was not well received by the public as thousands of complaints were submitted to the UK Advertising Standards Authority (Sweney, 2010). The focus of the commercial was predominantly geared towards the affective and experiential domain of human behaviour neglecting both normative influences on behaviour as well as important informational aspects. While the link between climate change and energy consumption was made explicit, no attention was paid to the psychological determinants of the behaviours in question.

Finally, more recently, the American "Wasting Water is Weird" (2011) video campaign was released featuring a character called "Rip". In the video, the Rip character visits people who are clearly wasting water (e.g. brushing their teeth with the tap running) and sarcastically illustrates that "they're weird" for wasting water. By trying to associate a negative social identity with wasting water, the commercial fully relies on normative influences on behaviour. Therefore, while zooming in on a particular behaviour (e.g. dishwashing), a major drawback of the campaign is that no attention is paid to cognitive or experiential processes and no contextual link is made to climate change. It is important to make the climate change context explicit, primarily because if people engage in conservation behaviours for hedonic or cost reasons, they are likely to stop doing so once the behaviour is no longer attractive or cost-effective, whereas environmental motives have proven more robust against such changes (Steg, 2008). Finally, no effort was made to research the psychological determinants of residential water usage. If research had been conducted, it perhaps would have been more apparent that providing information on alternative courses of action is an important aspect of trying to change unsustainable behaviours that have a strong habitual component (Gregory and Di Leo, 2003). In conclusion, climate change campaigns often adopt an either/or approach, appealing to only one aspect of human behaviour, thereby failing to consider other important psychological processes. Unsurprisingly, on the whole, evaluations of past public climate change campaigns have been disappointing at best (O'Neill and Hulme, 2009: Steg, 2008).

Figure 14.1 illustrates that in order for communication to be *persuasive*, it should take into account the interrelation between cognitive, experiential and normative influences on behaviour. Recent experimental evidence supports this notion. For example, Dolan and Metcalfe (2012) comment that little is known (empirically) about the interaction between social norms and basic information provision. Based on a large-scale energy conservation study, the authors conclude that, compared to only using a social norm prime, providing information *alongside* social norm messages is key to the success of behavioural change interventions – as it doubled the rate of energy conservation (Dolan and Metcalfe, 2012). Similarly, a recent field study by De Groot *et al.* (2013) showed that the combination of different normative appeals reduced the use of plastic bags in supermarkets significantly more compared to when the messages were administered individually. It is no surprise that integrating these theoretical dimensions can help guide the practical design of public climate change interventions. To illustrate, consider that it is well documented that human attitudes encompass both cognitive and affective dimensions (Albarracín *et al.*, 2005), especially in the context of climate change (Lorenzoni *et al.*, 2006). As a result, creating negative attitudes towards climate change draws on the interaction of both cognitive and affective processes. In addition, while knowledge about the potential consequences of climate change has been implicated in achieving behavioural change, this effect is enhanced when knowledge about consequences interacts with a feeling of personal and moral responsibility for those consequences (Bamberg and Möser, 2007; Joireman *et al.*, 2004; Wall, 2005). Furthermore, it is important that individuals believe that engaging in the target behaviour is the right thing to do (i.e. moral norm activation) but this feeling is more easily elicited when people are under the impression that the target behaviour is also being executed by important referent individuals (i.e. social norm activation). Because individual beliefs are often a function of the social group to which the individual belongs, an informational message is expected to be more persuasive if the right in-group source and context is provided (Mackie *et al.*, 1990; Van Knippenberg *et al.*, 1994). In sum, recent research is increasingly starting to validate the importance of exploring interactions between cognitive, experiential and normative influences on behaviour.

Conclusions

The aim of this chapter has been twofold. First, to evaluate the theoretical and empirical evidence for three major strategies to public change communication (cognitive, experiential and normative) and second, to provide a new communication model that is more likely to effectively encourage the behavioural shifts that climate change necessitates. It is concluded that in isolation, cognitive, experiential and normative approaches are unlikely to induce behavioural change. Instead, a new framework for communicating information about climate change is presented. It is argued that future interventions are more likely to reduce the gap between public communication and behavioural change when public campaigns: (1) effectively integrate cognitive, experiential and normative aspects of human behaviour in their message design; (2) make the climate change context explicit; and (3) foster a strong link between the behaviours that need to be changed and their psychological determinants.

Acknowledgements

This research has been supported by the Grantham Foundation for the Protection of the Environment, as well as the Centre for Climate Change Economics and Policy, which is funded by the UK's Economic and Social Research Council (ESRC) and by Munich Re.

References

Abbasi, D.R. (2006). *Americans and climate change: Closing the gap between science and action.* New Haven, CT: Yale School of Forestry and Environmental Studies.

Act on CO_2 Campaign (2009). Accessed on 20 May 2013. Retrieved from www.direct. gov.uk/en/Environmentandgreenerliving/Thewiderenvironment/index.htm.

Ajzen, I. (1991). The theory of planned behaviour. *Organizational Behaviour and Human Decision Processes, 50,* 179–211.

Albarracín, D., Johnson, B.T. and Zanna, M.P. (2005). *The handbook of attitudes.* Mahwah, NJ: Lawrence Erlbaum.

American Psychological Association (APA). (2010). Psychology and global climate change: Addressing a multi-faceted phenomenon and set of challenges. *Report of the American Psychological Association Task Force on the Interface Between Psychology and Global Climate Change.* Accessed on 6 March 2013. Retrieved from www.apa. org/science/about/publications/climate-change.aspx.

Anable, J., Lane, B. and Kelay, T. (2006). *An evidence base review of public attitudes to climate change and transport behaviour.* Report for the UK Department of Transport.

Anderson, C.R. (1977). The notion of schemata and the educational enterprise: General discussion of the conference. In C.R. Anderson, R.J. Spiro and W.E. Montague (eds), *Schooling and the acquisition of knowledge* (pp. 415–431). Hillsdale, NJ: Erlbaum.

Armitage, C.J., and Conner, M. (2001). Efficacy of the theory of planned behaviour: A meta-analytic review. *British Journal of Social Psychology, 40,* 471–499.

Bamberg, S. (2003). How does environmental concern influence specific environmentally-related behaviours? *Journal of Environmental Psychology, 23,* 21–32.

Bamberg, S. and Möser, G. (2007). Twenty years after Hines, Hungerford, and Tomera: A new meta-analysis of psychosocial determinants of pro-environmental behaviour. *Journal of Environmental Psychology, 27,* 14–25.

Bamberg, S. and Schmidt, P. (2001). Theory-driven, subgroup-specific evaluation of an intervention to reduce private car use. *Journal of Applied Social Psychology, 31,* 1300–1329.

Bamberg, S. and Schmidt, P. (2003). Incentive, morality or habit? Predicting students' car use for university routes with the models of Ajzen, Schwartz, and Triandis. *Environment and Behavior*, *35*, 1–22.

Bamberg, S., Hunecke, M. and Blöbaum, A. (2007). Social context, personal norm and the use of public transportation: Two field studies. *Journal of Environmental Psychology*, *27*, 190–203.

Baumeister, R.F. (1998). The self. In D.T. Gilbert, S.T. Fiske and G. Lindzey (eds), *The handbook of social psychology* (pp. 680–740). Boston, MA: McGraw-Hill.

Baumeister, R.F. and Bushman, B.J. (2008). *Social psychology and human nature*. San Francisco, CA: Wadsworth.

Becken, S. (2007). Tourists' perception of international air travel's impact on the global climate and potential climate change policies. *Journal of Sustainable Tourism*, *15*(4), 351–368.

Bicchieri, C. (2006). *The grammar of society*. New York: Cambridge University Press.

Blasch, J. and Farsi, M. (2012). *Retail demand for voluntary carbon offsets – a choice experiment among Swiss consumers*. Institute for Environmental Decisions Working Paper No. 18, Swiss Federal Institute of Technology (ETH), Zurich, Switzerland.

Böhm, G. (2003). Emotional reactions to environmental risks: Consequentialist versus ethical evaluation. *Journal of Environmental Psychology*, *23*, 199–212.

Böhm, G. and Pfister, H.-R. (2001). Mental representation of global environmental risks. In G. Böhm, J. Nerb, T. McDaniels and H. Spada (eds), *Environmental risks: Perception, evaluation and management* (pp. 1–30). New York: Elsevier Science/JAI Press.

Bolderdijk, J.W., Steg, L., Geller, E.S., Lehman, P.K. and Postmes, T. (2013). Comparing the effectiveness of monetary versus moral incentives in environmental campaigning. *Nature Climate Change*, *3*(4), 413–416.

Bord, R.J., O'Connor, R.E. and Fisher, A. (2000). In what sense does the public need to understand climate change? *Public Understanding of Science*, *9*, 205–218.

Boster, F.J. and Mongeau, P.A., (1984). Fear-arousing persuasive messages. In R.N. Bostrom (ed.), *Communication yearbook* (Vol. 8, pp. 330–375). Beverly Hills, CA: Sage.

Bostrom, A., Morgan, M.G., Fischhoff, B. and Read, D. (1994). What do people know about global climate change? Mental models. *Risk Analysis*, *14*, 959–970.

Boykoff, M.T. and Rajan, S.R. (2007). Signals and noise: Mass-media coverage of climate change in the USA and UK. *European Molecular Biology Organization*, *8*(3), 207–211.

Brody, S.D., Zahran, S., Vedlitz, A. and Grover, H. (2008). Examining the relationship between physical vulnerability and public perceptions of global climate change in the United States. *Environment and Behaviour*, *41*, 72–95.

Brönnimann, S. (2002) Picturing climate change. *Climate Research*, *22*, 87–95.

Bulkeley, H. (2000). Common knowledge? Public understanding of climate change in Newcastle, Australia. *Public Understanding of Science*, *9*, 313–333.

Cabinet Office (2011). *Behaviour change and energy use*. London: Department of Energy and Climate Change.

Camerer, C., Loewenstein, G. and Prelec, D. (2003). Neuroeconomics: How neuroscience can inform economics. *Journal of Economic Literature*, *43*, 9–64.

Carey, S. (1986). Cognitive science and science education. *American Psychologist*, *41*, 10, 1123–1130.

Centre for Research on Environmental Decisions (CRED). (2009). *The psychology of climate change communication: Translate scientific data into concrete experience*. Accessed on 29 April 2013. Retrieved from www.cred.columbia.edu/guide/guide/sec3.html.

Chaiken, S. and Trope, Y. (1999). *Dual-process theories in social psychology.* New York: Guilford.

Cialdini, R.B., Bator, R.J. and Guadagno, R.E. (1999). Normative influences in organizations. In L.L. Thompson, J.M. Levine and D.M. Messick (eds), *Shared cognition in organizations: The management of knowledge* (pp. 195–211). Mahwah, NJ: Lawrence Erlbaum.

Cialdini, R.B., Kallgren, C.A. and Reno, R.R. (1991). A focus theory of normative conduct. *Advances in Experimental Psychology, 24,* 201–234.

Climate Change Communication Advisory Group (CCCAG). (2010). *Communicating climate change to mass public audiences.* Accessed on 28 May 2013. Retrieved from www.pirc.info/projects/cccag/.

Cohen, S.A. and Higham, J.E.S. (2011). Eyes wide shut? UK consumer perceptions of aviation climate impacts and travel decisions to New Zealand. *Current Issues in Tourism, 14*(4), 323–335.

Corner, A. (2011, 16 December). *Social norm strategies do work – but there are risks involved.* Guardian Professional Network. Accessed on 29 May 2013. Retrieved from www.guardian.co.uk/sustainable-business/social-norm-behaviour-change.

Damasio, A.R. (1994). *Descartes' error: Emotion, reason, and the human brain.* New York: Grosset/Putnam.

Damasio, A.R. (1999). *The feeling of what happens.* New York: Harcourt-Brace and Company.

Darnton, A. (2005). *Public understanding of climate change.* Appendix 1 of *FUTERRA Sustainability Communications.*

Darnton, A. (2008). *Reference report: An overview of behavioural change models and their uses.* Government Social Research (GSR) Behaviour Change Knowledge Review. London: UK.

Das, E.H., de Wit, J.B. and Stroebe, W. (2003). Fear appeals motivate acceptance of action recommendations: Evidence for a positive bias in the processing of persuasive messages. *Personality and Social Psychology Bulletin, 29*(5), 650–664.

De Boer, J., Schösler, H. and Boersema, J.J. (2012). Climate change and meat eating: An inconvenient couple? *Journal of Environmental Psychology, 33,* 1–8.

De Groot, J.I.M., Abrahamse, W. and Jones, K. (2013). Persuasive normative messages: The influence of injunctive and personal norms on using free plastic bags. *Sustainability, 5,* 1829–1844.

De Hoog, N., Stroebe, W. and de Wit, B.F.J. (2005). The impact of fear appeals on processing and acceptance of action recommendations. *Personality and Social Psychology Bulletin, 31*(1), 24–33.

Devine-Wright, P. (2004). Towards zero-carbon: Citizenship, responsibility and the public acceptability of sustainable energy technologies. In. C. Buckle (ed.), *Solar energy society UK section of the International Solar Energy Society.* CIBSE: London.

Dolan, P. and Metcalfe, R. (2012). *Better neighbors and basic knowledge: A field experiment on the role of non-pecuniary incentives on energy consumption.* Department of Economics, Oxford University, UK.

Doyle, J. (2011). *Mediating climate change.* Farnham, UK: Ashgate Publishing.

Emmerich, R. (Director). (2004). *The day after tomorrow* [Motion picture]. United States: Twentieth Century Fox.

Epstein, S. (1994). Integration of the cognitive and the psychodynamic unconscious. *American Psychologist, 49,* 709–724.

Feinberg, M. and Willer, R. (2011). Apocalypse soon? Dire messages reduce belief in global warming by contradicting just-world beliefs. *Psychological Science, 22*(1), 34–38.

Ferguson, M.A. and Branscombe, N.R. (2010). Collective guilt mediates the effect of beliefs about global warming on willingness to engage in mitigation behavior. *Journal of Environmental Psychology, 30*(2), 135–142.

Festinger, L. (1954). A theory of social comparison processes, *Human Relations, 7,* 117–140.

Fischhoff, B., Slovic, P., Lichtenstein, S., Read, S. and Combs, B. (1978). How safe is safe enough? A psychometric study of attitudes towards technological risks and benefits. *Policy Sciences, 9,* 127–152.

Futerra (2005). *The rules of the game: The principles of climate change communication.* London. Accessed on 18 May 2013. Retrieved from www.futerra.co.uk/downloads/RulesOfTheGame.pdf.

GlobeScan (1999). Environics international environmental monitor survey dataset. GlobeScan, Inc. Toronto, CA. Accessed on 30 May 2013. Retrieved from http://130.15.161.246:82/webview/

GlobeScan (2000). Environics international environmental monitor survey dataset. GlobeScan, Inc. Toronto, CA. Accessed on 30 May 2013. Retrieved from http://130.15.161.246:82/webview/.

GlobeScan (2006, 3 May). BBC/Reuters/Media Center Poll: Trust in the Media. Accessed on 22 April 2013. Retrieved from www.globescan.com/news_archives/Trust_in_Media.pdf.

Goldstein, N.J., Cialdini, R.B. and Griskevicius, V. (2008). A room with a viewpoint: Using social norms to motivate environmental conservation in hotels. *Journal of Consumer Research, 35,* 472–482.

Gössling, S. and Peeters, P.M. (2007) "It does not harm the environment!" An analysis of industry discourses on tourism, air travel and the environment. *Journal of Sustainable Tourism, 15*(4), 402–417.

Gössling, S., Bredberg, M., Randow, A., Sandström, E. and Svensson, P. (2006). Tourist perceptions of climate change: A study of international tourists in Zanzibar. *Current Issues in Tourism, 9*(4&5), 419–435.

Government Communication Network (2009). *Communications and behaviour change.* Report. Accessed 26 August 2013. Retrieved from http://coi.gov.uk/documents/commongood/commongood behaviourchange.pdf.

Gregory, D.G. and Di Leo, M. (2003). Repeated behaviour and environmental psychology: The role of personal involvement and habit formation in explaining water consumption. *Journal of Applied Social Psychology, 33*(6), 1261–1296.

Grinde, B. (2002). Happiness in the perspective of evolutionary psychology. *Journal of Happiness Studies, 3*(4), 331–354.

Griskevicius, V., Cialdini, R.B. and Goldstein, N.J. (2008). Social norms: An underestimated and underemployed lever for managing climate change. *International Journal of Sustainability Communication, 3,* 5–13.

Grob, A. (1995). A structural model of environmental attitudes and behaviour. *Journal of Environmental Psychology, 15*(3), 209–220.

Guggenheim, D. (Director). (2006). *An inconvenient truth* [Motion picture]. United States: Paramount.

Hansen, J., Marx, S. and Weber, E.U. (2004). *The role of climate perceptions, expectations, and forecasts in farmer decision making: The Argentine pampas and South Florida. International Research Institute for Climate Prediction (IRI),* Palisades, NY: Technical Report 04–01.

Hargreaves, I., Lewis, J. and Speers, T. (2003). *Towards a better map: Science, the public and the media.* UK: ESRC.

Hastings, G., Stead, M. and Webb, J. (2004). Fear appeals in social marketing: Strategic and ethical reasons for concern. *Psychology and Marketing*, *21*(11), 961–986.

Heath, Y. and Gifford, R. (2002). Extending the theory of planned behavior: Predicting the use of public transportation. *Journal of Applied Social Psychology*, *32*(10), 2154–2189.

Heath, Y. and Gifford, R. (2006). Free-market ideology and environmental degradation: The case of belief in global climate change. *Environment and Behavior*, *38*(1), 48–71.

Helgeson, J., van der Linden, S. and Chabay, I. (2012). The role of knowledge, learning and mental models in perceptions of climate change related risks. In A. Wals and P.B. Corcoran (eds), *Learning for sustainability in times of accelerating change* (pp. 329–346). Wageningen, NL: Wageningen Academic Publishers.

Heskes, S. (1998). *Risicoperceptie broeikasprobleem. Rapportage van een kwalitatief onderzoek (Risk perception of the greenhouse problem. A qualitative study).* Amsterdam, The Netherlands: Heskes and Partners.

Hidalgo, C.M. and Pisano, I. (2010). Determinants of risk perception and willingness to tackle climate change. A pilot study. *Bilingual Journal of Environmental Psychology*, *1*(1), 105–112.

Higbee, K.L. (1969). Fifteen years of fear arousal: Research on threat appeals 1953–1968. *Psychological Bulletin*, *72*(6), 426–444.

Hines, J.M., Hungerford, H.R. and Tomera, A.N. (1986/87). Analysis and synthesis of research on responsible environmental behaviour: A meta-analysis. *Journal of Environmental Education*, *18*(2), 1–8.

Hounsam, S. (2006) *Painting the town green: How to persuade people to be environmentally friendly*. A report for everyone involved in promoting greener lifestyles to the public. Green-Engage communications. London.

Jacobson, G.D. (2011). The Al Gore effect: An Inconvenient Truth and voluntary carbon offsets. *Journal of Environmental Economics and Management*, *61*(1), 67–78.

Janis, I.L. (1967). Effects of fear arousal on attitude change: Recent developments in theory and experimental research. In L. Berkowitz (ed.), *Advances in experimental social psychology* (Vol. 3, pp. 166–224). San Diego, CA: Academic Press.

John, P., Cotterill, S., Moseley, A., Richardson, L., Smith, G., Stoker, G. and Wales, C. (2011). *Nudge, nudge, think, think: Experimenting with ways to change civic behaviour*. London: Bloomsbury Academic Publishing.

Joireman, J.A., Van Lange, P.A.M. and Van Vugt, M. (2004). Who cares about the environmental impact of cars? Those with an eye toward the future. *Environment and Behaviour*, *36*(2), 187–206.

Kahneman, D. (2003). Maps of bounded rationality: A perspective on intuitive judgment and choice. In T. Frangsmyr (ed.), *Les Prix Nobel: The Nobel Prizes 2002* (pp. 449–489). Stockholm: Nobel Foundation.

Kahneman, D., Slovic, P. and Tversky, A. (1982). *Judgment under uncertainty: Heuristics and biases*. New York: Cambridge University Press.

Kaiser, F.G. and Fuhrer, U. (2003). Ecological behavior's dependency on different forms of knowledge. *Applied Psychology: An International Review*, *52*(4), 598–613.

Kaiser, F.G., Wolfing, S. and Fuhrer, U. (1999). Environmental attitude and ecological behaviour. *Journal of Environmental Psychology*, *19*(1), 1–19.

Kearney, A.R. and Kaplan, S. (1997). Toward a methodology for the measurement of knowledge structures of ordinary people: The conceptual content cognitive map. *Environment and Behavior*, *29*(5), 579.

Kempton, W., Boster, J.S. and Hartley, J.A. (1995). *Environmental values in American culture*. Cambridge: MA: MIT Press.

Kollmuss, A. and Agyeman, J. (2002). Mind the gap: Why do people act environmentally and what are the barriers to pro-environmental behaviour? *Environmental Education Research*, 8(3), 239–260.

Krebs, D. and Janicki, M. (2002). Biological foundations of moral norms. In M. Schaller and C. Crandall (eds), *Psychological Foundations of Culture* (pp. 125–148). New Jersey: Lawrence Erlbaum Associates.

Krosnick, J.A., Holbrook, A.L., Lowe, L. and Visser, P.S. (2006). The origins and consequences of democratic citizens' policy agendas: A study of popular concern about global warming. *Climatic Change*, 77(1–2), 7–43.

LeDoux, J.E. (1996). *The emotional brain: The mysterious underpinnings of emotional life*. New York: Simon and Schuster, Inc.

Leiserowitz, A. (2004). Before and after *The Day After Tomorrow*: A US study of climate change risk perception. *Environment*, 46(9), 22–37.

Leiserowitz, A. (2006). Climate change risk perception and policy preferences: The role of affect, imagery and values. *Climatic Change*, 77(1), 45–72.

Leiserowitz, A. (2007). International public opinion, perception, and understanding of global climate change. *Human Development Report 2007/2008*.

Leiserowitz, A., Smith, N. and Marlon, J.R. (2010). *Americans' knowledge of climate change*. Yale University, New Haven, CT: Yale Project on Climate Change Communication.

Leventhal, H. (1970). Findings and theory in the study of fear communications. In L. Berkowitz (ed.), *Advances in experimental social psychology* (Vol. 5, pp. 119–186). San Diego, CA: Academic Press.

Lewicka, M. (1998). Confirmation bias: Cognitive error or adaptive strategy of action control? In M. Kofta, G. Weary and G. Sedek (eds), *Personal control in action: Cognitive and motivational mechanisms*. New York: Plenum Press.

Loewenstein, G.F., Weber, E.U., Hsee, C.K. and Welch, E. (2001). Risk as feelings. *Psychological Bulletin*, 127(2), 267–286.

Lorenzoni, I. and Langford, I.H. (2001). *Climate change now and in the future: A mixed methodological study of public perceptions in Norwich (UK)*. CSERGE Working Paper ECM 01–05.

Lorenzoni, I. and Pidgeon, N. (2005). *Defining dangers of climate change and individual behaviour: Closing the gap*. Avoiding Dangerous Climate Change Conference, Exeter, UK, February 2005.

Lorenzoni, I., Leiserowitz, A., De Franca Doria, M., Poortinga, W. and Pidgeon, N.F. (2006). Cross-national comparisons of image associations with "global warming" and "climate change" among laypeople in the United States of America and Great Britain. *Journal of Risk Research*, 9(3), 265–281.

Lorenzoni, I., Nicholson-Cole, S.A. and Whitmarsh, L. (2007). Barriers perceived to engaging with climate change among the UK public and their policy implications. *Global Environmental Change*, 17(3), 445–459.

Lowe, T., Brown, K., Dessai, S., de Franca Doria, M., Haynes, K. and Vincent, K. (2006). Does tomorrow ever come? Disaster narrative and public perceptions of climate change. *Public Understanding of Science*, 15(4), 435–457.

Mackie, D.M., Worth, L.T. and Asuncion, A.G. (1990). Processing of persuasive in-group messages. *Journal of Personality and Social Psychology*, 58(5), 812–822.

MacLean, P.D. (1990). *The triune brain in evolution: Role in paleocerebral functions.* New York: Plenum Press.

Maibach, E., Roser-Renouf, C. and Leiserowitz, A. (2008). Communication and marketing as climate change intervention assets: A public health perspective. *American Journal of Preventive Medicine, 35*(5), 488–500.

Malka, A., Krosnick, J.A. and Langer, G. (2009). The association of knowledge with concern about global warming: Trusted information sources shape public thinking. *Risk Analysis, 9*(5), 633–647.

Maloney, M.P., Ward, M.P. and Braucht, G.N. (1975). A revised scale for the measurement of ecological attitudes and knowledge. *American Psychologist, 30*(7), 787–790.

Manstead, A.S.R. (2000). The role of moral norm in the attitude–behaviour relation. In D.J. Terry and M.A. Hogg (eds), *Attitudes, behaviour and social context: The role of norms and group membership* (pp. 11–30). Mahwah, NJ: Erlbaum.

March, J. (1994). *A primer on decision-making: How decisions happen.* New York: Free Press.

Markowitz, E.M. and Shariff, A.F. (2012). Climate change and moral judgement. *Nature Climate Change, 2*, 243–247.

Marx, S., Shome D. and Weber, E.U. (2006). *Analytic vs. experiential processing exemplified through glacial retreat education module.* Center for Research on Environmental Decisions.

Marx, S.M., Weber, E.U., Orlove B.S., Leiserowitz, A., Krantz, D.H., Roncoli, C. and Philips, J. (2007). Communication and mental processes: Experiential and analytics processing of uncertain climate information. *Global Environmental Change, 17*(1), 47–58.

McGuire, W.J. (1969). The nature of attitudes and attitude change. In G. Lindzey and E. Aronson (eds), *Handbook of social psychology* (2nd edn, Vol. 3, pp. 136–314). Reading, MA: Addison-Wesley.

Meat Free Monday (2010). *Paul McCartney's meat free Monday campaign.* Accessed on 9 May 2013. Retrieved from: www.meatfreemondays.com/.

Meijnders, A.L. (1998) *Climate change and changing attitudes: Effect of negative emotion on information processing.* Unpublished doctoral dissertation, Eindhoven University of Technology, Eindhoven, NL.

Meijnders, A.L., Midden, C.J.H. and Wilke, A.M. (2001a). Role of negative emotion in communication about CO_2 risks. *Risk Analysis, 21*(5), 955–966.

Meijnders, A.L., Midden, C.J.H. and Wilke, A.M. (2001b). Communications about environmental risks and risk-reducing behavior: The impact of fear on information processing. *Journal of Applied Social Psychology, 31*(4), 754–777.

Meinhold, J.L. and Malkus, A.J. (2005). Adolescent environmental behaviours: Can knowledge, attitudes and self-efficacy make a difference? *Environment and Behavior, 37*(4), 511–532.

Morgan, M., Fischhoff, B., Bostrom, A. and Atman, C.J. (2002). *Risk communication: A mental models approach.* Cambridge, UK: Cambridge University Press.

Moser, S.C. (2006). Talk of the city: Engaging urbanites on climate change. *Environmental Research Letters, 1*(1), 1–10.

Moser, S.C. (2010). Communicating climate change: History, challenges, process and future directions. *Wiley Interdisciplinary Reviews: Climate Change, 1*(1), 31–53.

Moser, S.C. and Dilling, L. (2004) Making climate hot. *Environment: Science and Policy for Sustainable Development, 46*(10), 32–46.

Moser, S.C. and Dilling, L. (2007). *Creating a climate for change: Communicating climate change and facilitating social change.* New York: Cambridge University Press.

Myers, D.G. (2004). Theories of emotion. In D.G. Myers (ed.), *Psychology* (7th edn) (p. 500). New York: Worth Publishers.

Myers, G. and Macnaghten, P. (1998). Rhetorics of environmental sustainability: Commonplaces and places. *Environment and Planning A, 30*, 333–353.

Ngo, A.T., West, G.E. and Calkins, P.H. (2009). Determinants of environmentally responsible behaviours for greenhouse gas reduction. *International Journal of Consumer Studies, 33*(2), 151–161.

Nilsson, M. and Kuller, R. (2000) Travel behaviour and environmental concern. *Transportation Research Part D, 5*, 211–234.

O'Connor, R.E., Bord, R.J. and Fisher, A. (1999). Risk perceptions, general environmental beliefs, and willingness to address climate change. *Risk Analysis, 19*(3), 461–471.

O'Keefe, D.J. (2002). Guilt as a mechanism of persuasion. In J. P. Dillard and M. Pfau (eds), *The persuasion handbook: Developments in theory and practice* (pp. 329–344). Thousand Oaks, CA: Sage.

O'Neill, S. and Hulme, M. (2009). An iconic approach for representing climate change. *Global Environmental Change, 19*(4), 402–410.

O'Neill, S. and Nicholson-Cole, S. (2009). "Fear won't do it": Promoting positive engagement with climate change through visual and iconic representations. *Science Communication, 30*(3), 355–379.

Peta2 (2008). *Meat's not green campaign.* Accessed on 10 May 2013. Retrieved from http://features.peta2.com/meatsnotgreen/.

Pew Global Attitudes Project (2006). *No global warming alarm in the US.* Washington, DC: The Pew Research Center for the People and the Press.

Poortinga, W. and Pidgeon, N. (2003). *Public perceptions of risk, science and governance: Main findings of a British survey of five risk cases.* Centre for Environmental Risk, University of East Anglia.

Poortinga, W., Pidgeon, N. and Lorenzoni, I. (2006). Public perceptions of nuclear power, climate change and energy options in Britain. *Summary findings of a survey conducted during October and November 2005.* Understanding Risk. Working Paper 6 February. Norwich: Centre for Environmental Risk.

Read, D., Bostrom, A., Morgan, M.G., Fischhoff, B. and Smuts, T. (1994). What do people know about global climate change? Survey studies of educated laypeople. *Risk Analysis, 14*(6), 971–982.

Rogers, R.W. (1975). A protection motivation theory of fear appeals and attitude change. *Journal of Psychology, 91*(1), 93–114.

Rogers, R.W. (1983). Cognitive and physiological processes in fear appeals and attitude change: A revised theory of protection motivation. In J.T. Cacioppo and R.E. Petty (eds), *Social psychophysiology: A sourcebook* (pp. 153–176). New York: Guilford Press.

Sampei, Y. and Aoyagi-Usui, M. (2009). Mass-media coverage, its influence on public awareness of climate-change issues, and implications for Japan's national campaign to reduce greenhouse gas emissions. *Global Environmental Change, 19*(2), 203–212.

Schultz, P.W., Nolan, J.M., Cialdini, R.B., Goldstein, N.J. and Griskevicius, V. (2007). The constructive, destructive, and reconstructive power of social norms. *Psychological Science, 18*(5), 429–434.

Schwartz, S.H. (1977). Normative influences on altruism. In L. Berkowitz (ed.), *Advances in experimental social psychology* (pp. 222–280). New York: Academic Press.

Semenza, J.C., Hall, D.E., Wilson, D.J., Bontempo, B.D., Sailor, D.J. and George, L.A. (2008). Public perception of climate change: Voluntary mitigation and barriers to behaviour change. *American Journal of Preventive Medicine, 35*(5), 479–487.

Sharples, D.M. (2010). Communicating climate science: Evaluating the UK public's attitude to climate change. *Earth and E-environment, 5*, 185–205.

Sheeran, P. (2002). Intention-behaviour relations: A conceptual and empirical review. In W. Stroebe and M. Hewstone (eds), *European Review of Social Psychology* (Vol. 12, pp. 1–30). Chichester: Wiley.

Silver, N. (2009, 22 April). When hope is the enemy of change. *FiveThirtyEight.* Accessed on 29 May 2013. Retrieved from www.fivethirtyeight.com/2009/04/when-hope-is-enemy-of-change.html.

Sloman, S.A. (1996). The empirical case for two systems of reasoning. *Psychological Bulletin, 119*(1), 3–22.

Slovic, P. (1999). Trust, emotion, sex, politics and science: Surveying the risk-assessment battlefield. *Risk Analysis, 19*(4), 689–701.

Slovic, P., Finucane, M.L., Peters, E. and MacGregor, D. (2004). Risk as analysis and risk as feelings: Some thoughts about affect, reason, risk and rationality. *Risk Analysis, 24*(2), 311–322.

Slovic, P., Finucane, M.L., Peters, E. and MacGregor, D.G. (2006). The affect heuristic. *European Journal of Operational Research, 177*(3), 1333–1352.

Slovic, P., MacGregor, D.G. and Peters, E. (1998). *Imagery, affect and decision-making.* Eugene, OR: Decision Research.

Smith, N. and Leiserowitz, A. (2012). The rise of global warming skepticism: Exploring affective image associations in the United States over time. *Risk Analysis, 32*(6), 1021–1032.

Spence, A., Poortinga, W., Butler, C. and Pidgeon, N.F. (2011). Perceptions of climate change and willingness to save energy related to flood experience. *Nature Climate Change, 1*(1), 46–29.

Spence, A., Poortinga, W. and Pidgeon, N.F. (2012). The psychological distance of climate change. *Risk Analysis, 32*(6), 957–972.

Staats, H.J., Wit, A.P. and Midden, C.Y.H. (1996). Communicating the greenhouse effect to the public: Evaluation of a mass media campaign from a social dilemma perspective. *Journal of Environmental Management, 46*(2), 189–203.

Stamm, K., Clark, F. and Eblacas, P. (2002). Mass communication and public understanding of environmental problems: The case of global warming. *Public Understanding of Science, 9*(3), 219–237.

Steg, L. (2005). Car use: Lust and must. Instrumental, symbolic and affective motives for car use. *Transportation Research Part A, 39*, 147–162.

Steg, L. (2008). Promoting household energy conservation. *Energy Policy, 36*(12), 4449–4453.

Steg, L. and Vlek, C. (2009). Encouraging pro-environmental behaviour: An integrative review and research agenda. *Journal of Environmental Psychology, 29*(3), 309–317.

Sterman, J. (2008). Risk communication on climate: Mental models and mass balance. *Science, 322*, 532–533.

Sterman, J. and Booth Sweeney, L. (2002). Cloudy skies: assessing public understanding of global warming. *System Dynamics Review, 18*(2), 207–240.

Sterman, J. and Booth Sweeney, L. (2007). Understanding public complacency about climate change: Adults' mental models of climate change violate conservation of matter. *Climate Change, 80*(3–4), 213–238.

Stern, P.C., Dietz, T., Abel, T., Guagnano, G.A. and Kalof, L. (1999). A value belief norm theory of support for social movements: The case of environmental concern. *Human Ecology Review*, *6*(8), 1–97.

Stiff, B.J. and Mongeau, A.P. (2003). *Persuasive communication*. New York: Guilford Press.

Sutton, S.R. (1982). Fear-arousing communications: A critical examination of theory and research. In J.R. Eiser (ed.), *Social psychology and behavioural medicine* (pp. 303–337). London: John Wiley.

Sweney, M. (2010, 11 October). Government's £6m climate change ads cleared. *Guardian*. Accessed on 10 May 2013. Retrieved from: www.guardian.co.uk/media/2010/oct/11/government-climate-change-ad.

Tanner Jr, J.F., Day, E. and Crask, M.R. (1989). Protection motivation theory. An extension of fear appeals theory in communication. *Journal of Business Research*, *19*(4), 267–276.

Tonn, B., Hemrick, A. and Conrad, F. (2006). Cognitive representations of the future: Survey results. *Futures*, *38*(7), 810–829.

Truelove, H.B. (2009). *An investigation of the psychology of global warming: Perceptions, predictors of behaviour, and the persuasiveness of ecological footprint calculators*. Unpublished PhD dissertation, Washington State University, Washington, USA.

Ungar, S. (2000). Knowledge, ignorance and the popular culture: climate change versus the ozone hole, *Public Understanding of Science*, *9*(3), 297–312.

Van der Linden, S. (2011). Charitable intent: A moral or social construct? A revised theory of planned behaviour model. *Current Psychology*, *30*(4), 355–374.

Van der Linden, S. (2013). A response to Dolan. In A.J. Oliver (ed.) *Behavioural Public Policy* (pp. 209–215). Cambridge, UK: Cambridge University Press.

Van Knippenberg, D., Lossie, N. and Wilke, H. (1994). In-group prototypicality and persuasion: Determinants of heuristic and systematic message processing. *British Journal of Social Psychology*, *33*(3), 289–300.

Wall, R. (2005). *Psychological and contextual influences on travel mode choice for commuting*. Unpublished PhD thesis. Institute of Energy and Sustainable Development, De Montfort University, Leicester, UK.

Wasting Water is Weird Campaign (2011). Accessed on 29 May 2013. Retrieved from www.wastingwaterisweird.com/.

Weber, E.U. (1997). Perception and expectation of climate change: Precondition for economic and technological adaptation. In M. Bazerman, D. Messick, A. Tenbrunsel and K. Wade-Benzoni (eds), *Psychological perspectives to environmental and ethical issues in management* (pp. 314–341). San Francisco, CA: Jossey-Bass.

Weber, E.U. (2006). Evidence-based and description-based perceptions of long-term risk: Why global warming does not scare us (yet). *Climatic Change*, *77*, 103–120.

Weber, E.U. and Sonka, S. (1994). Production and pricing decisions in cash-crop farming: Effects of decision traits and climate change expectations. In B.H. Jacobsen, D.F. Pedersen, J. Christensen and S. Rasmussen (eds), *Farmers' decision making: A descriptive approach* (pp. 203–218). Copenhagen, Denmark: European Association of Agricultural Economists.

Weingart, P., Engels A. and Pansegrau, P. (2000). Risks of communication: Discourses on climate change in science, politics, and the mass media. *Public Understanding of Science*, *9*(3), 261–283.

Weinstein, N.D. (1980). Unrealistic optimism about future life events. *Journal of Personality and Social Psychology*, *39*(5), 806–820.

Whitmarsh, L. (2008). Are flood victims more concerned about climate change than other people? The role of direct experience in risk perception and behavioural response. *Journal of Risk Research*, *11*(3), 351–374.

Whitmarsh, L. (2009). Behavioural responses to climate change: Asymmetry of intentions and impacts. *Journal of Environmental Psychology*, *29*(1), 13–23.

Whitmarsh, L. O'Neill, S. and Lorenzoni, I. (2008). *Engaging the public with climate change: Behaviour change and communication*. London: Earthscan.

Witte, K. (1992). Putting the fear back into fear appeals: The extended parallel process model. *Communication Monographs*, *59*(4), 329–349.

Witte, K. (2000). A meta-analysis of fear appeals: Implications for effective public health campaigns. *Health Education and Behaviour*, *27*(5), 591–615.

Xiang, C. (2011). Why do people misunderstand climate change? Heuristics, mental models and ontological assumptions. *Climatic Change*, *108*(1–2), 31–46.

Zajonc, R. (1998). Emotions. In D. Gilbert, S. Fiske and G. Lindzey (eds), *The handbook of social psychology* (pp. 591–632). New York: McGraw-Hill.

15 Framing behavioural approaches to understanding and governing sustainable tourism consumption

Beyond neoliberalism, "nudging" and "green growth"?

C. Michael Hall

Introduction

Sustainable development, and the response to climate change in particular, has emerged as a central problem for tourism. The search to find "solutions" has led to the publication of numerous government, industry, institutional and academic reports as well as the adoption of "sustainable" tourism policies (Bramwell and Lane, 2012; Scott and Becken, 2010). Yet despite such actions tourism is empirically demonstrably less sustainable than ever, and continues to increase its absolute contribution to greenhouse gas emissions and therefore climate change (Gössling, 2009; Hall, 2011a; Peeters and Landré, 2011; Scott *et al.*, 2012a, 2012b).

At one level this can be explained by the growth in tourism numbers. In 2012, international tourist arrivals are expected to reach one billion for the first time, up from 25 million in 1950, 277 million in 1980 and 528 million in 1995 (UNWTO, 2012). The UNWTO predicts that the number of international tourist arrivals will increase by an average 3.3 per cent per year between 2010 and 2030 (an average increase of 43 million arrivals a year), reaching an estimated 1.8 billion arrivals by 2030 (UNWTO, 2011, 2012). Even with hoped-for per trip efficiency gains, the absolute contribution of tourism to climate change will continue to increase in the foreseeable future (Gössling *et al.*, 2010; Hall, 2010, 2011a; Peeters and Landré, 2011; Scott *et al.*, 2012b).

Scott *et al.* (2008) suggest that CO_2 emissions from tourism will grow by about 135 per cent to 2035 (compared with 2005), totalling approximately 3059 Mt. Most of this growth is associated with air travel (UNWTO, 2012). Owens *et al.* (2010) suggest that there will be a 2.0–3.6-fold increase in CO_2 emissions from aviation by 2050. Even the International Energy Agency's (2009) "Blue Shifts" scenario developed to describe a strong state interventionist stance with respect to policies that dampen air travel growth, the development of high-speed rail and the aggressive promotion of the use of telecommunications as a substitute for travel, still forecasts a tripling of air travel between 2005 and 2050. At the national level, the UK Department of Transport (2007) predicts that, taking radiative forcing into account, the 9 per cent contribution of aviation to total UK emissions in 2005 will have grown to approximately 15 per cent in 2020 and 29 per cent in 2050. It is, therefore, fundamentally unclear how more sustainable and absolute reductions in emissions and "carbon neutral" positions are to be achieved, given the growth rates expected in the forecasts and scenarios posited for tourism in the next 20–40 years (Gössling, 2009; Gössling and Schumacher, 2010; Peeters and Landré, 2011; Scott, 2011).

Despite forecast increases in absolute emissions from tourism, the notion of "green growth", i.e. that you can continue to have both increased economic growth at the same time as becoming more sustainable, has become an integral component of industry discourse on tourism and sustainability (Cabrini, 2012; UNWTO and UNEP, 2011). Nevertheless, the optimism of such a growth paradigm based on material/resource/energy efficiency, major changes in the energy mix towards renewables and continued increases in visitor numbers are extremely problematic, given the arithmetic constraints of growth and efficiency limits, governance and market limits, and system limits (Hall, 2009a; Hoffmann, 2011), as well as the general failure of many assessments of green economic growth and sustainable consumption to consider rebound effects (Arvesen *et al.*, 2011; Hall, 2009a, 2010; Santarius, 2012; Sorrell *et al.*, 2009), i.e. changed behaviours that may offset part of the environmental gain and/or efficiency improvements that also require resource use. No studies have been conducted on rebound effects specifically in relation to tourism. Barker *et al.* (2009) modelled the rebound effects resulting from the global energy efficiency measures incorporated into the IPCC's (2007) Fourth Assessment Report and estimated that for transport there would be a worldwide direct rebound of 9.1 per cent in 2020 and 9.1 per cent in 2030, and a macroeconomic rebound of 26.9 per cent in 2020 and 43.1 per cent in 2030, thus leading to a total economy wide rebound of 36.0 per cent in 2020 and 52.2 per cent in 2030. If this scale of rebound were applied to tourism then, even allowing for the estimated greater use of low-carbon fuels, the potential increase in tourism-related emissions would likely be more than 200 per cent by 2030 (Hall *et al.*, 2013). The situation becomes even more complicated because the business and policy strategies of almost all national and regional tourism organisations are predicated on increased visitor numbers.

In questioning the "green economy", Hoffmann (2011) takes the position that colossal de-carbonisation of the economy and society will only be achieved if current consumption patterns, methods and lifestyles are also subject to profound change. But how to achieve this in tourism has proven to be an extremely vexed question (Bramwell and Lane, 2012, 2013; Gössling *et al.*, 2012a; Hall, 2011a; Higgins-Desbiolles, 2010; Scott and Becken, 2010; Scott *et al.*, 2012b; Zeppel, 2012). This chapter takes a different tack in seeking to respond to the current policy-action gap that exists in relation to climate change and emissions reduction and sustainable tourism consumption and mobility; i.e. in order to reduce its demands on natural capital, tourism needs to become part of a circular economy, so that inputs of virgin raw material and energy and outputs in the form of emissions and waste requiring disposal are reduced (Hall, 2010).

The notion of a policy-action gap (Barrett and Fudge, 1981) is used in policy analysis not just to refer to a failure in implementation (Hall, 2009b) but also to the more fundamental framing of policy problems – with emissions reduction in tourism being a classic "wicked" or "messy" problem – given that how a problem is understood and defined predicates the range of solutions that are used to "solve" it (Hall, 2011a). Indeed, it is interesting to note that while there has been debate on the underlying disciplinary assumptions that have framed the problem of climate change, especially its presentation by the IPCC, and how this affects problem definition and solutions identification (Demeritt, 2001, 2006; Hall, 2013), the implications of this have not been fully appreciated in discussions of sustainable mobility and tourism consumption (Scott *et al.*, 2012b).

Many models, theories and studies of climate change and tourism, sustainable tourism, sustainable mobility and sustainable consumption in general contain underlying assumptions about an individual's capacity to act. These assumptions relate to both behaviour *and* governance, given that the range of policy measures that the state utilises to achieve its policy goals are based on assumptions regarding individual and collective behaviour (Hall, 2011b). Although usually not explicitly stated, such assumptions are contained not only in public policy positions but also in the recommendations of industry, institutions, consultants and academics on how to improve the sustainability of tourism because they implicitly suggest that by engaging policy setting A consumers will do B, i.e. by developing a code of conduct consumers will follow it. Therefore, this chapter aims to extend the work of Whitmarsh *et al.* (2011) with respect to the carbon capability of consumers, to also suggest that theories, paradigms and understandings of behaviour and how change might occur also frame the extent to which tourism governance is "carbon capable", especially in relation to policy learning for sustainable consumption (Hall, 2011a). The chapter first discusses three different approaches to consumer change before examining how the framing of behavioural capacity is connected to policy learning. Such framing has led not only to dominance of the paradigm of "ABC" – attitude, behaviour and choice – in favouring certain behavioural approaches to sustainable mobility but also the ignorance of other ways of framing consumer behaviour.

Approaches to behavioural change

In seeking to understand the different modes of policy intervention in changing behaviours with respect to climate change (or other concerns with sustainable consumption and mobility), there is a need to understand the situated meanings associated with climate change, sustainability, mobility, consumption and environmental behaviours. That is, how individuals translate and apply knowledge in their daily lives and decision-making (Whitmarsh *et al.*, 2011), including with respect to consuming travel and tourism (Gössling *et al.*, 2012a), as well as the different forms of governance (Hall, 2011b).

In the case of sustainable tourism mobility, the vast majority of individuals are not yet willing to give up international travel or change their travel markedly

(Cohen and Higham, 2011; Gössling *et al.*, 2012a; Kroesen, 2013; Mair, 2011; Scott *et al.*, 2012b), despite enthusiasm by governments and industry to encourage voluntary changes in consumer behaviour. As reported by the UK Sustainable Consumption Roundtable (2006, p. 35)

> consumers struggle with the idea of flying less. They candidly acknowledge that the prospect of cutting back on flights was extremely unattractive … flying abroad represents a major aspiration for consumers.… In addition, cutting back on flying is also very unattractive as consumers feel that there are few alternatives.

Even the most environmentally aware tourists might not be more willing to alter travel behaviour and may even be among the most active travellers (Barr *et al.*, 2010; Gössling *et al.*, 2009; McKercher *et al.*, 2010). Such a situation reflects the importance of recognising that relevant sustainable tourism consumption information is individually and socially contextualised and actioned. "Concepts and tools do not necessarily motivate behaviour change where individuals are not motivated to change or perceive barriers to doing so … amongst users of carbon calculators, many (though by no means all) use such tools to *offset* [original emphasis] their emissions rather than to *change* [emphasis added] their energy consumption behavior" (Whitmarsh *et al.*, 2011, p. 58).

For the majority of people, issues of climate change, energy, sustainable consumption and sustainable mobility have a relatively low salience in people's day-to-day activities, choices and actions (Sustainable Consumption Roundtable, 2006; Weber, 2006; Whitmarsh, 2009; Whitmarsh *et al.*, 2011). Awareness of environmental problems at an abstract level is usually not translated into personally relevant cognitions or motivating attitudes. For the majority of people, their understanding of environmental issues tends to be limited to abstract or vague concepts. Whitmarsh *et al.* (2011, p. 57) argue that climate change poses major challenges to communicators and educators as "It is a risk 'buried' in familiar natural processes such as temperature change and weather fluctuations … and has low salience as a risk issue because it cannot be directly experienced" (see also Weber, 2006). Interestingly, there is some evidence to support this observation, given that research suggests that the attitudes of tour operators as well as the general public change when they experience weather events that they believe are associated with climate change (Hall, 2006; Tervo-Kankare *et al.*, 2013). For example, in December 2012, the results of an US poll by the Associated Press-GfK found rising concern about climate change among Americans, with 80 per cent citing it as a serious problem for the United States, up from 73 per cent in 2009. According to Goldenberg (2012),

> Some of the doubters said in follow-up interviews that they were persuaded by personal experience: such as record temperatures, flooding of New York City subway tunnels, and news of sea ice melt in the Arctic and extreme drought in the mid-west.

The suggestion by Whitmarsh *et al.* (2009) that there are currently low levels of "carbon capability" among the UK population is arguably one that can be transferred to a wide range of individual and social contexts (see also Waitt *et al.*, 2012). Carbon capability is "the ability to make informed judgments and to take effective decisions regarding the use and management of carbon, through both individual behaviour change and collective action" (Whitmarsh *et al.*, 2009, p. 2). Carbon capability is not defined in a narrow individualistic sense of solely knowledge, skills and motivations (although these are important components). Instead, it suggests that a carbon capable individual, or family (Waitt *et al.*, 2012), have an understanding of the limits of individual action and where these encounter societal institutions, infrastructure and systems of provision that act as barriers to low carbon lifestyles, including travel and tourism, then they can recognise the need for collective action and other governance solutions (Whitmarsh *et al.*, 2009, 2011). Carbon capability is also regarded as an analogue to financial capability as applied to anthropogenic climate change, and similarly involved budget management, planning, staying informed and making choices (Lorenzoni *et al.*, 2011). Carbon capability has three core dimensions:

1   *Decision-making/cognitive/evaluative* (technical, material and social aspects of knowledge, skills, motivations, understandings and judgments);
2   *Individual behaviour or "practices"* (e.g. energy conservation, sustainable or slow travel); and

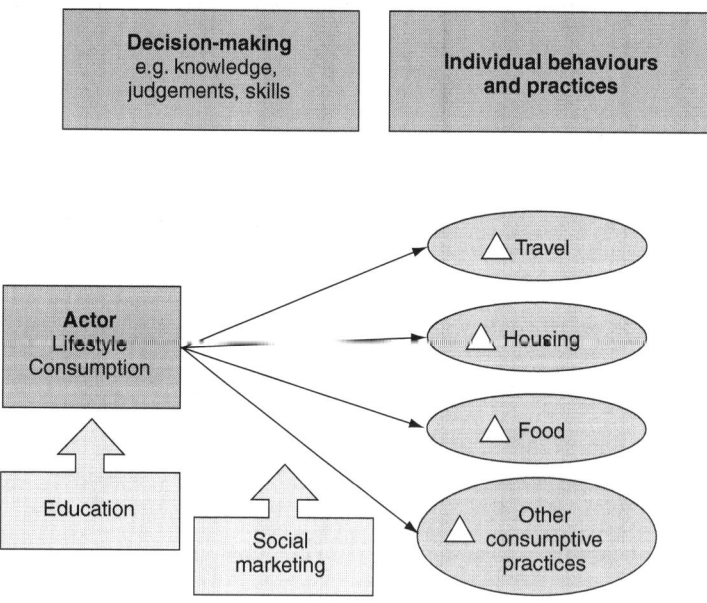

Figure 15.1 Dimensions of carbon literacy.

3   *Broader engagement with systems of provision and governance* (e.g. lobbying, voting, protesting, creating alternative social infrastructures of provision).

Under a carbon literacy approach the assumption is that, as individuals gain knowledge, perhaps as a result of educational and social marketing campaigns, they may change behaviours and practices (Figure 15.1). In contrast, a carbon capability approach argues that much consumption (and hence contribution to resource use and emissions) is inconspicuous, habitual and routine, rather than the result of conscious decision-making. This means that individual cognitive decisions about consumption are mediated through socially shaped lifestyle choices, resulting in sets of socio-environmental practices that are, in turn, delimited by social systems of provision and the rules and resources of macro-level structures and institutions (Giddens, 1984; Whitmarsh *et al.*, 2009) (Figure 15.2). Carbon capability does not imply that education and social marketing that tries to encourage behaviour change is worthless. Rather it highlights that the capacity for behavioural change needs to be understood in a much wider social, political and institutional context.

Approaches to consumer change

The notion of carbon capability inherently means looking beyond an understanding of environmental behavioural change that suggests that just providing information is sufficient for consumers to make "rational" or "appropriate" choices. However, it also raises the need to understand the underlying assumptions of different approaches to behaviour change. Three major contemporary approaches can be recognised in approaching issues of sustainability

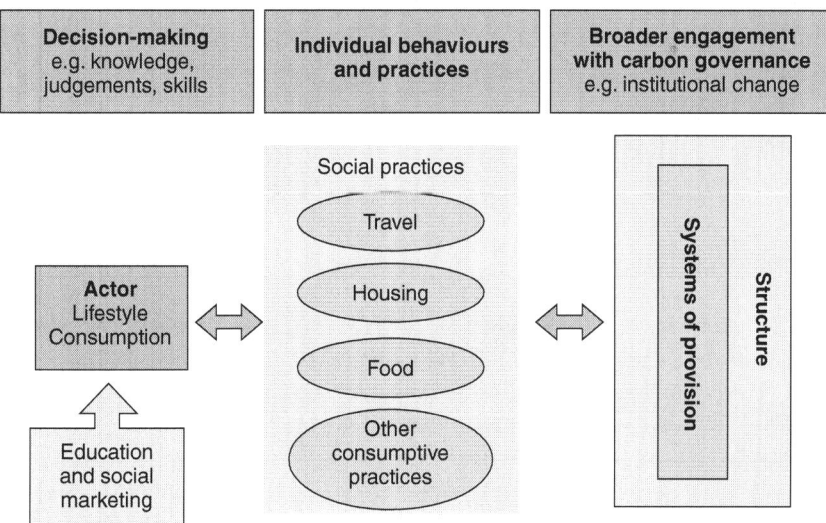

Figure 15.2 Dimensions of carbon capability.

tourism mobility, as well as other areas of sustainable consumption: utilitarian, social/psychological and systems of provision/institutional (Seyfang, 2011). An overview of these approaches is provided in Table 15.1 and is discussed in the following sections.

The utilitarian approach

The utilitarian approach to behavioural change utilises a conventional neo-classical microeconomic view of consumption by individuals as rational utility-maximisers. The approach assumes that individuals consume goods and services in free markets with perfect competition and information to decide a course of action that delivers the greatest utility to the individual. Even though the intellectual basis of neoclassical economics has been severely criticised, especially in terms of the relationship of its assumptions to real world economic behaviour (Keen, 2011), the approach underlies much contemporary neoliberal economic policy and has influenced the socio-political construction of rational utility maximising neoliberal consumer (Bone, 2010; McGuigan, 2006). From this perspective, efforts to promote sustainable mobility and consumption tend to rely on government intervention (or self-regulation) to correct "market failure" and ensure that private and corporate individuals have greater information on which to base their decisions. The approach aims to appeal to rational actors with information to overcome an "information deficit" and encourage "rational behaviour". However, access to information and education about climate change and more sustainable forms of consumption has not led to substantially improved sustainability behaviour (Christie, 2010; Gadenne *et al.*, 2011; Ockwell *et al.*, 2010). Addressing the issues of the "value-action gap" and consumer awareness has not been matched by behavioural change (Seyfang, 2011). Indeed, it is interesting to note that Maryon-Davis and Jolley's (2010) report on encouraging greater personal responsibility in the UK health system stated:

> The evidence from at least 30 years of health education and promotion has repeatedly demonstrated that appealing to people's sense of personal or social responsibility only works up to a point. Some people respond, but many others do not, and there are all sorts of reasons why not. Lack of information or knowledge is certainly an important barrier. So too are a lack of motivation to change or confidence to do so.
> But an even more fundamental barrier is the lack of opportunity to make healthier choices ... a set of factors described ... as "the causes of the causes". But it may also be due to supply-side factors such as lack of availability or choice, inadequate services or poor value-for-money – powerful commercial pressures.... Government policy can do much to reduce these barriers to healthy living. Sometimes legislation and regulation of the so-called choice architecture are important tools for nudging people into making healthier choices.
>
> (Maryon-Davis and Jolley, 2010, pp. 5–6)

Table 15.1 Approaches to consumer change

Approach	Scale	Understanding of decision-making	Consumption is …	Tools to achieve sustainable consumption	Dominant forms of governance
Utilitarian (green economics)	Individual	Cognitive information processing on basis of rational utility-maximisation	The means for increasing utility	Green labelling, tax incentives, pricing (including carbon trading), education	Markets (marketisation and privatisation of state instruments)
Social/psychological (behavioural economics/green consumerism/ABC model)	Individual	Response to psychological needs, behaviour and social contexts	Satisfier of psychological needs; cultural differentiator; marker of social meaning and identity	Nudging – making better choices through manipulating a consumer's environment	Markets (marketisation and privatisation of state instruments)
		Dominant paradigm of "ABC": attitude, behaviour, and choice		Social marketing in order to encourage behavioural change and promote sustainable lifestyles and behaviour	Networks (public–private partnerships)
Systems of provision/institutions (degrowth, steady-state tourism)	Community, society, network	Constrained/shaped by sociotechnical infrastructure and institutions	Routine habit, inconspicuous rather than conspicuous	Short-supply chains, local food, local tourism	Hierarchies (nation state and supranational institutions)

Sources: Burgess (2012), Gössling and Hall (2013), Hall (2009a, 2011a, 2011b, 2014), Seyfang (2011).

Social/psychological approaches: nudging and social marketing

The failure of neoclassical economic models to significantly increase levels of sustainability behaviour has led to the realisation that behaviour does not change simply because of better quality information (Whitmarsh, 2009; Whitmarsh *et al.*, 2009). The critique of neoclassical/rational models has primarily come from two social/psychological sources: behavioural economics and consumption studies.

Behavioural economics recognises that individuals have bounded rationality (Conlisk, 1996) and often engage in *satisficing* behaviour (Simon, 1959, 1965), i.e. an option that satisfies most needs but is not an optimal solution, as well as the role of social norms and routines that are not subject to rational cost-benefit calculations, including notions of community and fairness in economic outcomes (Folmer and Johansson-Stenman, 2011). Significantly, with respect to the manner of influencing behaviour it also stresses that too much choice in the market leads to information overload and subsequent difficulties in decision-making (Seyfang, 2011). The focus on satisficing is not new and has long been influential with respect to policymaking (Simon, 1965); however, it has recently assumed renewed significance with respect to climate change solutions (Thynne, 2008), including transition management in tourism (Gössling *et al.*, 2012b). It has also become an underlying dimension of the recent interest of behavioural economics in "nudging" (Amir and Lobel, 2008; Cialdini, 2007; Thaler and Sunstein, 2008).

The focus of nudging is in reconfiguring the "choice architecture" to encourage beneficial decision-making by consumers, such as reductions in emissions. The approach suggests that the goal of public policymaking should be to steer citizens towards making positive decisions as individuals and for society while preserving individual choice. Acting as "choice architects" policymakers organise the context, process and environment in which individuals make decisions and, in so doing, they exploit "cognitive biases" to manipulate people's choices (Alemanno, 2012). According to Alemanno (2012) and Bovens (2009), a policy instrument qualifies as a nudge when it satisfies the following:

1   The intervention must not restrict individual choice.
2   It must be in the interest of the person being nudged.
3   It should involve a change in the architecture or environment of choice.
4   It implies the strategic use of cognitive biases.
5   The action it targets does not stem from a fully autonomous choice (e.g. lack of full knowledge about the context in which the choice is made).

In the United Kingdom, nudging has become a centre-piece of the Cameron coalition government that came to power in 2010, with David Cameron reportedly making *Nudge* (Thaler and Sunstein, 2008) obligatory reading for his colleagues before the election, while in the US Sunstein became head of the Office of Regulatory Affairs in the Obama administration (Burgess, 2012). An example of the kind of solutions being explored are schemes that attempt to reduce electricity consumption by providing feedback to households on their own and neighbours' usage (Burgess, 2012), as well as changes in health-related behaviours (Maryon-Davis and Jolley, 2010). The influence of the concept of nudging on UK policy initiatives, including emissions reduction and changes in travel behaviour, is seen in the MINDSPACE report (Dolan *et al.*, 2010) which although published during the tenure of the Cameron government had been commissioned by the previous Labour government, thereby illustrating the approach's broad political appeal. In terms of its policy attractiveness, the report suggests "approaches based on 'changing contexts' – the environment within which we make decisions and respond to cues – have the potential to bring about significant changes in behaviour at relatively low cost" (Dolan *et al.*, 2010, p. 8). MINDSPACE is a mnemonic "which can be used as a quick checklist when making policy" (Dolan *et al.*, 2010, p. 8; see also Dolan *et al.*, 2012):

MESSENGER: We are heavily influenced by who communicates information.
INCENTIVES: Our responses to incentives are shaped by predictable mental shortcuts such as strongly avoiding losses.
NORMS: We are strongly influenced by what others do.
DEFAULTS: We "go with the flow" of pre-set options.
SALIENCE: Our attention is drawn to what is novel and seems relevant to us.
PRIMING: Our acts are often influenced by subconscious cues.
AFFECT: Our emotional associations can powerfully shape our actions.
COMMITMENTS: We seek to be consistent with our public promises, and reciprocate acts.
EGO: We act in ways that make us feel better about ourselves.

A key insight of such an approach is that people do not act as isolated individuals, and instead consumption is socially situated and is often deeply embedded with habits and norms being significant. This insight is shared with studies of consumption that recognise that consumption is a multilayered phenomena that is full of meaning including its role as a signifier of identity, cultural and social affiliations (consumption tribes) and relationships (Seyfang, 2011). The symbolic value of consumption and its function for social distinction as well as self-expression has long been recognised as important for tourism and leisure consumption (Chen and Hu, 2010; Miles, 1998), including specifically with respect to sustainable consumption (Barr *et al.*, 2011; Hall, 2011c).

The main focus of behavioural change from a consumption studies perspective is via the tool of social marketing. Social marketing is the application of commercial marketing techniques to influence the voluntary behaviour of target audiences and improve personal, environmental and societal well-being (Andreasen, 1994a, 1994b; Kotler and Lee, 2008; Troung and Hall, 2013). The potential of social marketing to influence sustainable behaviours in relation to climate change has become increasingly argued (Downing and Ballantyne, 2007; Peattie and Peattie, 2009), including with respect to tourism (Barr *et al.*, 2010,

2011; George and Frey, 2010; Gössling *et al.*, 2008), although the actual conduct of such tourism-specific campaigns is limited (Hall, 2014; Truong and Hall, 2013). However, although social/psychological approaches recognise the need to make major changes and while social norms are often cited as driving factors, there is little scope for wondering about how needs and aspirations come to be as they are, i.e. they do not fundamentally question structures and paradigms (Shove, 2010), which is not the case with the systems of provision approach.

Systems of provision/institutional approaches

The third approach takes a more structural perspective on the organisation of systems of consumption and provision by focusing on the contextual collective societal institutions, norms, rules, structures and infrastructures that constrain individual decision-making, consumption and lifestyle and social practices. These are referred to variously as "infrastructures of provision" (Southerton *et al.*, 2004a), institutions, and "systems of provision": the vertical commodity chains comprising production, finance, marketing, advertising, distribution, retail and consumption that "entail a more comprehensive chain of activities between the two extremes of production and consumption, each link of which plays a potentially significant role in the social construction of the commodity both in its material and cultural aspects" (Fine and Leopold, 1993, p. 33).

The significance of the systems of provision approach is that it highlights that particular socio-technical systems constrain choice to that available within the system of provision, and can therefore "lock-in" consumers to particular social practices of behaving and consuming (Lorenzoni *et al.*, 2007; Maréchal, 2007, 2010; Seyfang, 2011; Unruh, 2000), including potentially with respect to travel and tourism (Becken, 2007; Hares *et al.*, 2010; Verbeek and Mommaas, 2008). The approach provides a profound critique of the other two approaches for several reasons. First, it suggests that positioning the problem of sustainable mobility and consumption as a problem of personal choice fails to appreciate the socially situated and structured nature of consumption and is an "arguably suspect theory of choice" given that it assumes "that people could and would act differently if only they knew what damage they were doing. Such ideas inform programmes of research into the relationship between environmental belief and action, and the design of policy initiatives geared around the provision of more and better information" (Southerton *et al.*, 2004b, p. 4). Second, much research on consumption fails to recognise that consumers do not consume resources, they consume the services that are made possible by resources. Third, a focus on the end consumer obscures important questions about the design and production of options and relationships to demand and use. Fourth, the socially sustainable dimensions of consumption, such as distribution and equity effects, are ignored and conceptually "closed off" (Southerton *et al.*, 2004b, p. 5). Finally, it suggests that socio-technical systems, institutions and structures are not neutral, in that their formation and constitution is likely to sway behaviours and practices more in one direction than others.

The systems of provision approach has focused on alternative systems of provision, with one notable area for tourism being the development of alternative food networks such as farmers' markets and locally and spatially extended short supply chains, such as Fair Trade (Gössling and Hall, 2013; Sage, 2012). The emphasis on localism and short supply chains has been influential in the hospitality and restaurant sector together with the Slow Food movement and the growing interest in forms of slow tourism (Fullagar *et al.*, 2012; Gössling and Hall, 2013; Hall, 2012). Some aspects of the approach also focus on anti-consumption measures, such as voluntary simplicity or the more active notion of culture jamming – "the innovative and alternative ways in which people are offering a form of creative, non-violent resistance against the way we view the world, either for the sake of the interruption or for getting an alternative message across" (Woodside, 2001, p. 9) – which provides significant alternatives to contemporary consumer culture (Hall, 2011c).

Linking change approaches in consumption and governance

The three different approaches to behaviour change discussed above have different sets of assumptions and work within different paradigmatic frames. Acknowledging these assumptions may in itself be useful for improving dialogue with the academic community on the problems of achieving sustainable consumption and mobility. However, it is also important to recognise that the different approaches are also intimately related to different understandings of governance, the role of the state and policy learning.

Both the utilitarian (green labelling, tax incentives, pricing, education) and social/psychological (nudging, social marketing) approaches to encouraging sustainable consumption are grounded in the ABC model. Social change is thought to depend upon values and attitudes (A), which are believed to drive the kinds of behaviour (B) that individuals choose (C) to adopt. The ABC model resonates with current widely shared ideas about individual agency and media influence. "The combination of A and B and C generates a very clear agenda for effective policy, the conceptual and practical task of which is to identify and affect the determinants of pro-environmental behavior" (Shove, 2010, p. 1275). Shove reinforces the perspective of Southerton *et al.* (2004b), noted above, when she comments that present strategies of state intervention "presume that environmental damage is a consequence of individual action and that given better information or more appropriate incentives damaging individuals could choose to act more responsibly and could choose to adopt 'pro-environmental behaviours'". But in so doing they do not fundamentally question structures/paradigms and their role in influencing how and why environmentally problematic lifestyles and patterns of consumption are reproduced and how they change.

Instead, the third approach suggests that explanations of patterns of behaviour and consumption lie within socio-technical systems and structures. Such explanations "do not deny the possibility of meaningful policy action, but at a minimum they recognise that effect is never in isolation and that interventions go on within, not outside, the processes they seek to shape" (Shove, 2010, p. 1278). This also suggests that in order to encourage sustainable tourism consumption the system itself, and the "rules of the game", need to change (Hall, 2011a). However, the difficulty for this in policy terms is that it represents a direct challenge to the status quo.

The three approaches to consumer change and carbon capability have analogues in the policy learning literature (Hall, 2011a) (Table 15.2). First-order change tends to be characterised by incremental, routinised, satisficing behaviour that is based around government officials and policy experts that leads to a change in the "levels (or settings) of the basic instruments of ... policy" (Hall, 1993, p. 279). Second-order change is characterised by the selection of new policy instruments and techniques and policy settings due to previous policy experience but the overarching policy goals remain the same. Second-order change is more strategic, although officials and policy experts still remain relatively isolated from external political pressures. Third-order change, or a policy paradigm shift, takes place when a new goal hierarchy is adopted by policymakers because the coherence of existing policy paradigm(s) has been undermined, "where experiment and perceived policy failure has resulted in discrepancies or inconsistencies appearing which cannot be explained within the existing paradigm" (Greener, 2001, p. 135). In situations where existing institutions and policies cease to be relevant to policy problems, policy failure may also lead the state to search for policy advice outside of previous internal and external sources, including academia, think tanks and non-government organisations (Pierre and Peters, 2000).

The adoption of social/psychological approaches, such as nudging and social marketing, rather than simply providing education or relying on market mechanisms to encourage sustainable consumption, only represents a second-order change. It is a continuation of the ABC approach and not a fundamental reassessment or paradigm shift in thinking about policy intervention. As the MINDSPACE report states:

> Some leading proponents have portrayed the application of behavioural economics as a radical "third way" between liberal and paternalistic approaches to government. Others have tended to dismiss the approach as a distraction to the robust application of "normal" economics to policy. In crude terms, the first camp says the way to reduce carbon emissions is through harnessing the power of techniques such as comparisons with our neighbours' emissions; for the second camp, it is simply to get the price of carbon right, and then to let markets sort it out.
>
> Our position sits between the two camps. The application of behavioural economics does not imply a paradigm shift in policymaking. It certainly does not mean giving up on conventional policy tools such as regulation, price signals and better information.
>
> (Dolan *et al.*, 2010, p. 77)

Table 15.2 Relationship of dimensions of carbon capability to orders of change in policy learning

Policy learning	Characteristics	Carbon capability	Intervention
First-order change	Incremental, routine behaviour that is based around government officials and policy experts that leads to a change in the levels (or settings) of the basic instruments of policy, e.g. codes of conduct, voluntary instruments, new indicators	Focus on individual decision-making, e.g. knowledge, judgements, skills	Education
Second-order change	Selection of new policy instruments and techniques and policy settings as a result with previous policy experience but the overarching policy goals remain the same – more strategic, e.g. green growth	Individual behaviour/practices	Social marketing, nudging
Third-order change	Policy paradigm shift occurs when a new goal hierarchy is adopted by policymakers because the coherence of existing policy paradigm has been undermined, e.g. regulation and intervention, institutional change	Broader engagement with carbon governance	Lobbying, voting, protesting, regulatory change, creating alternative social infrastructures of provision

Source: after Hall (1993), Hall (2010, 2011b, 2014), Whitmarsh et al. (2009, 2011).

One of the appeals of behavioural economics approaches for policymaking, such as nudging, is the promise of cost-effectiveness, achieving "more for less", particularly in public services. The practical emphasis is on "smart" solutions that supposedly do not involve more resources; an imperative in recessionary times with many governments committed to reducing spending (Burgess, 2012). It also reinforces the interests of some stakeholders in continuing the process of deregulation, while Burgess (2012) also suggests that society-wide application of behavioural solutions assumes a fundamentally pliant and passive population that attaches limited value to individual liberty and autonomy in modern "risk society". Selinger and Powys White (2012) support Burgess' (2012) critique of nudging but go on to suggest that the cause-and-effect relations of even seemingly straightforward nudge proposals are not nearly as predictable as expected. Indeed, they are regarded as incapable of "solving" "wicked problems" such as sustainable consumption and climate change "occurring at a 'whole earth level' in which multiple systems, operating with different norms and at different scales, interact with one another to produce emergent behaviour that can be exceptionally difficult, if not impossible, to predict" (Selinger and Powys White, 2012, p. 29). As they go on to note,

> although discussions abound about how nudging can reduce energy consumption, nudge proponents do not even attempt to discuss at all the complex interventions that are needed to shift diverse citizens in diverse parts of the world away from a CO_2 intensive society or adapt to the changes that might accompany the perpetuation of CO_2 intensive industries, infrastructures, and lifestyles.
>
> (Selinger and Powys White, 2012, p. 29)

A further significant and complicating issue with respect to the social/psychological approach is that "Given that the ABC is the dominant paradigm in contemporary environmental policy, the scope of *relevant* social science is typically restricted to that which is theoretically consistent with it" (Shove, 2010, p. 1280, emphasis added). Policymakers fund and legitimise lines of enquiry that generate results that they can accept and manage, even if they do not necessarily provide the "solution" to the policy problem. The result is a self-fulfilling cycle of credibility (Latour and Woolgar, 1986) in which evidence of relevance and value to policymakers helps in securing additional resources for that approach. As Shove acknowledges the ABC paradigm is not just a theory of behavioural change:

> it is also a template for intervention which locates citizens as consumers and decision makers and which positions governments and other institutions as enablers whose role is to induce people to make pro-environmental decisions for themselves and deter them from opting for other, less desired, courses of action.
>
> (Shove, 2010, p. 1280)

Yet, it is also possible that the vocabulary of ABC is required in order to keep a very particular understanding of governance in place.

It is, nonetheless, clear that policymakers are highly selective in the models of change on which they draw and that their tastes in social theory are anything but random. An emphasis on individual choice has significant political advantages and in this context, to probe further, to ask how options are structured, or to inquire into the ways in which governments maintain infrastructures and economic institutions, is perhaps too challenging to be useful (Shove, 2010, p. 1283).

Conclusion: what would it take to go beyond the first three letters of the alphabet?

This chapter has highlighted that different approaches to behavioural change are based on different assumptions about the problem of sustainable consumption of tourism and mobility and the relative importance of the role of individual agency and structure. It has emphasised that while social/psychological approaches, including behavioural economics, do acknowledge the importance of social norms and routines, and identity and status in consumption behaviour, they do not then interrogate the systems of provision that give rise to the social practices of consumption of travel and tourism. The chapter does not deny the value of nudging and social marketing, which are at times indistinguishable (Burgess, 2012). However, it suggests that without facing up to the implications of structure and institutions then the likelihood of tourism activities and behaviours being "locked-in" to particular unsustainable socio-technical systems of provision is greatly increased (Gössling *et al.*, 2012b). If one wants to develop appropriate tourism behaviours then it is vital that socio-technical systems are changed.

Just as significantly the chapter has emphasised that different ways of understanding behavioural change, and the approaches used to implement them, relate to different forms of governance. Nudging and social marketing are closely related to what is often referred to as "new governance". Yee (2004, p. 487) provided a very basic definition of this approach by describing new modes of governance as "new governing activities that do not occur solely through governments". In the European context, Heritier (2002) focuses on new modes of governance that include private actors in policy formulation and/or while being based on public actors are only marginally based on regulative powers, or not based on regulation at all. Some of the characteristics of the key elements of these so-called new modes of governance were also indicated in Table 15.1. Each of the conceptualisations of governance structures is related to the use of particular sets of policy instruments (Hall, 2011b). Given the artificiality of any policy–action divide (Barrett and Fudge, 1981), these instruments and modes of governance can also be connected to different conceptual approaches to implementation (Hall, 2009b). In something of a chicken and the egg situation it is not really possible to say what came first – the mode of governance or the intervention mechanism for behavioural change that implements public policies. What is important to stress, therefore, is that the mode of governance and the manner of intervention become mutually reinforcing. One cannot be adequately understood without the other, even if this is often the case in the many academic, institutional and governmental papers that propose methods to improve sustainable behaviours by tourists (usually while still travelling in increasing numbers to tourist destinations) (Hall, 2011a). Yet the mutual reinforcement between modes of governance and intervention also creates a path dependency in which solutions to sustainable tourism mobility are only identified within "green growth" arguments for greater efficiency and market-based solutions yet with continued focus on growth in both GDP and tourism numbers, and an ideology that frames the problem of sustainable consumption in terms of individual consumption and responsibility (Bailey and Wilson, 2009; Shove, 2010; Whitmarsh *et al.*, 2011).

"Lock-in" to a particular socio-technical system of provision has been recognised as a major constraint to emissions reduction and avoidance of dangerous climate change (Maréchal, 2007, 2010), except perhaps by those that believe in and promote the neoliberal, technocentric and ecological modernisation values underpinning the carbon economy (Bailey and Wilson, 2009). In such a situation policy learning is extremely difficult (Hall, 2011a), with changes occurring only at the margin or in limited terms, such as the development of nudging, but without challenging the basic paradigm of economic and tourism growth (Table 15.2). Such an observation has implications for academic work as well as for policy problems and their solution. As Shove suggests, paradigms and approaches that lie beyond the pale of the ABC are doomed to be forever marginal no matter how interactive or how policy-engaged their advocates might be. To break through this log jam, it would be necessary to reopen a set of basic questions about the role of the state, the allocation of responsibility, and in very practical terms the meaning of manageability, within climate-change policy (Shove, 2010, p. 1283).

Yet, the public policy system and much of the research that feeds into it is not necessarily conducive to breakthroughs and rapid paradigm shift. Hall (2011a), for example, noted that in research on tourism governance far too much attention has been given to the assumption that a well-designed institution is "good" because it facilitates cooperation and networking rather actually focusing on the norms and institutionalisation of institutional arrangements and their potential outcomes. What, after all, is the point of encouraging governance mechanisms such as partnerships, network development, self-regulation and individual responsibility if they continue to have no practical effect on the sustainability of tourism and consumption? If the ethical value of "individual choice" leads to increased emissions from lifestyle and travel actions and worsening environmental change then how ethical is it?

This is not to suggest that there is no value in nudging and social marketing, and even education and market-based solutions. They are part of the suite of approaches that need to be used together *with* a fundamental examination of the socio-technical system itself if there is to be a sustainable post-carbon transition (Gössling *et al.*, 2012b; van den Bergh and Kemp, 2008). Change can occur within the first and second orders of policy learning and with limited regulatory shifts. As the MINDSPACE discussion paper reported:

> ... generally the broader sweep of policy history suggests that [culture] change is driven by a mix of both broad social argument and small policy steps. Smoking is perhaps the most familiar example. Over several decades the behavioural equilibrium has shifted from widespread smoking to today's status as an increasingly minority activity. Better information; powerful advertising (and the prohibition of pro-smoking advertising); expanding bans; and changing social norms have formed a mutually reinforcing thread of influence to change the behavioural equilibrium. There is every reason to think that is a pattern that we will see repeated in many other areas of behaviour too, from sexual behaviour to carbon emissions.
>
> (Dolan *et al.*, 2010, p. 77)

In the case of smoking, education, social marketing and limited regulation has had a substantial impact on lowering the proportion of the population that smokes on a regular basis. The problem is that it has taken generations to achieve "cultural change" in developed countries; is almost unknown in developing countries where the majority of smokers can now be found; has been challenged by vested industrial and political interests along the way; has taken many years for counter-institutional research to be accepted; and, even then, smoking still kills millions of people a year within an uneven pattern of regulatory control across the world. In the case of sustainable tourism mobility, emissions and climate change we do not have the luxury to wait that long for such an imperfect set of behavioural "solutions".

Acknowledgements

Papers on the different approaches to consumer behaviour and implications for sustainable tourism were presented at the *Psychological and Behavioural Approaches to Understanding and Governing Sustainable Tourism Mobility* Workshop, Freiburg, Germany, July 2012, and the *21st Nordic Symposium in Tourism and Hospitality Research*, Umeå , Sweden, November 2012. The author is grateful for comments received at these meetings, the valuable comments of the anonymous referees, as well as financial support from the Freiburg Institute for Advanced Studies and the University of Eastern Finland.

References

Alemanno, A. (2012). Nudging smokers – the behavioural turn of tobacco risk regulation. *European Journal of Risk Regulation*, *1/2012*, 32–42.
Amir, O. and Lobel, O. (2008). Stumble, predict, nudge: How behavioral economics informs law and policy. *Columbia Law Review*, *108*, 2098–2139.

Andreasen, A.R. (1994a). *Marketing social change: Changing behaviour to promote health, social, development, and the environment*. San Francisco: Jossey-Bass.

Andreasen, A.R. (1994b). Social marketing: Its definition and domain. *Journal of Public Policy and Marketing*, *3*, 108–114.

Arvesen, A., Bright, R.M. and Hertwich, E.G. (2011). Considering only first-order effects? How simplifications lead to unrealistic technology optimism in climate change mitigation. *Energy Policy*, *39*, 7448–7454.

Bailey, I. and Wilson, G.A. (2009). Theorising transitional pathways in response to climate change: Technocentrism, ecocentrism, and the carbon economy. *Environment and Planning A*, *41*, 2324–2341.

Barker, T., Dagoumas, A. and Rubin, J. (2009). The macroeconomic rebound effect and the world economy. *Energy Efficiency*, *2*, 411–427.

Barr, S., Gilg, A. and Shaw, G. (2011). "Helping people make better choices": Exploring the behaviour change agenda for environmental sustainability. *Applied Geography*, *31*, 712–720.

Barr, S., Shaw, G., Coles, T. and Prillwitz, J. (2010). "A holiday is a holiday": Practicing sustainability, home and away. *Journal of Transport Geography*, *18*, 474–481.

Barrett, S. and Fudge, C. (1981). *Policy and action*. London: Methuen.

Becken, S. (2007). Tourists' perception of international air travel's impact on the global climate and potential climate change policies. *Journal of Sustainable Tourism*, *15*, 351–368.

Bone, J.D. (2010). Irrational capitalism: The social map, neoliberalism and the demodernization of the West. *Critical Sociology*, *36*, 717–740.

Bovens, L. (2009). The ethics of nudge. In T. Grüne-Yanoff and S. Ove Hansson (eds), *Preference change: Approaches from philosophy, economics and psychology* (pp. 207–209). Berlin: Springer.

Bramwell, B. and Lane, B. (2012). Towards innovation in sustainable tourism research? *Journal of Sustainable Tourism*, *20*, 1–7.

Bramwell, B. and Lane, B. (2013). Getting from here to there: Systems change, behavioural change and sustainable tourism, *Journal of Sustainable Tourism*, *21*, 1–4.

Burgess, A. (2012). "Nudging" healthy lifestyles: The UK experiments with the behavioural alternative to regulation and the market. *European Journal of Risk Regulation*, *1/2012*, 3–16.

Cabrini, L. (2012). *Tourism in the UN Green Economy Report*. UNWTO High-level Regional Conference on Green Tourism, 3 May 2012, Chiang Mai, Thailand. Retrieved from asiapacific.unwto.org/sites/all/files/.../2012may chiangmai lc 0.pdf.

Chen, P.T. and Hu, H.H. (2010). How determinant attributes of service quality influence customer perceived value: An empirical investigation of the Australian coffee outlet industry. *International Journal of Contemporary Hospitality Management*, *22*, 535–551.

Christie, L.H. (2010). *Understanding New Zealand homeowners' apparent reluctance to adopt housing-sustainability innovations* (Unpublished thesis). Victoria University of Wellington.

Cialdini, R. (2007). *Influence: The psychology of persuasion*. New York: Harper Business.

Cohen, S.A. and Higham, J.E.S. (2011). Eyes wide shut? UK consumer perceptions on aviation climate impacts and travel decisions to New Zealand. *Current Issues in Tourism*, *14*, 323–335.

Conlisk, J. (1996). Why bounded rationality? *Journal of Economic Literature*, *34*, 669–700.

Demeritt, D. (2001). The construction of global warming and the politics of science. *Annals of the Association of American Geographers*, *91*, 307–337.

Demeritt, D. (2006). Science studies, climate change and the prospects for constructivist critique. *Economy and Society*, *35*, 453–479.

Department of Transport (2007). *Air passenger demand and CO$_2$ forecasts*. London: Department of Transport.

Dolan, P., Hallsworth, M., Halpern, D., King, D., Metcalfe, R. and Vlaev, I. (2012). Influencing behaviour: The mind space way. *Journal of Economic Psychology*, *33*, 264–277.

Dolan, P., Hallsworth, M., Halpern, D., King, D. and Vlaev, I. (2010) *MINDSPACE: Influencing behaviour through public policy*. London: Cabinet Office and Institute for Government.

Downing, P. and Ballantyne, J. (2007). *Tipping point or turning point? Social marketing & climate change*. London: Ipsos MORI Social Research Institute.

Fine, B. and Leopold, E. (1993). *The world of consumption*. London: Routledge.

Folmer, H. and Johansson-Stenman, O. (2011). Does environmental economics produce aeroplanes without engines? On the need for an environmental social science. *Environmental and Resource Economics*, *48*, 337–361.

Fullagar, S., Markwell, K. and Wilson, E. (eds) (2012). *Slow tourism*. Bristol: Channel View Publications.

Gadenne, D., Sharma, B., Kerr, D. and Smith, T. (2011). The influence of consumers' environmental beliefs and attitudes on energy saving behaviours. *Energy Policy*, *39*, 7684–7694.

George, R. and Frey, N. (2010). Creating change in responsible tourism management through social marketing. *South African Journal of Business Management*, *41*(1), 11–23.

Giddens, A. (1984). *The constitution of society: Outline of the theory of structuration*. Cambridge: Polity Press.

Goldenberg, S. (2012). Extreme weather more persuasive on climate change than scientists. *Guardian*, 14 December. Retrieved from www.guardian.co.uk/environment/2012/dec/14/extreme-weather-climate-change-scientists.

Gössling, S. (2009). Carbon neutral destinations: A conceptual analysis. *Journal of Sustainable Tourism*, *17*, 17–37.

Gössling, S. and Hall, C.M. (2013). Sustainable culinary systems: An introduction. In C.M. Hall and S. Gössling (eds), *Sustainable culinary systems: Local foods, innovation, and tourism & hospitality* (pp. 3–44). London: Routledge.

Gössling, S. and Schumacher, K.P. (2010). Implementing carbon neutral destination policies: Issues from the Seychelles. *Journal of Sustainable Tourism*, *18*, 377–391.

Gössling, S., Hall, C.M., Ekström, F., Brudvik Engeset, A. and Aall, C. (2012b). Transition management: A tool for implementing sustainable tourism scenarios? *Journal of Sustainable Tourism*, *20*, 899–916.

Gössling, S., Hall, C.M., Lane, B. and Weaver, D. (2008). The Helsingborg statement on sustainable tourism. *Journal of Sustainable Tourism*, *16*, 122–124.

Gössling, S., Hall, C.M., Peeters P. and Scott D. (2010). The future of tourism: Can tourism growth and climate policy be reconciled? A climate change mitigation perspective. *Tourism Recreation Research*, *35*, 119–130.

Gössling, S., Hall, C.M. and Scott, D. (2009). The challenges of tourism as a development strategy in an era of global climate change. In E. Palosou (ed.), *Rethinking development in a carbon-constrained world: Development cooperation and climate change* (pp. 100–119). Helsinki: Ministry of Foreign Affairs.

Gössling, S., Scott, D., Hall, C.M., Ceron, J. and Dubois, G. (2012a). Consumer behaviour and demand response of tourists to climate change. *Annals of Tourism Research*, *39*, 36–58.

Greener, I. (2001). Social learning and macroeconomic policy in Britain. *Journal of Public Policy*, *21*, 133–152.

Hall, C.M. (2006). New Zealand tourism entrepreneur attitudes and behaviours with respect to climate change adaption and mitigation. *International Journal of Innovation and Sustainable Development*, *1*, 229–237.

Hall C.M. (2009a). Degrowing tourism: Décroissance, sustainable consumption and steady-state tourism. *Anatolia: An International Journal of Tourism and Hospitality Research*, *20*, 46–61.

Hall, C.M. (2009b). Archetypal approaches to implementation and their implications for tourism policy. *Tourism Recreation Research*, *34*, 235–245.

Hall, C.M. (2010). Changing paradigms and global change: From sustainable to steady-state tourism. *Tourism Recreation Research*, *35*, 131–145.

Hall, C.M. (2011a). Policy learning and policy failure in sustainable tourism governance: From first and second to third order change? *Journal of Sustainable Tourism*, *19*, 649–671.

Hall, C.M. (2011b). A typology of governance and its implications for tourism policy analysis. *Journal of Sustainable Tourism*, *19*, 437–457.

Hall, C.M. (2011c). Consumerism, tourism and voluntary simplicity: We all have to consume, but do we really have to travel so much to be happy? *Tourism Recreation Research*, *36*, 298–303.

Hall, C.M. (2012). The contradictions and paradoxes of slow food: Environmental change, sustainability and the conservation of taste. In S. Fullagar, K. Markwell and E. Wilson (eds), *Slow tourism: Experiences and mobilities* (pp. 53–68). Bristol: Channel View Publications.

Hall, C.M. (2013). The natural science ontology of environment. In A. Holden and D. Fennell (eds), *The Routledge handbook of tourism and the environment* (pp. 6–18). Abingdon: Routledge.

Hall, C.M. (2014). *Tourism and social marketing*. London: Routledge.

Hall, C.M., Scott, D. and Gössling, S. (2013). The primacy of climate change for sustainable international tourism. *Sustainable Development*, *21*, 112–121.

Hall, P.A. (1993). Policy paradigms, social learning, and the state: The case of economic policymaking in Britain. *Comparative Politics*, *25*, 275–296.

Hares, A., Dickinson, J. and Wilkes, K. (2010). Climate change and the air travel decisions of UK tourists. *Journal of Transport Geography*, *18*, 466–473.

Heritier, A. (2002). New modes of governance in Europe: Policy-making without legislating? In A. Heritier (ed.), *Common goods: Reinventing European and international governance* (pp. 186–206). Oxford: Rowman & Littlefield.

Higgins-Desbiolles, F. (2010). The elusiveness of sustainability in tourism: The culture-ideology of consumerism and its implications. *Tourism and Hospitality Research*, *10*, 116–115.

Hoffmann U. (2011). *Some reflections on climate change, green growth illusions and development space*, UNCTAD Discussion Paper No. 205. Geneva: UNCTAD.

Intergovernmental Panel on Climate Change (IPCC). (2007). In M.L. Parry, O.F. Canziani, J.P. Palutikof, P.J. van der Linden and C.E. Hanson (eds), *Climate change 2007: Impacts, adaptation and vulnerability, contribution of working group II to the fourth assessment report of the Intergovernmental Panel on Climate Change*. Cambridge: Cambridge University Press.

International Energy Agency (IEA). (2009). *Transport, energy and CO₂: Moving towards sustainability*. Paris: International Energy Agency.

Keen, S. (2011). *Debunking economics – Revised and expanded edition: The naked emperor dethroned?* London: Zed Books.

Kotler, P. and Lee, N.R. (2008). *Social marketing: Influencing behaviours for good.* Thousand Oaks, CA: Sage.

Kroesen, M. (2013). Exploring people's viewpoints on air travel and climate change: Understanding inconsistencies. *Journal of Sustainable Tourism, 21,* 271–290.

Latour, B. and Woolgar, S. (1986). *Laboratory life: The construction of scientific facts.* Princeton: Princeton University Press.

Lorenzoni, I., Nicholson-Cole, S. and Whitmarsh, L. (2007). Barriers perceived to engaging with climate change among the UK public and their policy implications. *Global Environmental Change, 17,* 445–459.

Lorenzoni, I., Seyfang, G. and Nye, M. (2011) Carbon budgets and carbon capability: Lessons from personal carbon trading. In I. Whitmarsh, S. O'Neill and I. Lorenzoni (eds), *Engaging the public with climate change: Behaviour change and communication* (pp. 31–46). London: Earthscan.

Mair, J. (2011). Exploring air travellers' voluntary carbon-offsetting behaviour. *Journal of Sustainable Tourism, 19,* 215–230.

Maréchal, K. (2007). The economics of climate change and the change of climate in economics. *Energy Policy, 35,* 5181–5194.

Maréchal, K. (2010). Not irrational but habitual: The importance of "behavioural lock-in" in energy consumption. *Ecological Economics, 69,* 1104–1114.

Maryon-Davis, A. and Jolley, R. (2010). *Healthy nudges: When the public wants change but the politicians don't know it.* London: Faculty of Public Health.

McGuigan, J. (2006). The politics of cultural studies and cool capitalism. *Cultural Politics, 2,* 137–158.

McKercher, B., Prideaux, B., Cheung, C. and Law, R. (2010). Achieving voluntary reductions in the carbon footprint of tourism and climate change. *Journal of Sustainable Tourism, 18,* 297–318.

Miles, S. (1998). *Consumerism: As a way of life.* London: Sage.

Ockwell, D., O'Neill, S. and Whitmarsh, L. (2010). Behavioural insights: Motivating individual emissions cuts through communication. In C. Lever-Tracey (ed.), *Routledge handbook of climate change and society* (pp. 341–350). London: Routledge.

Owens, B., Lee, D.S. and Lim, L. (2010). Flying into the future: Aviation emissions scenarios to 2050. *Environmental Science Technology, 44,* 2255–2260.

Peattie, K. and Peattie, S. (2009). Social marketing: A pathway to consumption reduction? *Journal of Business Research, 62,* 260–268.

Peeters, P. and Landré, M. (2011). The emerging global tourism geography – an environmental sustainability perspective. *Sustainability, 4*(1), 42–71.

Pierre, J. and Peters, G.B. (2000). *Governance, politics and the state.* London: Palgrave Macmillan.

Sage, C. (2012). *Environment and food.* London: Routledge.

Santarius, T. (2012). *Green growth unravelled – how rebound effects baffle sustainability targets when the economy keeps growing.* Wuppertal Institute for Climate, Environment and Energy. Retrieved from www.boell.de/downloads/WEB_12,1022_The_Rebound_Effect-Green_Growth_Unraveled_TSantarius_V101.pdf.

Scott, D. (2011). Why sustainable tourism must address climate change. *Journal of Sustainable Tourism, 19,* 17–34.

Scott, D. and Becken, S. (2010). Adapting to climate change and climate policy: Progress, problems and potentials. *Journal of Sustainable Tourism*, *18*, 283–295.

Scott, D., Amelung, B., Becken, S., Ceron, J., Dubois, G., Gössling, Peeters, P. and Simpson, M. (2008). *Climate change and tourism: Responding to global challenges* (Technical report; pp. 23–250). Madrid: UNWTO and UNEP.

Scott, D., Gössling, S. and Hall, C.M. (2012a). International tourism and climate change. *WIRES Climate Change*, *3*(3), 213–232.

Scott, D., Gössling, S. and Hall, C.M. (2012b). *Tourism and climate change: Impacts, adaptation and mitigation*. Abingdon: Routledge.

Scott, D., Peeters, P. and Gössling, S. (2010). Can tourism deliver its "aspirational" emission reduction targets? *Journal of Sustainable Tourism*, *18*, 393–408.

Selinger, E. and Powys White, K. (2012). Nudging cannot solve complex policy problems. *European Journal of Risk Regulation*, *1/2012*, 26–31.

Seyfang, G. (2009). *Low-carbon currencies: The potential of time banking and local money systems for community carbon-reduction*. CSERGE Working Paper EDM 09–04, CSERGE, School of Environmental Sciences, University of East Anglia.

Seyfang, G. (2011). *The new economics of sustainable consumption: Seeds of change*. Basingstoke: Palgrave Macmillan.

Shove, E. (2010). Beyond the ABC: Climate change policy and theories of social change. *Environment and Planning A*, *42*, 1273–1285.

Simon, H.A. (1959). Theories of decision making in economics. *American Economic Review*, *49*, 253–283.

Simon, H.A. (1965). *Administrative behavior* (2nd edn). New York: Free Press.

Sorrell, S., Dimitropoulos, J. and Sommerville, M. (2009). Empirical estimates of the direct rebound effect: A review. *Energy Policy*, *37*, 1356–1371.

Southerton, D., Chappells, H. and Van Vliet, B. (eds) (2004a). *Sustainable consumption: The implications of changing infrastructures of provision*. Cheltenham: Edward Elgar.

Southerton, D., Van Vliet, B. and Chappells, H. (2004b). Introduction: Consumption, infrastructures and environmental sustainability. In D. Southerton, H. Chappells and B. Van Vliet (eds), *Sustainable consumption: The implications of changing infrastructures of provision* (pp. 1–14). Cheltenham: Edward Elgar.

Sustainable Consumption Roundtable (2006). *Shifting opinions and changing behaviours*. A Consumer Forum report by Opinion Leader Research, May 2006. London: Sustainable Development Roundtable.

Tervo-Kankare, K., Hall, C.M. and Saarinen, J. (2013). Christmas tourists' perceptions of climate change in Rovaniemi, Finnish Lapland. *Tourism Geographies*, *15*, 292–317.

Thaler, R.H. and Sunstein, C.R. (2008). *Nudge: Improving decisions about health, wealth and happiness*. London: Yale University Press.

Thynne, I. (2008). Symposium introduction: Climate change, governance and environmental services: Institutional perspectives, issues and challenges. *Public Administration and Development*, *28*, 327–339.

Troung, V.D. and Hall, C.M. (2013). Social marketing and tourism: What is the evidence? *Social Marketing Quarterly*, *19*, 110–135.

Unruh, G.C. (2000). Understanding carbon lock-in. *Energy Policy*, *28*, 817–830.

UNWTO. (2011). *Tourism towards 2030: Global overview*. UNWTO General Assembly, 19th Session, Gyeongju, Republic of Korea, 10 October 2011. Madrid: UNWTO.

UNWTO. (2012). *UNWTO tourism highlights. 2012 edition*. Madrid: UNWTO.

UNWTO and UNEP. (2011). Tourism: Investing in the green economy. In *Towards a green economy* (pp. 409–447). Geneva: United Nations Environmental Programme.

Van den Bergh, J. and Kemp, R. (2008). Transition lessons from economics. In J. van den Bergh and F. Bruinsma (eds), *Managing the transition to renewable energy: Theory and practice from local, regional and macro perspectives* (pp. 81–128). Cheltenham: Edward Elgar.

Verbeek, D. and Mommaas, H. (2008). Transitions to sustainable tourism mobility: The social practices approach. *Journal of Sustainable Tourism, 16*, 629–644.

Waitt, G., Caputi, P., Gibson, C., Farbotko, C., Head, L., Gill, N. and Stanes, E. (2012). Sustainable household capability: Which households are doing the work of environmental sustainability? *Australian Geographer, 43*, 51–74.

Weber, E.U. (2006). Experience-based and description-based perceptions of long-term risk: Why global warming does not scare us (yet). *Climatic Change, 77*, 103–120.

Whitmarsh, L. (2009). Behavioural responses to climate change: Asymmetry of intentions and impacts. *Journal of Environmental Psychology, 29*, 13–23.

Whitmarsh, L., O'Neill, S., Seyfang, G. and Lorenzoni, I. (2009). *Carbon capability: What does it mean, how prevalent is it, and how can we promote it?* Tyndall Working Paper No. 132. Tyndall Centre for Climate Change Research, University of East Anglia.

Whitmarsh, L., Seyfang, G. and O'Neill, S. (2011). Public engagement with carbon and climate change: To what extent is the public "carbon capable"? *Global Environmental Change, 21*, 56–65.

Woodside, S. (2001). *Every joke is a tiny revolution: Culture jamming and the role of humour* (Unpublished Master's thesis). University of Amsterdam, Media Studies, Amsterdam.

Yee, A.S. (2004). Cross-national concepts in supranational governance: State-society relations and EU-policy making. *Governance – An International Journal of Policy and Administration, 17*, 487–524.

Zeppel, H. (2012). Climate change and tourism in the Great Barrier Reef Marine Park. *Current Issues in Tourism, 15*, 287–292.

16 New governance models for behaviour change in tourism mobilities

A research agenda

Stefan Gössling, Paul Peeters, James E.S. Higham and Scott A. Cohen

There is a groundswell of opinion in tourism, transport and cognate academic fields, that the travel and tourism industry is profoundly environmentally flawed (Gössling *et al.*, 2010; Wheeller, 2012). Deeply embedded in neoliberal consumer society and entrenched in the structures of late-capitalism (Harvey, 2011), efforts to address the environmental failures of global tourism have, for the time being, rested largely with the consumer. This edited book has interrogated the behavioural and psychological dimensions of (tourist) mobility consumption, highlighted the complexity of consumer decision-making and drawn into question the efficacy of a consumer-led industry response to the climate crisis.

The chapters in the first part of the book explored psychological understandings of climate change and tourism mobilities. These chapters unpack some of the key barriers to behaviour change in sustainable mobility, focusing on the attitude-behaviour gap as a significant hurdle to actualising behavioural change, the importance of identity and emotions to consumer decision-making in tourism and transport contexts, and how the hedonic and affective representations surrounding tourism spaces make them particular tricky settings for enacting sustained positive behaviour change. The chapters show that the barriers to unlocking behavioural change amongst consumers are considerable, and that the travelling public is unlikely to change "spontaneously" on the basis of environmental awareness alone. The socio-psychological insights in this part instead point towards increased governance as paramount in developing more sustainable mobility practices, if these changes are to be significant and in line with global climate policy.

Part II of the book turned to behavioural aspects of climate change and tourism mobilities, and dealt with issues such as how carbon offsetting can ironically induce *more* travel rather than deter it, and the multiple ways in which time and distance are implicated in mobility decisions, including how changing information technologies can redefine these concepts. Longer-term planning horizons, and the impacts of individual lifestyles on demand modelling are explored, as well as how public transport can be promoted to visitors in urban destinations. The chapters in this part span a range of behavioural issues as they relate to (un)sustainable mobility, from localised ground transport and real-time travel information, to mega-events and the perceived cultural value of long-distance travel.

The final part of the book focused on governance and policies based upon psychological, behavioural and social mechanisms. It commences with a comprehensive review of the cognitive, experiential and normative approaches to climate change communication before proposing an integrative conceptual framework for enhanced communication interventions. This aims to narrow the gap between awareness and attitudes on one hand, and behaviour on the other, that is evidenced in many of the other chapters. The part concludes with a challenge to move beyond socio/psychological approaches that attempt to foster sustainable mobility behaviour, such as nudging and social marketing, and question more seriously the systems of provision that perpetuate these practices. Significant structural change will require more radical approaches to governance, but the wheels of change turn slowly and in the case of anthropogenic climate change time is in limited supply.

Overall, the chapters support earlier insights that increasing climate awareness and environmental concern has little bearing upon tourism consumption (Cohen *et al.*, 2011; Eijgelaar *et al.*, 2010; Hares *et al.*, 2010; Higham and Cohen, 2011; McKercher *et al.*, 2010), but they provide new perspectives as to why this might be the case. Travel decisions, the book shows, are deeply embedded socially and culturally, and intimately related to emotions, identity, time, happiness, performances of self or the attainment (or avoidance) of "possible selves", all of which represent subconscious and little investigated psychological factors that bear upon travel decisions. The wide disparities that are apparent in domestic ("home") and tourism ("away") decision-making and behavioural contexts (Barr *et al.*, 2010) cement the conclusion that the autonomy of individual pro-environmental response, when set within the systems of provision in late-capitalist consumer society, is fraught with challenge.

A research agenda on governing behaviour change

Transport growth, including both passengers and freight, is one of the key challenges of the twenty-first century. If environmental problems related to transportation are to be solved – including climate change, but also issues of land use, noise and population growth – there can be little doubt that this will require a massive change in the way humans travel and consume mobility. One important finding of several chapters in this volume is that the low sustainability of the current tourism system is embedded in structures that make it easy and often cheaper to travel unsustainably, and difficult to travel sustainably, raising a wide range of questions regarding transport infrastructures, taxation, management and governance.

Within this situation, behavioural change is likely to become increasingly important, as technology-based solutions have remained inadequate to catch up with the rapid growth in global mobility. In light of the issues outlined in the chapters of this book, a comprehensive research initiative is needed to better understand the way the public will react to command-and-control as well as market-based policies, in order to design acceptable and effective policies. In this context, it will be necessary to better understand the social, cultural and psychological structures underlying the reasoning of consumers – as well as policy makers. For this purpose, a broad psychological research focus will be required, including fundamentally new research approaches rooted in neurosciences, as well as social, clinical and evolutionary psychology. This research agenda has been spelled out to some extent in the various chapters of the book, which explore new directions to an understanding of mobility consumption with the purpose to inform discussions about new models of mobility governance. Research may in particular focus on:

1 Policy-making and transport governance

There is widespread consensus that transport governance has to play a more prominent role in steering sustainable transport choices. Yet, it is clear that policy makers have shown limited interest in creating legislative frameworks that would achieve significant changes in transport behaviour. Specifically, the taxation of particularly carbon-intense transport modes, the support of slow tourism and transport, or infrastructure development to encourage modal shifts to bicycle, train and bus are key issues that would have to be resolved by policy makers. While this general need for governance is widely acknowledged, the notable lack of political initiatives to support such changes deserves further study: are there specific reasons for inaction? In this context, there is also a notable lack of communication between politicians and scientists. Evidence would suggest that politicians have far more links to industry than science, and in particular insights from the social sciences. The importance of the (non)existence of such linkages, as well as the structure of current spheres of influence represents another important field deserving further study to understand the lack of progress on transport governance.

2 Taxation, rationing and trading

One of the key challenges in changing mobility patterns is that these are highly unevenly distributed between and within societies. In many industrialised countries, energy-intense mobility patterns are a norm, while in the least developed countries, a large proportion of daily mobility will be covered on foot, by bike or bus. Similar differences can be observed within societies, where a small share of highly (aero)-mobile travellers is responsible for a comparably large share of the overall distances consumed. As current measures – to the extent that these even exist – usually focus on economic instruments to initiate behavioural change, questions arise as to whether such measures will or will not be able to change the consumption patterns of the most mobile, who also tend to represent the most wealthy share of society. Yet, even if a general shift to transport modes with lower environmental impacts and a reduction in the distances travelled was initiated through taxation on the basis of environmental impacts – as an example, international aviation remains untaxed – it is questionable whether this would affect the disproportionally more wealthy. If this paradox remains unresolved, only more radical measures, such as carbon rationing and trading, may prevent rapid further growth of emissions from the transport system. Further research is thus needed to better understand perceptions of abundance and scarcity in a world that may need more radical measures to cope with increasingly limited resources.

3 De-marketing, choice editing and desirable futures

Consumer perceptions and behaviour will remain crucial to achieve significant change in transport systems, both because of individual actions adding up to global problems/solutions, but as well in terms of consumer influence on industry and government. Industry and government create the institutional frameworks that foster mobility, with widespread options to de-market products and patterns, to edit consumer choices and to nurture visions of desirable low-carbon futures. As consumers are heavily influenced by opinion leaders, the question of how to make "hypo-mobility" fashionable, i.e. to generate dissociative identities associated with hypermobility, deserves further study. Specifically, ongoing trends in urban environments, which move, for example, from car-ownership to car-sharing, but also towards air travel as routine, require a better understanding. The role of social media in supporting such trends needs to be investigated, as well as the role of peers and social media in shaping trends, visions and what is considered desirable forms of mobility.

Concluding points

This book has focused specifically on social, cultural and psychological aspects of tourist transport behaviour. The chapters provided critical and fine-grained insights into the behavioural and psychological dimensions of travel decision-making. As we drew out in developing the above research agenda, the behaviour and psychology of the travelling public represents only one piece of the emissions reduction puzzle. While a comprehensive understanding of tourist psychology is necessary to inform policy makers, it will not be sufficient to achieve emission reductions, and bring tourism onto a climatically sustainable pathway, *if treated in isolation*. The individual consumer cannot be held accountable for the environmental failures of the tourism industry, in the absence of radical change in the systems of provision. Placing the burden of responsibility on the individual will continue to influence some, frustrate or infuriate others and be treated with disregard (out of a sense of futility) by most (Higham *et al.*, 2014). Radical structural change may take the form of infrastructure planning, including financial and economic infrastructure (i.e. taxation regimes and emission trading schemes) for sustainable mobility. These points raise timely and equally important questions regarding the climate psychology of decision makers in government and industry.

It should also be noted in conclusion that the focus in this book has been largely western-centric. An exception was Chapter 7 by Malhado, Araujo and Rothfuss which addressed the role of information in shaping mobility choices in the context of mega-events in South Africa, reminding us of the importance of considering geographical context when discussing the manifold ways in which societies may respond to the transport emissions challenge. These lines of inquiry need to be further extended to emergent world regions, where rapidly expanding middle classes are fuelling increases in aero-consumption and the replication of the low-cost carrier model (Freire-Medeiros and Name, 2013). Randers' (2012) global forecast for the 40 years to 2052 foresees the emerging markets of the world (i.e. Brazil, the United Arab Emirates, Nigeria, India, Indonesia and China among others) driving relentlessly to close the consumption gap that exists between developing and developed (consumer) societies. The efficacy of individual consumers bearing the costs (social, economic) and responsibilities (psychological, behavioural) of profoundly unsustainable consumer societies is clearly open to question. While the rampant accumulation of capital marches on through the privatisation of profit (Harvey, 2011), it is evident that the individual consumer will not bear sole responsibility for the fundamental failures of global capitalism (Urry, 2011). Changes to the systems of provision (Hall, 2013, also Chapter 15 this volume) are clearly required, given the environmental failures of highly mobile western consumer societies (Harvey, 2011). This will be an important focus of our continuing academic endeavours.

References

Barr, S., Shaw, G., Coles, T. and Prillwitz, J. (2010). "A holiday is a holiday": Practicing sustainability, home and away. *Journal of Transport Geography, 18*(3), 474–481.

Cohen, S.A., Higham, J.E.S. and Cavaliere, C.T. (2011). Binge flying: Behavioural addiction and climate change. *Annals of Tourism Research, 38*(3), 1070–1089.

Eijgelaar, E., Thaper, C. and Peeters, P. (2010). Antarctic cruise tourism: The paradoxes of ambassadorship, "last chance tourism" and GHG emissions. *Journal of Sustainable Tourism 18*(3), 337–354.

Freire-Medeiros, B. and Name, L. (2013). Flying for the very first time: Mobilities, social class and environmental concerns in a Rio de Janeiro favela. *Mobilities, 8*(2), 167–184.

Gössling, S., Hall, C. M., Peeters, P. and Scott, D. (2010). The future of tourism: Can tourism growth and climate policy be reconciled? A climate change mitigation perspective. *Tourism Recreation Research, 35*(2), 119–130.

Hall, C.M. (2013). Framing behavioural approaches to understanding and governing sustainable tourism consumption: Beyond neoliberalism, "nudging" and "green growth"? *Journal of Sustainable Tourism, 21*(7), 1091–1109.

Hares, A., Dickinson, J. and Wilkes, K. (2010). Climate change and the air travel decisions of UK tourists. *Journal of Transport Geography, 18*(3), 466–473.

Harvey, D. (2011). *The enigma of capital and the crises of capitalism.* Oxford: Oxford University Press.

Higham, J.E.S. and Cohen, S.A. (2011). Canary in the coalmine: Norwegian attitudes towards climate change and extreme long-haul air travel to *Aotearoa*/New Zealand. *Tourism Management, 32*(1), 98–105.

Higham, J.E.S., Cohen, S.A. and Cavaliere, C.T. (2014). Climate change, discretionary air travel and the "flyers' dilemma". *Journal of Travel Research*, doi: 10.1177/0047287 513500393.

McKercher, B., Prideaux, B., Cheung, C. and Law, R. (2010). Achieving voluntary reductions in the carbon footprint of tourism and climate change. *Journal of Sustainable Tourism*, *18*(3), 297–317.

Randers, J. (2012). *2052: A global forecast for the next forty years*. Vermont: Chelsea Green Publishing.

Urry, J. (2011). *Climate change and society*. Cambridge: Polity.

Wheeller, B. (2012). Sustainable mass tourism: More smudge than nudge – the canard continues. In T.V. Singh (ed.), *Critical Debates in Tourism* (pp. 39–43). Bristol: Channel View Publications.

Index

Page numbers in *italics* denote tables, those in **bold** denote figures.